The Future of the Law of the Sea

Gemma Andreone
Editor

The Future of the Law of the Sea

Bridging Gaps Between National, Individual and Common Interests

Springer Open

Editor
Gemma Andreone
Institute for International Legal Studies
Italian National Research Council
Rome, Italy

ISBN 978-3-319-51273-0 ISBN 978-3-319-51274-7 (eBook)
DOI 10.1007/978-3-319-51274-7

Library of Congress Control Number: 2017936385

Printed on acid-free paper

This Springer imprint is published by Springer Nature
The registered company is Springer International Publishing AG
The registered company address is: Gewerbestrasse 11, 6330 Cham, Switzerland

Foreword

The present book is the final product of the work of Marsafenet, which is a network of experts on maritime safety and security, sponsored by the COST funding scheme under H2020. Marsafenet is a collaborative partnership that works together in addressing recent law of the sea issues. The work of Marsafenet has led to a number of symposia held in various countries, the publication of several books and the creation of the *MarSafeLaw Journal*, an open access, peer-reviewed journal. It must be envisaged that this remarkable collective effort will not vanish with the end of Marsafenet—the network of friendships and scholarly collaboration must continue, as must the *MarSafeLaw Journal*.

The goal of the present volume is to address various law of the sea issues not only through the lens of State interests, as in traditional international law, but also from the point of view of common values and the interests of individuals. This is achieved in the two main parts of the book, which discuss respectively 'the equitable and sustainable exploitation of marine environment and of its resources' and 'the national and international response to maritime crimes'.

Common values are already present in the United Nations Convention on the Law of the Sea (UNCLOS), namely, but not exclusively, in the notion of the 'common heritage of mankind'. The interests of individuals are less evident in the Convention, but nonetheless present, as pioneering studies on the law of the sea and human rights have shown. The present volume, much like those already published by Marsafenet, makes interesting contributions to both points of view.

All the essays contained in the book at hand centre on maritime safety and security problems that have arisen in the last few decades, after the entry into force of UNCLOS. These problems and the responses thereto not only necessitate enhanced international cooperation; they also require action at the domestic level. Questions arise concerning the adoption of domestic legislation and regulations to implement systematically (and not only episodically) the rules of international law within domestic legal systems. Questions concerning the implementation of international judgments in domestic legal systems also arise. Recent cases brought before the International Tribunal for the Law of the Sea (most recently, the *Virginia*

G and the *Nordstar*) and before an Arbitral Tribunal (the *Duzgit Integrity*) suggest that these matters are becoming more pressing.

In light of this, the continuation of the endeavours of Marsafenet—be it by a reborn Marsafenet or by individual scholars or by new collective research projects—would be best served by focusing on the domestic law impact of UNCLOS. How have State Parties implemented the rules of UNCLOS in their domestic systems? How are the mechanisms employed for the implementation of UNCLOS assessed from the point of view of compliance under international law with the Convention? Do the means for implementing UNCLOS include remedies for individuals? How does legislation implementing UNCLOS interact at the domestic level with legislation implementing the European Human Rights Convention? How has UNCLOS been implemented in the EU legal system? What is the best way to prevent divergent implementation legislation or practices from jeopardising the unifying impact of UNCLOS?

All these questions, and others germane to them, are open to debate. They may not be the only questions currently arising as regards the law of the sea, but they are of particular practical and scientific interest and well suited for a comparative study by a multinational team of scholars such as Marsafenet.

Emeritus Professor of International Law
State University of Milan, Italy

Former Judge of the International
Tribunal for the Law of the Sea

Tullio Treves

Scope and Aim of the Volume

This volume is the final publication of the research carried out within the framework of Cost Action IS1105 Marsafenet (i.e. Network of legal experts on the legal aspects of maritime safety and security), which has brought together more than 80 researchers from 23 countries between 2012 and 2016.

As the main goals of international law in regulating marine spaces are the protection of the marine environment and the sustainable and equitable exploitation of its resources, in addition to the peaceful use of the oceans, this publication proposes some reflections on both maritime safety and security issues.

The principles and norms of the law of the sea are primarily codified in the 1982 United Nations Convention on the Law of the Sea (UNCLOS) and related instruments, which constitute the current legal framework for ocean governance.

As is widely known, this field of international law has long been characterised by a tension between the concept of the freedom of the seas (*mare liberum*) and that of the closed sea (*mare clausum*). Over time, the pivotal issue has always been the extent to which the sea is an international space, with resources freely available to all, and the extent to which it can be claimed by each State. The juxtaposition of these two positions continues to lie at the core of contemporary international law of the sea and international practice, and it is largely focused on national interests in maritime spaces and resources.

Nevertheless, in line with ongoing developments in the international legal order, recognition and protection of the rights of individuals, as well as the common interests of the international community, have become increasingly important in the regulation of maritime spaces and resources. The emergence of such interests, the need to protect them effectively and the growing interest in maritime activities by non-State actors, such as private companies and NGOs, are all elements that are gradually changing the nature of the law of the sea. As a result, there is an emerging inclination for a structural change that may reconcile these objectives with the perpetual importance of State sovereignty.

In several UNCLOS provisions, there are already a number of direct and indirect references that highlight the need to protect the common values of the international community. The rights and freedoms of the individual are also considered and protected by some UNCLOS provisions.

The future challenge of the law of the sea is to strengthen the protection of all of the (sometimes conflicting) interests at stake—national, individual and common interests—and to achieve a fair balance among them, in order to foster the realisation of a just and equitable international economic order. The development of legal instruments and mechanisms intended to take into account and balance the diverse interests of States, international organisations, non-State actors (including individuals, groups of individuals, private entities and NGOs) and the international community as a whole is following a tortuous and asynchronous pattern—existing instruments are frequently used and adapted to face issues and challenges that they were not created for.

Against this background, the volume at hand strives to address this dynamic development of the law of the sea, focusing on a few key issues that are under the spotlight of the current international agenda, and which also lie at the heart of the conflict between competing actors and interests.

This volume explores the diverse phenomena that are challenging the international law of the sea today, using a unique perspective, which involves simultaneous analysis of the national, individual and common interests at stake. This perspective can constitute a useful element in the effort to bring today's legal complexity and fragmentation to a homogenous vision of the sustainable use of the marine environment and its resources, as well as the international and national responses to maritime crimes. These two areas of investigation have been chosen because they represent an interesting research laboratory for identifying and analysing the evolving nature of the international law of the sea.

Accordingly, this volume is divided into two sections: one devoted to equitable and sustainable exploitation of the marine environment and its resources, the other to national and international responses to maritime crimes.

The first section addresses the legal instruments and mechanisms concerned with regulating the use of the sea and the exploitation of marine living and non-living resources (e.g. marine protected areas, fisheries agreements, marine scientific research, biodiversity protection and blue energy at sea).

These issues are at the forefront of this discussion due to the increasing use of ocean spaces and resources to meet global demands for energy, food and the mobility of people and goods. They lie in the common ground between different overlapping legal regimes such as the law of the sea, environmental law, energy law, climate change law and EU law, and involve conflicts between the competing economic and environmental interests of States, individuals and private actors, at both the local and global levels.

The second group of issues at stake is concerned with maritime crimes and the changing international security landscape, focusing on human trafficking and smuggling at sea, piracy, private maritime security and the proliferation security initiative. These issues reveal the interplay between the law of the sea, human rights law and international criminal law, highlighting the need for effective instruments for the protection of individuals at sea under a number of different, conflicting perspectives.

Like all previous Marsafenet publications, this volume is open access and has been peer reviewed by anonymous reviewers actively involved in the Marsafenet network.

My deepest gratitude goes to all the authors and reviewers for their important scholarly contributions.

Special acknowledgment is owed to Valentina Rossi (Institute for Research on Innovation and Services for Development of the Italian National Research Council, Italy) and Claudia Cinelli (KG Jebsen Centre for the Law of the Sea, University of Tromso, Norway and University of Pisa, Italy) for their support and assistance in the preparation of this volume.

Chair of the Cost Action IS1105 Gemma Andreone
Marsafenet
Institute for International Legal Studies of the Italian
National Research Council
Rome, Italy

Peer Review Process

Each chapter of this book has been assessed by anonymous reviewers.

A list of the reviewers is available from Springer International Publishing.

Funding Institutions

COST is supported by
EU Framework Programme Horizon 2020

COST (European Cooperation in Science and Technology) is a pan-European intergovernmental framework. Its mission is to enable breakthrough scientific and technological developments leading to new concepts and products and thereby contribute to strengthening Europe's research and innovation capacities.

It allows researchers, engineers and scholars to jointly develop their own ideas and take new initiatives across all fields of science and technology while promoting multi- and interdisciplinary approaches. COST aims at fostering a better integration of less research intensive countries to the knowledge hubs of the European Research Area.

The COST Association, an international not-for-profit association under Belgian Law, integrates all management, governing and administrative functions necessary for the operation of the framework. The COST Association has currently 36 Member Countries.

More information about COST is available at http://www.cost.eu.

COST is supported by the EU Framework Programme Horizon 2020.

COST Action IS 1105

This publication is based upon work from COST Action IS 1105, supported by COST (European Cooperation in Science and Technology).

MARSAFENET—the acronym for NETwork of experts on the legal aspects of MARitime SAFEty and security—aims to bring together experts in international law of the sea in order to increase the knowledge on maritime security and safety and to develop a common conceptual and methodological framework with the goal of contributing to fill the legal gaps and of transforming scientific results into feasible solutions. The network is intended to foster the identification and exploitation of synergies between EU policies on maritime safety and security. In terms of societal implications, it is aimed at facilitating the detection of solutions for old and new issues and criticalities, which may be implemented within the public realm (decision-makers, international institutions, international and national tribunals, EU institutions, etc.) and within the private sector (shipping sector, civil society, NGOs, etc.).

This Cost Action takes an in-depth look at current urgent maritime matters focusing on four main issues: shipping and marine environmental protection, new developments of economic activities at sea, international maritime security and border surveillance and, finally, protection of fragile and semi-enclosed seas.

MARSAFENET is currently composed of more than 80 legal experts from 23 different countries.

More information about COST Action IS1105 is available at www.marsafenet.org.

National Research Council of Italy

ISGI

Institute for International Legal Studies

The Grant Holder of the Cost Action IS1105 is the Institute for International Legal Studies (ISGI). The Institute for International Legal Studies is a scientific organ of the Italian National Research Council (CNR), founded as a Centre in 1986 in cooperation with the Italian Society for International Organization (SIGI). Key areas of activity include also training and expert consultancy in cooperation with national and international institutions such as the Ministry of Foreign Affairs, the Ministry of Economic Development and the Ministry of Welfare, Health and Social Policies; the Asser Institute of the Hague; ESA; UNIDROIT; and UNESCO.

More information about ISGI is available at www.isgi.cnr.it.

Contents

List of Contributors

Montserrat Abad Castelos Faculty of Social Science and Law, University Carlos III, Madrid, Spain

Gemma Andreone Institute for International Legal Studies, Italian National Research Council, Rome, Italy

Giorgia Bevilacqua Institute for Research on Innovation and Services for Development, Italian National Research Council, Naples, Italy

Emmanuella Doussis Institute of European Integration and Policy, National and Kapodistrian University of Athens, Athens, Greece

Magne Frostad Faculty of Law, K.G. Jebsen Centre for the Law of the Sea, University of Tromsø, The Arctic University of Norway, Tromsø, Norway

Kamrul Hossain Northern Institute for Environmental and Minority Law, Arctic Center, University of Lapland, Rovaniemi, Finland

Pirjo Kleemola-Juntunen Northern Institute for Environmental and Minority Law, Arctic Centre, University of Lapland, Rovaniemi, Finland

Jasenko Marin Department for Maritime and Transport Law, Faculty of Law, University of Zagreb, Zagreb, Croatia

Robert Mikac Department of International Politics and Diplomacy, Faculty of Political Sciences, University of Zagreb, Zagreb, Croatia

Kathleen Morris Department of International Relations, University of Southern California, Los Angeles, CA, USA

Mišo Mudrić Department for Maritime and Transport Law, Faculty of Law, University of Zagreb, Zagreb, Croatia

Gabriela A. Oanta University Institute of European Studies "Salvador de Madariaga", University of A Coruña, A Coruña, Spain

Enrique J. Martínez Pérez Faculty of Law, University of Valladolid, Valladolid, Spain

Marta Chantal Ribeiro Faculty of Law, University of Porto, Porto, Portugal

Sarra Sefrioui Faculty of Law, Abdel Malek Essaâdi University, Tangier, Morocco

Marta Sobrido University Institute of European Studies "Salvador de Madariaga", University of A Coruña, A Coruña, Spain

José Manuel Sobrino University Institute of European Studies "Salvador de Madariaga", University of A Coruña, A Coruña, Spain

Tullio Treves Department of Italian and Supranational Public Law, State University of Milan, Milan, Italy

Part I
The Equitable and Sustainable Exploitation of Marine Environment and of Its Resources

Adapting to Sea Level Rise: A Law of the Sea Perspective

Sarra Sefrioui

1 Introduction

A fresh look at the latest satellite data from 2002 and 2014 shows that seas are rising by around 1.4 mm a year due to thermal expansion rather than 0.7–1 mm as was expected.[1] Many geographical coastal features and low-lying island countries appear to be at risk of being deleted from the map in the next decades due to global warming and sea level rise.[2]

The reality of climate change and sea level rise does not only have geographical impacts. It may also generate legal implication of changing baselines, which in turn influences the outer limits of maritime zones. The potential submergence of important base points may potentially lead to the loss of maritime jurisdiction subject to maritime claims and to the loss of jurisdictional rights over valuable resources within these maritime spaces. This could consequently have disastrous economic consequences.[3]

It is important to recall that a coastal State's maritime claims to maritime zones—territorial sea, contiguous zone, exclusive economic zone and continental shelf under the United Nations Convention on the Law of the Sea (UNCLOS)—are measured from baselines except for one of the situations where the outer limits of

[1]The Guardian. http://www.theguardian.com/environment/2016/jan/26/sea-level-rise-from-ocean-warming-underestimated-scientists-say. Accessed 1st Jul 2016.

[2]Kiribati where 32 islets have already disappeared under the sea, Maldives, Nauru, Kosrae, Marshall islands, Salomon islands, Tuvalu and some areas in the US, For more details about which areas within the contiguous U. S. are most at risk and the potential impact of sea level rise, see https://coast.noaa.gov/digitalcoast/stories/population-risk.html. 25/09/2016.

[3]Schofield (2009a), p. 70.

S. Sefrioui (✉)
Faculty of Law, Abdel Malek Essaâdi University, Tangier, Morocco
e-mail: sarra.sefrioui@gmail.com

G. Andreone (ed.), *The Future of the Law of the Sea*,
DOI 10.1007/978-3-319-51274-7_1

continental shelf exceed 200 nautical miles.[4] Baselines are located at the interface between the land area and sea for the purpose of maritime jurisdiction. They also divide the internal waters of a coastal State from the territorial sea—the most landward of the belts of offshore jurisdiction. The international rights and duties of coastal States and flag States differ substantially between internal waters and the territorial sea as shown above.[5]

Moreover, baselines are quite important to the delimitation of boundaries. In the bilateral delimitation of maritime boundaries, baselines form the starting point in delimitations between adjacent and opposite States with overlapping claims to maritime area—the role of baselines.[6]

However, baselines are facing sea level rise effects and at the same time the silence of UNCLOS to the question whether these baselines and therefore maritime zones—or one of them—shift or remain stable and effective. Case law on maritime delimitation provides little responses.

This article examines the potential effects of sea level rise on baselines, the outer limits of maritime zones, and maritime boundary. It will discuss in the first section the question on how the law of the sea can adapt to sea level rise and what measures can be adopted to address the implications of sea level rise on baselines and the establishment of maritime zones. Therefore, the second section of this chapter will focus on the analysis of the effects of sea level rise on baselines from which the maritime limits and boundaries are determined. The third section will provide the potential legal responses to mitigate the effects of sea level rise regarding baseline alteration and disappearance. It tries to answer the question of whether baselines should be ambulatory or permanently fixed. The result of this study will be presented in the conclusion in the fourth section.

2 Legal Implication of Sea Level Rise on Baselines from Which Maritime Limits and Boundaries Are Determined

This section mainly discusses the legal framework of maritime zones and the current legal regime of baselines to contextualize the study before analyzing the legal implication of sea level rise on baselines, which in turn influences the outer limits of maritime zones.

[4]International Law Association (2008).

[5]Ibid., p. 4.

[6]Ibid., p. 5. International Law Association referring to the International Court of Justice (ICJ) case ICJ (2009).

2.1 *General Description of Maritime Zones*

According to the United Nations Convention of 10 December 1982 on the Law of the Sea (UNCLOS), maritime zones can be divided mainly into six areas as follows. The internal waters are located on the landward side of the baselines and comprise the maritime waters adjacent to the land territory of the coastal State (article 8 § 1 UNCLOS); the territorial sea is adjacent to the internal waters, and it measures 12 nautical miles from the baseline (articles 2 § 1 and 3 UNCLOS); the contiguous zone measured 24 nautical miles from the baseline where coastal State has, notably, policing powers in relation to its customs, fiscal, sanitary and immigration laws, and regulations (article 33 UNCLOS); the exclusive economic zone is 200 nautical miles wide (article 57 UNCLOS) where coastal State has sovereign rights on this zone in respect to environmental protection, scientific research, exploration, and the use of natural resources (article 56 UNCLOS). The continental shelf is the prolongation of the coastal State's land territory submerged for 200 miles from the baselines when the outer edge of the continental margin is less or up to 350 nautical miles (or 100 nautical miles from the 2500 m isobath) if it is wider (article 76 UNCLOS). However, since the coastal State's right to outer limits of continental shelf relies not only on the 200 nautical miles rule but also on the "natural prolongation" criterion, it implies that the outer limits of the continental shelf must not always be measured from baselines. The coastal State has sovereign rights over this area in respect to the exploration and exploitation of natural resources (article 77 § 1 UNCLOS). Finally, the high seas are not subject to the State's sovereignty and are located beyond the external limit of the EEZ (as a maximum of 200 miles from the baselines) (article 86 UNCLOS). The outer limit of all these zones are determined and delimited from baselines except for one of the situations where outer limits of continental shelf exceed 200 nautical miles.

2.2 *Each Maritime Zone Is Measured from Lines Joining Appropriate Points on Land: Baselines*

UNCLOS establishes the legal framework of baselines. However, in some cases, it appears that a distinction between baselines serving for measuring the limits of the maritime zones and baseline serving for establishing maritime boundary can be drawn.

2.2.1 The Establishment of Baselines Under UNCLOS: Relevant Provisions

The establishment of baselines is a fundamental operation for a coastal State wishing to have jurisdiction over maritime zones adjacent to the continental coasts.

These lines are the basis of the coastal State's appropriation of the maritime zones in the sense that it constitutes the lines from which maritime zones are measured. Consequently, baselines are the starting line of the outer limits of maritime zones, and then the significant change on these lines will accordingly affect the jurisdiction of outer limits of maritime zones. They aim to correct the curves of the coast and to prevent their reproduction by enveloping the coast as an "envelope line."[7]

There are different types of baselines (normal baselines, straight baselines and other bay closing lines, straight line across the mouth of the river) that mainly depend on the general configuration of the coast. A State unilaterally determines the base points that are relevant according to UNCLOS. This national operation[8] of the unilateral determination of the relevant baselines has, however, international aspects.[9]

Normal Baselines

The relevant regulation concerning baselines was included in the Geneva Convention 1958[10] and is currently in UNCLOS 1982. Normal baselines are defined by article 5 of UNCLOS as follows: "Except where otherwise provided in this Convention, the normal baseline for measuring the breadth of the territorial sea is the low-water line along the coast as marked on large-scale charts officially recognized by coastal State." Thus, the normal baseline is the low-water line[11] drawn also according to UNCLOS, including article 6 (reefs), article 8 (internal waters), article 9 (mouths of rivers), article 13 (low-tide elevations). It appears that the coastal State would try to choose the lowest line to establish its baseline far from the coast seaward.

A reference to the method of drawing baselines is made in article 14 of UNCLOS, which underlines that "the coastal State may determine baselines in turn by any of the methods provided for in the foregoing articles to suit different conditions." Therefore, only States have the right to choose the reference level of its

---

[7]Kapoor and Kerr (1986), p. 58.

[8]See, ICJ, Judgment (1951).

[9]ICJ, Judgment (2009), para. 137.

[10]The Geneva Convention in 1958 is one of the four first conventions on the law of the sea matters that has codified, in many provisions, the customary international law.

[11]The low-water line is defined by the International Hydrographic Organization as the line until which water is retreated at low water along the coast in particular in a beach ("[la ligne] *jusqu'à laquelle se retire l'eau à basse mer le long de la côte, en particulier sur une plage*." It is identified by a "*ligne mince et ondulée formée de sable fin, de débris de coquilles, de petits morceaux d'algues, de détritus divers, etc., laissée par les vagues, qui marque la limite supérieure atteinte par leur mouvement sur la plage*." Organisation Hydrographique Internationale (1998), p. 41. The different level used to establish a low water line are Lowest Astronomical Tide, LAT, Mean Low Water Springs, MLWS; Mean Lower Low Water, MLLW; Mean Sea Level, MSL. See, United Nations Office for Oceans Affairs and the Law of the Sea (1989), p. 47.

low-water line, and it is submitted to any evaluation.[12] The majority of States in the world have established normal baselines in a sense that they are considered as the "default" baselines.[13] However, a straight baseline has a particular regime.

Straight Baselines

Straight baselines are drawn where a coastline is deeply indented and cut into, or in the presence of a fringe of islands along the coast in its immediate vicinity. They are drawn by joining the appropriate base points on land in accordance with article 7 (1) of UNCLOS.[14] The straight baselines have the objective to smoothen the curves of the coast. If these baselines are drawn seaward in the coast, their purpose is not, however, to extend the territorial sea unduly.[15] The International Court of Justice in the *Qatar/Bahrain* case generally highlighted that "the method of straight baselines, which is an exception to the normal rules for the determination of baselines [...] must be applied restrictively."[16]

The Convention determines some "rules" for the establishment of the straight baselines in that they are drawn in some situations, including in the presence of "a delta and other natural conditions, the coastline is *highly unstable*" (article 7 (2) emphasis added). Also, straight baselines must not depart from any appreciable extent from the general direction of the coast, and the sea areas lying within the lines must be sufficiently closely linked to the land domain in order to be subject to the regime of internal waters (article 7(3) UNCLOS). Article 7(4) also stipulates that straight baselines "shall not be drawn to and from low-tide elevations unless lighthouses or similar installations which are permanently above sea level have been built on them or except in instances where the drawing of baselines to and from such elevations has received general international recognition."

The provisions of article 7 of UNCLOS give rise to several comments related to the exact meaning of the terms used. Of particular interest here is that the term "highly unstable" is not clear. The Convention fails to provide any rule related to the change of geography and seems to give a "stable" solution of straight baselines to avoid fluctuation in case of use of normal baselines.

---

[12]See, Calerton and Schofield (2001), pp. 21–23.

[13]Prescott and Schofield (2000), pp. 94–97.

[14]Straight baselines may also be employed across mouths of rivers (Article 9 UNCLOS) and bays (Article 10 UNCLOS), which is of less interest here.

[15]United Nations Office for Oceans Affairs and the Law of the Sea (1989), p. 39.

[16]ICJ, Judgment (2001), para. 212.

2.2.2 Baselines in Establishing Maritime Limits and in Drawing Maritime Boundary

The distinction between maritime limits and maritime boundaries shapes the potential response to sea level rise and therefore whether ambulatory or fixed baselines will have any implication.[17] To claim jurisdiction over maritime areas, a State may unilaterally establish maritime limits that mark the outer limit of its national jurisdiction measured from baselines.[18] However, where there are overlapping claims, maritime delimitation boundary is established from baselines.

Maritime Limits

A coastal State's maritime zone limit is usually the outer limit of this maritime zone and the inner limit of another zone until the 200-nautical-mile limit that makes the end of the exclusive economic zone and the beginning of the high seas.[19] Maritime limits determine the extent of maritime zones establishing the ending line of the maritime zones seaward. A State can unilaterally determine its maritime limits when they do not overlap with the neighboring State's maritime limits.

According to UNCLOS, coastal States must deposit charts and geographical coordinates that show straight baselines or the outer limits of the territorial sea, the exclusive economic zone, and the continental shelf derived therefrom with the United Nations Secretary-General. However, UNCLOS does not require that baselines must be published in charts and lists of geographical coordinates.[20] Baselines "shall be shown on charts of a scale or scales adequate for ascertaining their position."[21] When determining its baselines, a coastal State must take into consideration that it is a national operation that has international aspects.[22] The validity of baselines can be challenged by other states as it was argued by the ICJ in the *Anglo-Norwegian Ficheries* case that it is "a unilateral act, because only the coastal State is competent to undertake it, the validity of the delimitation with regard to other States depends upon international law."[23] Baselines that serve for identifying maritime limits have international implications since they can be used for maritime

---

[17]Lisztwan (2012), p. 171.

[18]The outer limit is defined as "limite jusqu'à laquelle un État côtier revendique ou peut revendiquer une juridiction spécifique conformément aux dispositions de la Convention. Les limites extérieures de la mer territoriale, de la zone contiguë et de la zone économique exclusive sont constituées par des lignes dont chaque point est séparé du point le plus proche de la ligne de base par une distance égale à la largeur de la zone mesurée (art. 4; art. 33, par. 2 et art. 57)". Division des affaires maritimes et du droit de la mer (2001), p. 142.

[19]See, Caflisch (1985), p. 376.

[20]Article 16 (2) of UNCLOS.

[21]Ibid. (1) of UNCLOS.

[22]ICJ, Judgment (2009), para. 137.

[23]ICJ, Judgment (1951), p. 132.

delimitation.[24] Therefore, the alteration of baselines from which limits are determined may also influence the maritime delimitation boundary.

Maritime Delimitation

Where claims to maritime areas overlap, a maritime boundary is measured from a selection of base points that form the starting point from which the maritime boundary between adjacent and opposite States is measured.[25] States must negotiate and agree on a maritime boundary or reach delimitation through submission to third-party dispute resolution (including the International Tribunal on the Law of the Sea (ITLOS), the ICJ, or an arbitral tribunal[26]). The methodology for determining baselines is not provided by UNCLOS, neither by jurisprudence. Moreover, States do not address the potential shift of their baselines that might be caused by sea level rise. It is left to the agreement through negotiation between the States concerned. States, in their agreement, may agree to fix their baselines and maritime limits regardless of any potential change because of sea level rise.

If they fail to find an agreement, article 15 of UNCLOS provides that "where the coasts of two States are opposite or adjacent to each other, neither of the two States is entitled, failing agreement between them to the contrary, to extend its territorial sea beyond the median line every point of which is equidistant from the nearest points on the baselines from which the breadth of the territorial seas of each of the two States is measured." This selection would give to these points the legal status as such and therefore the legal validity. Nevertheless, the Court or the Tribunal most likely "select base points by reference to the physical geography of the relevant coasts."[27] The relevant coasts would be the projections of which overlap with that of another State. They are determined by the Court in the first step of the judicial maritime boundary delimitation that consists of drawing a provisional line. The International Court of Justice in *Romania/Ukraine* case held that those base points on the relevant coasts should be chosen that "mark a significant change in the direction of the coast, in such a way that the geometrical figure formed by the line connecting all these points reflects the general direction of the coastlines".[28] Articles of the Convention dealing with maritime delimitation do not, however, address the impact, if any, of shifting coastal geography or any corresponding change in equities.

---

[24]"The delimitation of sea areas has always an international aspect", ICJ, Judgment (1951), p. 132. PCA Award (2006), para. 365, and explicitly in ICJ, Judgment (2009), para. 137.

[25]ICJ, Judgment (2009), para. 137.

[26]Article 287 (1). State Parties may, however, except boundary delimitations from such compulsory procedures. See ibid. article 298(1)(a)(i).

[27]ICJ, Judgment (2009), para. 137.

[28]*Ibid.*, p. 127.

2.3 Baselines Would Naturally Change Because of Sea Level Rise

Legal consequences of the sea level rise are difficult to predict with precision. The change in the coast can be in many ways. However, potential examples and situations can be examined. The first situation to address is that when base points and baselines shift (retreat) landward, and the second is when base points and baselines situated on islands, rocks, and low-tide elevations disappear.

2.3.1 When Baselines Shift Landward

When coastal States' baselines retreat landward with no overlapping maritime claims, the coastal State would lose part of its territory, and the baseline from which the breath of the maritime zones is measured would shift landward. With regard to the maritime zone limits unilaterally established, they would also retreat in the same way as the baseline. Therefore, the legal status of the maritime zones would change: part of the territorial sea landward becomes internal water, and seaward becomes EEZ. Therefore, part of the EEZ becomes high seas. This has implications on sovereign rights: innocent passage, freedom of navigation, fishing rights, etc.

When the coastal State has a maritime delimitation agreement with an opposite or adjacent State, this would have two implications:

(a) If the boundary agreement divides their exclusive economic zones, in most cases coastline retreat will only increase the exclusive economic zones of the two States. As such, coastline shift will not affect the types of zones delimited[29] when the total area of the two EEZs does not exceed 400 nautical miles.
(b) If the total area exceeds 400 nautical miles after the coast retreats, a new area of high seas is created.[30]

Moreover, the shift landward of the baseline may change the initial direction of the coast. In this case, if the retreat is considerable and the distance from the base point and the new base point is significant, this "former base point" would not be "replaced" by a new one because the latter would draw a baseline, which would depart to an appreciable extent from the general direction of the coast, against the spirit of the provision of UNCLOS that "the drawing of baselines must not depart to any appreciable extent from the general direction of the coast" (article 7 (3)). However, the ICJ has faced, in the *Nicaragua/Honduras* case a highly unstable coastline in Cape Gracias a Dios, where the Nicaragua-Honduras land boundary ends. In this case, if the Delta shifted landward, it would actually lead to the

[29] Lisztwan (2012), p. 176.
[30] *Ibid.*

baseline more closely following the overall shape of the coastline. The Court held that "[g]iven the close proximity of these base points to each other, any variation or error in situating them would become disproportionately magnified in the resulting equidistance line."[31] The land boundary along the Rio Coco ends in a prominent delta—Cape Gracias a Dios—created by sediment transported down the river. The parties to the case agreed that the sediment transported by the River Coco has "caused its delta as well as the coastline to the north and south of the Cape, to exhibit a very active morpho-dynamism."[32] The Court has underlined that "continued accretion at the Cape might render any equidistance line so constructed today arbitrary and unreasonable in the near future."[33] Therefore, the Court did could not determine any base point for the construction of the equidistance line and concluded that "where [. . .] any base points that could be determined by the Court are inherently unstable, the bisector method may be seen as an approximation of the equidistance method."[34] However, sea level rise does not only create a shift of baseline landward; it can also submerge islands and low-tide elevations on which base points are established.

2.3.2 When Base Points Are Situated on Disappeared Island and Low-Tide Elevation

Islands and low-tide elevations would serve to establish base points and baselines for the purpose of drawing maritime limits and/or maritime boundaries. Therefore, in this section, we will discuss how in some cases the legal regime of an island may change to low-tide elevation regime due to the submergence of the island. Moreover, the distinction of these features implies that coastal States' maritime rights may alter depending on the category into which the maritime feature falls. This could mean a huge loss of coastal States' rights, maritime areas, and resources.

A low-tide elevation is defined by article 13(1) of UNCLOS as an area of land "above water at low-tide but submerged at high tide." The Convention specifies that straight baselines may be drawn to and from low-tide elevations if lighthouses or similar permanently uncovered installations have been constructed on them or if there has been general international recognition (article 7 § 4 UNCLOS). However, low-tide elevation may only be used for measuring the breadth of the territorial sea where the low-tide elevation is situated wholly or partly at a distance not exceeding the breadth of the territorial sea from the mainland or an island (article 13 (2) UNCLOS). Therefore, the effect of its disappearance by permanent submergence depends on its geographical situation with respect to the outer limits of the

[31]ICJ, Nicaragua/Honduras Case: para. 277.

[32]Ibid.

[33]Ibid.

[34]Ibid.

territorial sea.[35] It creates a loss of the 12 miles that it generates if it is situated wholly or partly within the territorial sea area. Where a low-tide elevation (or former island) lies at a distance exceeding the breadth of the territorial sea from the mainland or an island, it has no territorial sea of its own (article 13(2) UNCLOS).

It is important to underline that islands are distinguished to be low-tide elevations (article 13(1) UNCLOS). A low-tide elevation is "a naturally formed area of land which is surrounded by and above water at low-tide but submerged at high tide" (article 13(1) UNCLOS), and an island, however, is "a naturally formed area of land, surrounded by water, which is above water at high tide" (article 121 (1) UNCLOS). Islands under article 121 paragraph 1 of UNCLOS remain above water at high tide since it "is a naturally formed area of land, surrounded by water, which is above water at high tide." With regard to the generation of maritime zone, low-tide elevations "literally do not rise to the status of islands."[36] Islands generate maritime zones such as territorial sea, contiguous zone, EEZ, and continental shelf. Coastal islands are used as baselines when situated within 12 nautical miles and would enlarge the extent of the territorial sea seaward since it generates its own territorial sea. Article 7(4) of UNCLOS recognizes similar rights to low-tide elevations only within its limited circumstances. If sea level rises, some islands may become submerged at least at high tide. This consequently led to a different legal situation with regard to maritime entitlement since islands are different from low-tide elevations.

Therefore, when an island has become a mere low-tide elevation, coastal States would surely try to strengthen against further submersion in order to maintain the straight baseline.[37] To preserve its emergence above water at low tide, coastal States may engage in some activities of construction of artificial installations on the low-tide elevation. The question that is raised here is to know if these artificial works are legally accepted and do not change the status of the low-tide elevation. To these issue, some scholars (Prescot and Brid[38]) have argued that these activities are not against the provisions of article 7 of the Convention since a low-tide elevation has to be internationally recognized. Nevertheless, in a jurisdictional maritime delimitation, even though the general recognition of a low-tide elevation is one of the conditions to use it to draw straight baselines, it seems difficult to accept that the Court of the Tribunal would still consider a disappeared low-tide elevation. For instance, the International Court of Justice accepted the use by Norway of a nonconstructed low-tide elevation to draw straight baselines,[39] but these two situations cannot be assimilated.

This would lead to conclude that if the land features from which baselines may be drawn retreat into each other and disappear, there will be no baselines from

[35]See, Calerton and Schofield (2001), p. 38.

[36]Roach and Smith (1996), p. 73.

[37]See, Freestone and Pethick (1994).

[38]Bird and Prescott (1989), pp. 177–196.

[39]ICJ, Fisheries Case: 116.

which to define the internal waters, territorial sea, contiguous zone, exclusive economic zone, and continental shelf zone.[40] With the disappearance of this zones, the maritime area would be subject to the regime of high seas since this regime applies "to all parts of the sea that are not included in the exclusive economic zone, in the territorial sea or in the internal waters of a State, or in the archipelagic waters of an archipelagic State."[41]

3 Potential Responses to Baseline Alteration and Base Point Disappearance

It is necessary to recall that the legal framework of the United Nations Convention on the Law of the Sea of 1982 is based on maritime geography. However, unlike the Convention, maritime geography is by its nature unstable and evolving. In this context, the question that can be raised is how a lawyer can find legal answers to the questions raised by the instability of the low-water line and some base points and their potential impact on maritime baselines.

The answer to this question is not easy. Two theories have emerged generating different consequences: the approach that encourages the use of ambulatory baselines and the opposite approach, which has opted for the stability and preservation of baselines vis-à-vis the change in geography.

3.1 The Practical Implications of the Use of Shifting Baselines

Referring to the above analysis of low water normal baselines of article 5 of UNCLOS, it is understood that there is a common uncertainty related to the coastal State having the choice regarding which one is the low-water line, which is inherently changing by sea level rise. It is to recall that the lower water line is the starting line of the outer limits of maritime zones.

In interpreting article 5 of UNCLOS, the International Law Association' (ILA) Committee on Baselines under the International Law of the Sea has argued that the charted low-water line is the legal normal baseline and the chart itself is the legal document that determines the position of that baseline irrespective of the physical realities of the coast.[42] As discussing above, the low-water line depends on the choice of vertical datum that is the level of reference for vertical measurements of a tide.

---

[40]See Hestetune (2010).

[41]Article 86 of UNCLOS.

[42]ILA's interpretation of article 5 of the UNCLOS, pp. 1, 2.

Article 5 of LOSC presents another uncertainty in that it does not specify a particular vertical datum and thus low-water line to be used. "With respect to the changes in the location of the low-water line caused by the tidal cycle, this line can be fixed by identifying the single vertical, or tidal, datum (among several used in the hydrographic community) to represent low tide. This vertical datum is the 'zero level' to which elevation and depth measurements are reduced. The intersection of the sea—when it is at that chosen level—with the coast is the low-water line. The low-water line thus defined is an elusive feature if not a purely conceptual construct."[43] Therefore, the choice is left to the coastal State since there is no "wrong" answer.[44]

Once the selected low-water line is shown on the charts officially, there is therefore recognition by coastal States and normal baseline could remain in place, irrespective of sea level rise. This would ensure safety and prevent navigation from uncertainty.

However, another interpretation was given to article 5 of UNCLOS by the ILA in the case of unstable coasts due to sea level rise. Normal baselines could adapt to physical realities, and therefore they could be dynamic.[45] This means that it would create a baseline system that reflects the actual geographical conditions by being ambulatory.[46]

The ambulatory approach considers that the Convention does not provide any provision on the consequences of sea level rise on the baselines, islands, and low-tide elevations, and consequently nothing can require from a coastal State to permanently fix its limits and boundaries. The very few provisions that might be seen as dealing with stability of maritime limits are related to the continental shelf and with the deltas' baseline provisions, but they are far from being sufficient. Thus, with regard to the continental shelf, UNCLOS requires from the coastal State to "deposit with the Secretary-General of the United Nations charts and relevant information, including geodetic data, permanently describing the outer limits of its continental shelf."[47] As to the baselines of the deltas, UNCLOS provides that "the appropriate points may be selected along the furthest seaward extent of the low-water line and, notwithstanding subsequent regression of the low-water line, the straight baselines shall remain effective until changed by the coastal State in accordance with this Convention."[48]

The ambulatory or shifting baseline approach has been developed by some scholars[49] who consider that with sea level rise uncertainty in maritime boundaries is created in that the baseline from which the boundary is drawn is ambulatory. In

[43]Ibid., p. 6.

[44]Calerton and Schofield (2001), p. 14.

[45]*Supra* note 3, p. 2.

[46]Ibid., p. 6.

[47]Article 76 paragraph 9 of UNCLOS.

[48]Article 7 paragraph 2 of UNCLOS.

[49]See Caron (1990), p. 635; Di Leva and Morita (2008); Reed (2000).

this sense, when feature from or on which the baseline is drawn disappears, the baseline must move and the maritime boundary generated from it has to be redrawn and calculated from the new baseline. Therefore, the maritime boundary generated from the previous disappeared baseline is not valid anymore and is reestablished from the new baseline. In this case, the outer limits of maritime zones are ambulatory in that they will move with the baselines from which they are measured and normal baselines may change with the change of the low-water line.[50]

The disappearance of baseline point implies the disappearance of the boundary generated by that point. The boundary though follows respectively the "movement" base point and the baseline. It results that maritime limits and boundaries shift when baselines shift, islands, or low-tide elevations disappear.

Some scholars have suggested that the implications of sea level rise on maritime boundaries could lead to "renegotiation of maritime boundary agreements based on the principle of equidistance to correspond with new geographic realities; re-evaluation of both equity and equidistance principles by international courts and tribunals in settling boundary disputes; or finally, reversion of highly disputed exclusive economic zone claims to the legal status of high seas."[51] This was, for example, the case of the two coastal State neighbors Switzerland and Italy.[52] The maritime boundary between these States shifts because of the height of the glaciers. Therefore, it was not permanently determined and led the parties between 2008 and 2009 to negotiate the definition of a new maritime boundary.[53]

This shift may generate some critical consequences. It would create uncertainty in maritime boundaries that would not be appreciated by a law that aims to generate stability between States in their relation. Modifying maritime boundaries regularly would create legal insecurity for States that have a constant unstable coast. They would have to constantly review their maritime limits and boundaries, and this would also create conflicts and instability for any neighboring state, even those that have more stable baselines.

In fact, the change of baseline could create conflicts between adjacent or opposite coastal States over the exploitation of natural resources.[54] If the shift of baseline is to be applied, some States that would lose part of their territory, islands, or low-tide elevations may invest huge financial efforts to maintain them even

---

[50]Alexander (1983), p. 535.

[51]Houghton et al. (2010), pp. 813–814.

[52]The original proposal to move the Swiss-Italian border comes from a member of Italy's centre-left opposition party and the final border will be agreed by a commission of experts from Switzerland's Federal Office of Topography and Italy's Military Geographic Institute. See, https://www.newscientist.com/article/dn16854-climate-changes-europes-borders-and-the-worlds/. See also, http://www.independent.co.uk/news/world/europe/melting-snow-prompts-border-change-between-switzerland-and-italy-1653181.html. Accessed on 01/10/2016.

[53]However, the result of this negotiations have not been provided.

[54]Caron (1990), pp. 640–641.

"artificially."[55] On the other side, by adjusting and correcting baselines, coastal States must take into consideration the costs of adaptation and the time that has to be spent in the long process of modifying maritime borders. Moreover, by applying the ambulatory baseline approach and if baselines are not marked on large-scale charts, navigation charts would not be precise in determining the maritime limits and boundaries and ships would not know exactly in which zone they navigate and to which rights they are subject (right of innocent passage, fishing rights, etc.).

In State practice, some States having an unstable baseline have made reference, in their national legislation, to the point on which the base point is situated without indicating the exact geographical coordinates of the point. To avoid the risk of establishment of a baseline that would not be stable, some States did not register their geographical coordinates but are content in the publication of the marine charts, which are formally easier to update with more flexibility in determining base points. For example, the Mexican legislation, in determining base points in the Mexican Golf, indicated that the departure point of the baseline is situated in the middle of the point in the mouth of Rio Grande without adding any precision about the geographical coordinates about this point.[56]

Taking into consideration all these implications of the shifting baseline approach, the ILA argued that the actual low-water line is the legal normal baseline and charts, and it should be considered as the evidence of the physical coastal realities or the actual coastal configuration.[57] The interpretation of article 5 of UNCLOS by the coastal State is fundamental for addressing the potential impacts of sea level rise with regard to maritime zones.

3.2 *Toward the Preservation of Baselines and Its Practical Implications*

As discussed above, neither in the case of normal baselines nor in straight baselines does UNCLOS provide that the maritime zone limits and boundaries can move with baselines. It "permanently" fixes the outer limit of the continental shelf to every State since they have to deposit to the Secretary-General of the United Nations

[55]Ibid., pp. 639–640; Soons (1990), pp. 222–223. Examples can be cited of some states. Indonesia that was planning to construct giant dikes around twelve islands in order to protect its territorial sea. Also, the case of Okinotorishima Island can be cited where the Japan is spending colossal sums to prevent its erosion and thus claim an EEZ. Approximately 163,000 miles of seabed and fishing zone were threatened. It is an island that is isolated from the coastal State, uninhabited and does not have fresh water. The island is a strategic point in the crossroad of the maritime roads converging to the centers of the world development. In 1977, Japan have declared a 200 nautical miles around Okinotorishima (Law No 30 of 2nd May 1977. http://www.un.org/Depts/los/LEGISLATIONANDTREATIES/PDFFILES/JPN_1977_Law.pdf. Accessed 3rd Mar 2016. See also, Song (2009), pp. 145–176.

[56]Federal Act relating to the Sea, 8 Jan 1986. http://www.un.org/Depts/los/LEGISLATIONANDTREATIES/PDFFILES/MEX_1986_Act.pdf. Accessed 3rd July 2016.

[57]Supra note 3, p. 2.

charts and relevant information, including geodetic data, permanently describing the outer limits of their continental shelf.[58] It also fixes the baselines for deltas and other natural conditions that make coastlines highly unstable. Since UNCLOS does not fix the outer boundary of the territorial sea, the contiguous zone, and the EEZ, one may think that these maritime zone boundaries can be ambulatory. However, the unique provision of UNCLOS to the question of instability of geography is illustrated in article 7(2).[59] It indicates that despite the possible shift of the coast landward, the appropriate points and the straight baselines joining them "shall remain effective until changed by the coastal State." This article would present some help in our contest because it concerns, according to the Convention, the case of "the presence of a delta and other natural conditions the coastline is *highly unstable* (emphasis added)" (article 7 (2)). Although it is unclear in which case a coast can be considered as highly unstable, the ICJ in the *Nicaragua/Honduras* case, noting the highly unstable nature of the mouth of the River Coco at the Nicaragua-Honduras land boundary terminus, decided that fixing base points on either bank of the river and using them to construct a provisional equidistance line would be "unduly problematic."[60]

However, the Convention is silent about the legal solution for changes of coasts or disappearance of features on which baselines and base points are established. The preservation of baselines and base points approach have been proposed in 1990 by A.H.A. Soons and was followed by several scholars:[61] "[C]oastal states are entitled, in the case of landward shifting of the baseline as a result of sea level rise, *to maintain* the outer limits of the territorial sea and of the [exclusive economic zone] where they were located at a certain moment in accordance with the general rules in force at that time."[62] Following this idea, other scholars (Prescott and Schofield) have underlined that some States, such as the United Kingdom and the Netherlands, have considered the nautical chart as the only legal document that defines baselines. In fact, by recognizing that coastline change by the time, the nautical chart or the straight baseline geographical coordinates as deposited in the Secretary-General must remain the reference legal document regardless of the coastline changes.

By fixing permanently the baselines, resource conflicts between States are avoided. It could appear that the coastal State that had less than 200 nautical mile EEZ and has lost part of its coast would gain more maritime resources because its coastline retreats, but all States would not have more than they are entitled to under the Convention. It is important to understand that since the breath of the maritime zones is fixed by UNCLOS, equity considerations impose States to recall that

[58]Article 76 paragraph 9 of UNCLOS. See, Freestone and Pethick (1994), pp. 73–90.

[59]Article 7 paragraph 2 UNCLOS indicated that "Where because of the presence of a delta and other natural conditions the coastline is highly unstable, the appropriate points may be selected along the furthest seaward extent of the low-water line and, notwithstanding subsequent regression of the low-water line, the straight baselines shall remain effective until changed by the coastal State in compliance with this Convention."

[60]ICJ, Judgment (2007), para. 273.

[61]See also, Rayfuse (2010).

[62]Soons (1990), p. 225 (emphasis added).

choosing to fix the boundaries or to adjust them with the ambulatory baselines and base points will not allow States to gain more than what they presently possess.[63]

Changing boundaries to adapt to the coastline changes would lead the state to protect the baselines by artificial costly installations. However, fixing baselines would avoid costs of adjustment to constant uncertain changes of the baselines[64] and costs of nautical maps modifications.[65]

By fixing the boundaries, the principles governing the oceans and those agreed upon related to the maritime zones and maritime boundaries by the States to the Convention in their negotiations are preserved. Fixing (freezing) baselines would promote stability in the location of limits of maritime zones and also in maritime delimitation boundaries—bilateral and multilateral—agreements. And as the ICJ stated in the *Temple of Preah Vihear* case, "when two countries establish a frontier between them, one of the primary objects is to achieve stability and finality."[66] The Division of Ocean Affairs' Handbook on Maritime Delimitation underlines that maritime boundary delimitation agreements "have a vocation for permanence and stability."[67]

Baselines, because they have legal meaning and not only a geographical meaning, are characterized by legal stability and should not be moving with the geography. The law of the sea in general is the law that governs relations between States in their maritime affairs.[68] In this sense, what fundamentally interests the law of the sea, including the Convention, is the stability and security of the relations between States, including in their international boundary regime. Even though the particularity of this law is that it is based on geography—in which change and instability are inherent—law is considered the priority because it provides stability and security and answers perfectly to the objectives of the Convention. Thus, in stating factors that States should consider in boundary negotiations, the UN Division for Ocean Affairs and the Law of the Sea advises States not to take into consideration any future geographical or geological shifts and the corresponding impacts on resource distribution or equities.[69]

Moreover, the Law of the Sea Convention is not the sole source of law governing maritime affairs and entitlements. Reference is made to other complementary Conventions that regulate general international law aspects in the law of the sea such as boundary agreements. According to the Vienna Convention for the Law of

[63]Caron (1990), p. 16.

[64]Rayfuse (2009), p. 6.

[65]Caron (1990), p. 647.

[66]See, ICJ Merits (1962).

[67]Division for Ocean Affairs and the Law of the Sea (2000), para. 322.

[68]The Preamble of the United Nations Convention for the Law of the Sea underlines that: it establishes a legal order for the seas and oceans which will facilitate international communication, and will promote the peaceful uses of the seas and oceans, the equitable and efficient utilization of their resources, the conservation of their living resources, and the study, protection and preservation of the marine environment.

[69]United Nations (2014), pp. 239–322.

Treaties,[70] stability of boundary agreements is achieved regardless of coastline movement. Even though geography changes and baseline shifts, maritime boundary agreements and their geographical coordinates remain secure and stable.[71]

Article 62 of the Vienna Convention, however, underlines an exception by which a State can unilaterally terminate an agreement because of a fundamental change in circumstances (*rebus sic stantibus*[72] or "things standing thus"). In the light of this article, some scholars have considered that a change in the geography would be a fundamental change that justifies the termination of an agreement and therefore its revision or replacement by a new agreement that would take into consideration the new situation. The question that can arise here is as follows: is an involuntary change of circumstances based on geography considered a fundamental change? Can one of the parties invoke article 62 to unilaterally terminate the maritime boundary agreement? The answer must be preceded by a clarification. Both parties know, at the time of conclusion of their maritime boundary agreement, that change of geography is inherent to this kind of agreements and can initially be expected; thus, stable geography is not the "circumstance" that forms the ground of their consent. Therefore, article 62 of the Vienna Convention cannot be invoked, and coastline changes will not affect the maritime boundary agreement. International Courts have not accepted the recognition of the right of unilateral termination, given the importance of the stability of the treaty regime.[73] Some States like Argentina and Chile have expressly rejected the application of this theory.[74] Moreover, the terms of article 62(2) of the Vienna Convention explicitly excludes boundary agreements; although it is still debatable by the doctrine if it also applies to maritime boundaries, it stipulated that "[a] fundamental change of circumstances may not be invoked as a ground for terminating or withdrawing from a treaty [...] if the treaty establishes a boundary [...]."[75] The ICJ in the *Aegean Sea* case implied that maritime boundaries fall within the Article 62(2) exception: "Whether it is a land frontier or a boundary line in the continental shelf that is in question, the process is essentially the same, and inevitably involves the same element of stability and permanence, and is subject to the rule excluding boundary agreements

[70] Vienna Convention on the Law of Treaties (1969), p. 331.

[71] See, Lisztwan (2012), pp. 154–200.

[72] Villiger (2009), p. 766.

[73] Gabčíkovo-Nagymaros Project (Hung./Slovk (1997)), p. 7.

[74] *Argentina*, United Nations Treaty Collection, Accessed on https://treaties.un.org/pages/ViewDetailsIII.aspx?src=TREATY&mtdsg_no=XXIII-1&chapter=23&Temp=mtdsg3&clang=_en#EndDec (Argentina reservation) "The Argentine Republic does not accept the idea that a fundamental change of circumstances which has occurred with regard to those existing at the time of the conclusion of a treaty, and which was not foreseen by the parties, may be invoked as a ground for terminating or withdrawing from the treaty [...]"; *Chile,* United Nations Treaty Collection, Accessed on https://treaties.un.org/doc/Publication/UNTS/Volume%201223/volume-1223-A-18232-English_French.pdf (Chile reservation) "The Republic of Chile declares its adherence to the general principle of the immutability of treaties [...] and [...]formulates a reservation relating to the provisions of article 62, paragraphs 1 and 3, of the Convention, which it considers inapplicable to Chile.".

[75] Article 62(2)(a) of the Vienna Convention on the Law of Treaties (1969).

from fundamental change of circumstances."[76] A state would therefore be unlikely to succeed in unilaterally terminating a maritime boundary treaty by invoking the principle of *rebus sic stantibus* under article 62 of the Vienna Convention.[77]

However, the stability of boundaries and the legal stability are defended by the legal reasoning[78] even though the approach of fixing baselines is criticized by being inappropriate and insufficient.[79] Avoiding instability and insecurity in maritime limits and boundaries would lead to fix them as they are situated at the time of agreement between States and at the time of deposit to the UN Secretary-General. Article 76 paragraph 9 of UNCLOS can be applied analogically to fix baselines and boundaries even though the technic of fixing them is not established yet.

4 Conclusion

Climate change and sea level rise create important challenges for the international community in general and for coastal States in particular. The consequences of climate change are affecting every State in the world regardless of their level of richness, size of the territory, the power of the State. It is clear that sea level rise will affect coastal States, in the first place and more than the others.

However, it creates also challenges for international law. The different consequences of sea level rise are showing that law has to take this issue into consideration to find a legal response of adaptation. In this sense, to figure out these solutions, States have to realize the effects of climate change and sea level rise, determine their maritime limits and boundaries, and implement the legal approach of stabilizing them.

This article examined how sea level rise is being a threat to baselines and base points from which the maritime limits and boundaries are determined and the "absence" of response of the provisions of the United Nations Convention for the Law of the Sea even though it is the "Constitution of the Oceans."[80] But this Constitution could not predict all the different situations. Facing sea level rise, baselines can either retreat or lose base points established on low-tide elevations or islands giving light to complex legal consequences and questions of whether to change the coordinates of baselines and therefore of the limits and boundaries and adapt them to the potential new ones or to freeze baselines allowing the stabilization of the limits and maritime boundary agreement.

Despite some critical views over the preservation of baseline approach, it remains the approach that can be applicable and efficient in a way that it responds to the purpose of law and agreement in the sense of being stable.

---

[76]ICJ (1978), para. 85.

[77]Lisztwan (2012), p. 192.

[78]Kamto (2009), p. 492.

[79]Lisztwan (2012), pp. 154–200.

[80]Tommy (1983).

However, it must be noted that the legal approach to cope with the consequences of sea level rise would not change these consequences; it is only an adaptation theory promoting the stabilization of international legal agreements. More efforts have to be engaged to give importance to public and private sectors in the protection of the seas and oceans to reduce implications of sea level rise and to encourage the important role of the contributions of international courts to the determination of some balancing of national, individual, and common interests.

References

Alexander L (1983) Baseline delimitations and maritime boundaries. V J Int Law 23:503–536

Bird E, Prescott JRV (1989) Rising global sea levels and national maritime claims. Marine Policy Rep 177:177–196

Caflisch L (1985) La délimitation des espaces entre États dont les côtes se font face ou sont adjacentes. In: Dupuy R-J, Vignes D (dir) Traité du nouveau droit de la mer. Paris, Economica, Brulyant, Bruxelles, pp 375–440

Calerton CM, Schofield C (2001) Developments in the technical determination of maritime space: charts, datums, baselines, maritime zones and limits. Maritime Briefing, No 3, I.B.R.U, University of Durham, United Kingdom

Caron D (1990) When law makes climate change worse: rethinking the law of baselines in light of a rising sea level. Ecol Law Q 17:621–653

Di Leva C, Morita S (2008) Maritime rights of coastal states and climate change: should states adapt to submerged boundaries? World Bank Law and Development Working Paper Series No. 5, p 17. http://siteresources.worldbank.org/INTLAWJUSTICE/Resources/L&D_number5.pdf

Division des affaires maritimes et du droit de la mer (2001) Manuel sur la délimitation des frontières maritimes. United Nations, New York

Division for Ocean Affairs and the Law of the Sea (2000) Handbook on the delimitation of maritime boundaries. United Nations, New York

Dorst L (2008) Hydrographic service of the Royal Netherlands Navy. The effects of changing baselines on the limits of the Netherlands in the North Sea. In: Fifth ABLOS conference: difficulties in implementing the provisions of UNCLOS, 15–17 October 2008. http://www.gmat.unsw.edu.au/ablos/ABLOS08Folder/Session6-Paper3-Dorst.pdf. Accessed 15 July 2016

Federal Act relating to the Sea, 8 Jan 1986. http://www.un.org/Depts/los/LEGISLATIONANDTREATIES/PDFFILES/MEX_1986_Act.pdf. Accessed 3 July 2016

Freestone D, Pethick J (1994) Sea level rise and maritime boundaries: international implications of impacts and responses. In: Blake GH (ed) Maritime boundaries. World boundaries, vol 5. - London, pp 73–90

Hestetune J (2010) The invading waters: climate change dispossession, state extinction, and international law. California Western School of Law

Houghton KJ, Vafeidis AT, Neumann B, Proelss A (2010) Maritime boundaries in a rising sea. Nat Geosci 3:813–816

ICJ, Gabčíkovo-Nagymaros Project (1997) (Hungary/Slovkia)

ICJ, Judgment (1951) Fisheries case (United Kingdom v. Norway)

ICJ, Judgment (1978) Aegean Sea Continental Shelf (Greece v. Turkey)

ICJ, Judgment (2001) Maritime Delimitation and Territorial Question between Qatar and Bahrain (Qatar v. Bahrain)

ICJ, Judgment (2007) Case Concerning Territorial And Maritime Dispute Between Nicaragua and Honduras in The Caribbean Sea (Nicaragua v. Honduras)

ICJ, Judgment (2009) Maritime Delimitation in the Black Sea (Romania v. Ukraine)

International Law Association (2008) Committee on baselines under the law of the sea. Internal Discussion Document

Kamto M (2009) Sur quelques questions techniques liées à la détermination du tracé d'une frontière maritime délimitée. In: Vignes D, Raigón RC, Cataldi G (dir) L'évolution et l'État actuel du droit international de la mer: Mélanges de droit de la mer offerts. Bruxelles, Bruylant, pp 481–505
Kapoor DC, Kerr AJ (1986) A guide to maritime boundary delimitation. Carswell, Toronto Law No 30 of 2nd May 1977. http://www.un.org/Depts/los/LEGISLATIONANDTREATIES/PDFFILES/JPN_1977_Law.pdf. Accessed 3 Mar 2016
Lisztwan J (2012) Stability of maritime boundary agreements. Yale J Int Law 37(1):154–200
Organisation Hydrographique Internationale (1998) Dictionnaire hydrographique. Monaco
PCA Award (2006) Arbitration between Barbados and the Republic of Trinidad and Tobago (Barbados/Trinidad and Tobago)
Prescott JRV, Schofield CH (2000) The maritime political boundaries of the world. Martinus Nijhoff, Leiden
Rayfuse R (2009) W(h)ither Tuvalu? International law and disappearing states. In: Symposium International sur les îles et les océans (revised version of the presented document), Paper presented international symposium of Islands and Oceans, Tokyo, Japan, 22–23 January 2009, pp 70–77
Rayfuse R (2010) International law and disappearing states: utilising maritime entitlements to overcome the statehood dilemma, Univ. N.S.W. Faculty of Law Research Series, Paper 52
Reed MW (2000) Shore and sea boundaries. The development of international maritime boundary principles through United States practice, vol 3, p 185
Roach JA, Smith RW (1996) United States responses to excessive maritime claims, 2nd edn
Schofield C (2009a) Against a rising tide: ambulatory baselines and shifting maritime limits. In: International symposium on Islands and Oceans, Tokyo, p 70
Schofield CH (2009b) Against a rising tide: ambulatory baselines and shifting maritime limits in the face of sea level rise. Paper presented at the proceedings of the international symposium of Islands and Oceans, Terashima, Tokyo, Ocean Policy Research Foundation, pp 70–77
Song Y-H (2009) Okinotorishima: a "Rock" or an "Island"?: recent maritime boundary controversy between Japan and Taiwan/China. In Hong S-Y, Van Dyke JM (eds) Maritime boundary disputes, settlement processes and the law of the sea, Publications on ocean development series. Martinus Nijhoff, Leiden, pp 145–176
Soons AHA (1990) The effects of a rising sea level on maritime limits and boundaries. Netherlands Int Law Rev 37(2):207–232
Tommy KTB (1983) Third United Nations Conference on the Law of the Sea, A Constitution for the Oceans (Dec. 6 & 11, 1982), The Law of the Sea: Official Text of The United Nations Convention on the Law of the Sea with Annexes and Index, at xxxiii, xxxiii
United Nations (2014) Handbook on the delimitation of maritime boundaries
United Nations Office for Oceans Affairs and the Law of the Sea (1989) Vienna Convention on the law of treaties, 23 of May 1969, 1155 U.N.T.S. 331
Villiger ME (ed) (2009) Commentary on The 1969 Vienna Convention on the Law of Treaties. Brill/Nijhoff

The Common Fisheries Policy: A Difficult Compromise Between Relative Stability and the Discard Ban

José Manuel Sobrino and Marta Sobrido

1 Introduction

The fishing opportunities of the European Union (EU)[1] vary annually, mainly in response to biological considerations. To ensure the greatest possible stability, for over 30 years the allocation of EU fishing opportunities to its Member States (MSs) has been based on a predictable share of the stocks for each MS, known as *relative stability* (RS).

In December 2013, Regulation (EU) No 1380/2013 of the European Parliament and the Council on the Common Fisheries Policy was adopted.[2] The new regulation keeps RS as a criterion for allocating fishing opportunities to MS and also bans *discards*, which are catches returned to the sea.[3] Implementing the discard ban is a major challenge for mixed fisheries in which more than one species is present and where different species are likely to be caught in the same fishing operation,[4] e.g., cod, haddock, whiting, and saithe in Northwest Atlantic waters.

[1]For easier reading we shall refer to European Union (EU). But for EU we also mean the European Economic Community (EEC, 1958–1993) and the European Community (EC, 1993–2009). The EEC came into being in 1958. With the entry into force of the Treaty on European Union in November 1993, the EEC became the EC. And with the introduction of the Lisbon Treaty in December 2009, the EU replaced and succeeded the EC.

[2]Regulation (EU) No 1380/2013 of the European Parliament and of the Council of 11 December 2013 on the Common Fisheries Policy, amending Council Regulations (EC) No 1954/2003 and (EC) No 1224/2009 and repealing Council Regulations (EC) No 2371/2002 and (EC) No 639/2004 and Council Decision 2004/585/EC. OJ 2013 L 354/22–61.

[3]Definition of *discards*: article 4.1.10; ibid.

[4]Definition of *mixed fisheries*: article 4.1.36; ibid.

J.M. Sobrino • M. Sobrido (✉)
University Institute of European Studies "Salvador de Madariaga", University of A Coruña, Campus de Elviña, s/n, A Coruña 15071, Spain
e-mail: j.sobrino@udc.es; marta.sobrido@udc.es

G. Andreone (ed.), *The Future of the Law of the Sea*,
DOI 10.1007/978-3-319-51274-7_2

The aim of this paper is to examine the compromise between RS and the discard ban.

2 Origin and Enshrining of Relative Stability

The origin and evolution of the Common Fisheries Policy (CFP) is closely linked to the evolution of international fisheries law (in particular, the creation of the exclusive economic zone)[5] and the enlargement of the EU (mainly the first and third extension, which included the major European fishing states). Knowledge of both processes is also essential for understanding the rationale behind and the resulting form of RS.

The first EU fisheries legislation, which among other things established the "equal access principle," was adopted at the beginning of the 1970s as a preliminary step for the negotiations that would result in the first enlargement of the EU,[6] involving the major fishing states of the United Kingdom (UK), Ireland, Denmark (Danish accession included Greenland but not the Faroe Islands),[7] and Norway, although the latter did not finally join. RS emerges shortly after, in connection with the outcome of the accession negotiations that enabled that first enlargement. But it is also closely linked to two other factors: the establishment of the exclusive economic zone and the third enlargement of the EU. In relation to the first factor we must recall how, in reaction to certain non-EU countries asserting jurisdiction

---

[5]See Rey Aneiros (2001) and Sobrino Heredia (2003).

[6]The EU has had competence to adopt legislation on fisheries from the outset (EEC Treaty, 1957), but fishing was not a priority for MSs then (France, Germany, Italy, Belgium, Netherlands and Luxembourg). Moreover, fishing mostly took place in what were then waters of the high seas. The first step was taken in 1966, with a report from the Commission. In 1968, three draft regulations appeared that eventually led to the adoption in 1970 of two regulations—Council Regulations (EEC) Nos 2141/70 and 2142/70—which, among other things, introduced the "equal access principle". In accordance with this principle the fishing regulation applied by each MS in its maritime waters—waters under its sovereignty or within its jurisdiction—must not lead to differences in the treatment of other MSs. Churchill and Owen (2010), pp. 4–6, state that MSs wanted an *acquis* for fisheries before starting negotiations with the four candidate states and, in fact, the two regulations cited were adopted the day before formal negotiations started with these states.

[7]Treaty of Accession of Denmark, Ireland and the United Kingdom (1972); OJ 1972 L 73; MSs of the EU since 1 January 1973. Both Greenland and the Faroe Islands are part of Denmark, but when Denmark joined the EU in 1973, the Faroe Islands decided to remain outside precisely because of fishing: "the Faroese have not found it their interest to become subject to the Common Fisheries Policy" (The Government of the Faroe Islands, http://www.government.fo/foreign-relations/missions-of-the-faroe-islands-abroad/the-mission-of-the-faroes-to-the-european-union/the-faroe-islands-and-the-european-union/). The Faroe Islands is like a third country with respect to the EU. As for Greenland, it joined the EU in 1973 as part of Denmark but withdrew from it in 1985 as result of a referendum held in 1982. Since then, Greenland is part of the OCT (Overseas Countries and Territories; articles 198–204 TFEU).

over their waters out to 200 miles, the EU adopted very similar fishing areas from 1977.[8] Moreover, the EU took responsibility for managing fishing rights in these new waters and also the fishing rights of EU vessels in the waters of third states. With regard to the second factor, the negotiations concerning what would become the third enlargement of the EU involved states with major fishing interests such as Spain and Portugal.[9] After the first enlargement, the UK and Ireland were by far the largest EU fishing powers,[10] a position that would later be occupied by Spain, although it, like Portugal, did not have great resources in its waters. Against this backdrop, most MSs wanted to have a European system of fisheries management established before starting negotiations with these two countries.[11]

2.1 Origin of Relative Stability

In November 1976, the Council adopted the "Hague Resolution," which deals with the external aspects (affirmation of the competence of the EU to negotiate with third countries)[12] as well as internal aspects (affirmation of competence of the EU for the

[8]Joint action by MSs concerning the waters of the North Sea and North Atlantic. For the evolution of the EU position, and that of its members, on the establishment of the exclusive economic zone within the framework of the III United Nations Conference on the Law of the Sea (1973–1982), see Treves (1976).

[9]Portugal submitted a formal application for membership in the EU on March 28, 1977. Spain did the same 4 months later, on 28 July 1977. With regard to Spain, the letters exchanged in July 1977 between the Spanish Prime Minister and the Chairman of the European Communities on Spain's request to start negotiating its integration can be found in *Revista de Instituciones Europeas* 4 (1977) 1031–1036. The negotiations culminated in 1985 in the Treaty of Accession of Spain and Portugal (1985); OJ 1985 L 302. Spain and Portugal have been MSs of the EU since 1 January 1986.

[10]In 1981, when the EU was formed by 10 countries, almost 90% of EU resources were captured by the UK (64%) and Ireland (25%); see Lostado i Bojo (1985), p. 41.

[11]As stated by Churchill and Owen (2010), pp. 11–14, Spain and Portugal had large fleets (the Spanish fleet was nearly ¾ the size of the entire EU fleet, at the time composed by ten MSs), and the waters under the jurisdiction of these two states did not have many resources because their continental shelves—not in the legal but the geological sense—are narrow, and waters located on the continental shelves are the richest in fishery resources. The Commission then spoke of an "imbalance in the fisheries sector between the tonnage of the Spanish fleet and the fishing zones available to Spain"; European Commission, "Opinion on Spain's application for membership", sent to the Council by the Commission on 29 November 1978; available at *Bulletin of the European Communities*, Supplement 9/78 (1978) 16. On the other hand, Portugal and Spain had sufficient fisheries access agreements with third countries with which the EC had no agreements; see Sobrino Heredia (1990).

[12]Council Resolution of 3 November 1976 on certain external aspects of the creation of a 200-mile fishing zone in the Community with effect from 1 January 1977; OJ 1981 C 105/1.

adoption of conservation measures in EU waters)[13] that result from the creation, in 1977, of the 200-mile fishing zone. In this resolution, which was published late[14] and incomplete,[15] the Council states that the CFP must take into account the vital needs of regions where local populations depend greatly on fishing and related industries. Known as the "Hague Preferences," this provision is specifically directed towards Ireland and northern parts of the UK.[16] Despite its open wording, these "preferences" are a recognition that must be understood in the context of the negotiations with the UK and Ireland. Therefore, they cannot be automatically extended to other areas with similar needs; it became clear during the subsequent accession of new states.

The full implementation of the Hague Resolution via a specific distribution among MSs required years of difficult negotiations.[17] Finally, in 1983, the Council adopted a regulation[18] and, based on it, made the first allocation among the MSs[19]

---

[13]Annex VI to the Hague Resolution of 3 November 1976; text reproduced in Opinion of Advocate General Reischl delivered on 11 September 1979, *France v UK*, 141/78, ECLI:EU:C:1979:202, p. 2945. Regarding EU fishing regime and third states, see Del Vecchio (1982) and Meseguer Sánchez (1981).

[14]Not published until 1981 (OJ 1981 C 105/1). In 2001, the Advocate General Alber drew attention to its late and incomplete publication; see Opinion of Advocate General Alber delivered on 13 November 2001, *Spain v Council*, joined cases C-61/96, C-132/97, C-45/98, C-27/99, C-81/00 and C-22/01, ECLI:EU:C:2001:602, footnote 7.

[15]The Hague Resolution had eight annexes. However, in the Official Journal of the European Union (OJ) only Annex I, on the external aspects, was published. In 1998 the Court of Justice of the European Union (CJEU) drew attention to this incompleteness. See Judgment of the Court of 19 February 1998, *NIFPO and Northern Ireland Fishermen's Federation v Department of Agriculture for Northern Ireland*, C-4/96, ECLI:EU:C:1998:67, paragraph 5. The Advocate General in this same case noted that the explanation given by the Council to justify this defect was that some of its annexes contained confidential material regarding instructions given by the Council to the Commission concerning future negotiations by the EU with non-member countries and international organisations (see Opinion of Advocate General La Pergola issued on 30 September 1997, case C-4/96, cit., ECLI:EU:C:1997:444, point 7). In that judgment, the CJEU reproduces the text of Annex VII (see the judgment in case C-4/96, cit., paragraph 4). And many years before, the Advocate General Reischl had reproduced the text of Annex VI (see Opinion in case C-141/78, cit., p. 2945).

[16]"(. . ..) the expression 'northern parts of the United Kingdom' for the purposes of the Hague Preferences comprises Scotland, Northern Ireland, the Isle of Man, and that part of England between the ports of Bridling- ton and Berwick"; Opinion in case C-4/96, cit., footnote 9.

[17]See Holden (1985), point 6 "Allocation Between Member States of the EEC". Holden was at that moment Directorate General for Fisheries Commission of the European Communities.

[18]Council Regulation (EEC) No 170/83 of 25 January 1983 establishing a Community system for the conservation and management of fishery resources; OJ 1983 L 24/1–13. This regulation establishes that the volume of the catches available to the EU must be shared across the MSs in a manner which assures each MS *relative stability* in fishing activities for each of the stocks considered (art. 4.1) with the notion of relative stability understood in accordance with the Hague Preferences (recitals 6–7).

[19]Council Regulation (EEC) No 172/83 of 25 January 1983 fixing for certain fish stocks and groups of fish stocks occurring in the Community's fishing zone, total allowable catches for 1982, the share of these catches available to the Community, the allocation of that share between the Member States and the conditions under which the total allowable catches may be fished; OJ 1983 L 24/30–67.

stating that three criteria were considered for doing this: the traditional fishing activities, the specific needs of areas particularly dependent on fishing and its dependent industries (the Hague Preferences), and the loss of fishing potential in the waters of third countries.[20] With regard to the Hague Preferences for Ireland, Northern Britain, and Greenland (the latter until it left the EU in 1985), Ireland, the UK, and Denmark were assured that their fishing opportunities for certain species would reach a minimum threshold, on the understanding that this threshold would be in absolute terms and not a percentage. As noted above, these preferences were not the result of an objective study of the special needs of European fishing territories but the result of a negotiation.[21]

2.2 *Enshrining Relative Stability*

This allocation formula laid down in 1983[22] continues to take place; it appeared in subsequent reforms (1992, 2002, 2013)[23] and is in the current regulation, despite the proposed amendment put forward by the European Commission's Green Book on the reform of the CFP (2009).[24]

---

[20]Ibid. recital 3. These criteria, which were set by the Council in 1980 without further details on their application or how much weight was to be placed on each criterion (Council declaration of 30 May 1980 on the common fisheries policy, OJ 1980 C 158/2) were interpreted by the Commission, which presented a mathematical model that took into account the three criteria and served as a starting point for the allocation of quotas, stock by stock. As Holden (1985) states, the Commission interpreted what should be understood by traditional fishing activities ("average catches in the period 1973–78, less industrial by-catches beyond permitted limits and human consumption species caught directly for reduction to meal and oil"), the specific needs of areas particularly dependent on fishing and its dependent industries ("For Greenland: a major share of the catch possibilities in Greenland waters; For Ireland: doubling of the 1975 catches by 1978; For north Britain: maintaining a minimum catch possibility equal to the landings in 1975 by vessels less than 24 m long at ports in northern Ireland, Scotland and along the east coast of England as far south as Bridlington"), and the loss of fishing potential in the waters of third countries ("the difference between what a Member State is actually allowed to catch and what it would have caught if there had been no extension to 200-mile limits. What it would have caught is calculated as its average percentage share of the particular stock for the period 1973–76 multiplied by the TAC, if known, or the estimated TAC, otherwise").

[21]Penas Lado (2016), p. 28, points out that, in exchange for this guarantee, the UK and Ireland agreed to lower TACs—understood as the percentage applied in the context of the RS—than they wanted.

[22]Allocation formula laid down in Regulation 172/83, cit.

[23]Council Regulation (EEC) No 3760/92 of 20 December 1992 establishing a Community system for fisheries and aquaculture; OJ 1992 L 389/1–14; see recitals 12–14. Council Regulation (EC) No 2371/2002 of 20 December 2002 on the conservation and sustainable exploitation of fisheries resources under the Common Fisheries Policy; OJ 2002 L 358/59–80; see recitals 16–18. Regulation 1380/2013, cit., see recitals 35–37.

[24]European Commission, *Green Paper of 22 April 2009—Reform of the Common Fisheries Policy*, COM(2009) 163 final; Green Paper 2009, hereinafter. See point 5.3.

However, it has been completed as a result of new accessions and distribution of more species. On the one hand, since 1983 new states have acceded to the EU, including major fishing nations such as Spain and Portugal. When joining the EU, each new state accepts the *acquis communautaire* and therefore also RS, not as an underlying principle to ensure a distribution that tends to provide stability but as the specific percentages determined in 1983.[25] The incorporation of a new state does not entitle that state to require a review of the percentages set when it was not part of the EU. With regard to these species, the new MS only gets what negotiates in its accession treaty.[26] On the other hand, in the distribution of 1983, not all species from all areas were included, and over time it has been necessary to offer new fishing opportunities. This decision is made by the Council, taking into account the "interests" of the states,[27] a notion that does not necessarily take into account, if any, the historical catches of the MSs for that stock zone.[28]

In brief, the current allocation percentages are established as follows: first, by taking into account the percentages established in 1983, if necessary with the application of the Hague Preferences,[29] e.g., the allocation of cod and whiting quotas in ICES division VIIa (Irish Sea);[30] second, based on the percentages set forth in the Acts of Accession of states that adhered after 1983, e.g., the allocation of anchovy quotas in ICES division VIII (Bay of Biscay);[31] and third, in accordance with the

[25]As stated expressly by the CJEU. Judgment of the Court of 30 March 2006, *Spain v Council*, C-87/03 and C-100/03, ECLI:EU:C:2006:207, paragraphs 28–29.

[26]For new states, the application of RS is not based on the three criteria taken into account in the regulations of 1983; rather, it is only based on historical catches. Penas Lado (2016), p. 27.

[27]Art. 16.1 of Regulation 1380/2013: "...The interests of each Member State shall be taken into account when new fishing opportunities are allocated". And in the same sense, formerly, the regulations from 1992 and 2002; see article 8.4.iii) of Regulation 3760/92, and article 20.2 of Regulation 2371/2002.

[28]As stated expressly by the CJEU, the allocation of new fishing opportunities among MSs requires the assessment of a complex economic situation for which the Council enjoys a wide discretionary power (Judgments in cases C-87/03 and C-100/03, cit., paragraph 38). That is not always harmonious process.

[29]Penas Lado (2016), p. 348, points out that the Hague Preferences have been applied continuously although not always harmoniously owing to the resistance of negatively affected MSs. This author explains that its application is not automatic but established within a negotiating framework, and often only 50% is applied (i.e., the average between what the UK and/or Ireland would receive by applying RS and what they would obtain under the Hague Preferences).

[30]The distribution of these quotas and the application of the Hague Preferences were specifically addressed in case C-4/96. Judgment of case C-4/96, cit., paragraph 18: "Under that mechanism, Ireland and the United Kingdom are granted annual quotas calculated on the basis of the mid-point between the notional quotas resulting from the application of the 1983 allocation keys alone and the notional quotas corresponding to their Hague Preferences".

[31]The Act concerning the conditions of accession of Spain (cit., arts. 161.1.f and 162) established that 90% of the quota for anchovy in the Bay of Biscay correspond to Spain and the remaining 10% to France. For example, the TAC of 22,500 tonnes for the EU in the Gulf of Biscay in 2016 is distributed as follows: 22,500 tonnes for Spain and 2500 for France; Council Regulation (EU) 2016/72 of 22 January 2016, OJ 2016 L 22/42.

percentages for new fishing opportunities, e.g., the angler fish quota in ICES division IV (Norwegian waters), which was allocated for the first time in 2005.[32]

The resulting percentages of these three processes constitute the RS distribution key.

3 The Nature of Relative Stability

RS is a criterion of distribution that, as its name suggests, aims to ensure the greatest possible stability. Given the constraints surrounding the setting of quotas, it is a *relative* stability as opposed to an *absolute* stability. Furthermore, its implementation as a distribution key does not guarantee the relative stability of the situation prior to its implementation but a *future* relative stability (1).

The mechanisms for allowing flexibility in this distribution key exclude fishing communities as direct beneficiaries, and RS is established as a guarantee of stability only for states (2). RS is a mandatory distribution key, albeit one susceptible to modification (3).

3.1 Future Predictability

Due to the difficult situation in the fisheries sector (structural problems of great socioeconomic impact), from the beginning of the CFP it was decided that this area would not be subject to the rules governing the single market in the EU for economic activities.[33]

To ensure the maximum stability, it was agreed that the EU itself would distribute the fishing opportunities among states (leaving the distribution of national quotas among fishing operators to each state, according to the criteria deemed appropriate). It was also decided that such a distribution would be the same every year, not in absolute terms but at least in relative terms. The distribution cannot be the same in absolute terms because the EU fishing opportunities (in EU waters, in waters subject to the sovereignty or jurisdiction of third countries, and in

[32]This allocation took place with the Council Regulation (EC) No 27/2005 of 22 December 2004, OJ 2005 L 12/1. Its character of "new fishing opportunity" was confirmed by the CJEU in Case C 141/05; 5; Judgment of the Court of 8 November 2007, *Spain v Council*, C-141/05, ECLI:EU: C:2007:653, paragraph 90.

[33]See European Commission, *Green Paper on the future of the Common Fisheries Policy*, COM (2001) 135 final; Green Paper 2001, hereinafter; section 5.1.4.1. This exception to the single market has received great criticism from many quarters (authors, authorities, etc.). For instance, by a region highly dependent on fishing, Galicia (Spain); see Xunta de Galicia, *Declaration by the Autonomous Government (Xunta) of Galicia on the Principles of the European Union in the Future Common Fisheries Policy* (Santiago de Compostela: Xunta de Galicia, 2002).

international waters) are not the same every year. It varies depending on the biological status of fisheries and on the outcome of agreements with third parties.

Furthermore, its implementation as a distribution key does not guarantee the relative stability of the situation prior to its implementation—not for those who were MSs in 1983 and not for new states that have joined since then. What is guaranteed is the stability of the original terms agreed upon, which amounts to a future stability.

In this respect, the Hague Preferences deserve special mention, although some states—such as Belgium, Denmark, Germany, the Netherlands, and France—are manifestly against them. The EU regulation refers to the Hague Preferences as an element that integrates RS. It was like this at the beginning (1983)[34] and has continued to be so in subsequent regulations (1992, 2002, 2013).[35] The Hague Preferences guarantee preferential treatment to Ireland and the UK for certain fisheries. When the application of RS would result in total allowable catches (TACs) for the UK and Ireland in certain fisheries below the minimum originally agreed upon, these states may request the application of the Hague Preferences. This system, which is applied by the Council, results in higher TACs for Ireland and the UK than they would otherwise receive. Inevitably, this occurs at the expense of other states that have a share in the fisheries in question.[36]

3.2 *The States as Beneficiaries of Relative Stability*

RS is a guarantee, a guarantee of stability. But it is a guarantee for states,[37] not a guarantee for the economic actors involved in fishing. For this reason, the situation of the latter is irrelevant from a legal point of view.

Despite being presented as a criterion for safeguarding the interests of the fishing industry and fishing communities,[38] the fact is, as the European Commission warned in 2009, RS no longer provides a guarantee that fishing rights remain

[34]Regulation 170/83, cit., recitals 35–37.

[35]Regulation 3760/92, cit., recitals 12–14. Regulation 2371/2002, cit., recitals 16–18. Regulation 1380/2013, cit., recitals 35–37.

[36]See Parliamentary questions (European Parliament); Answer given by the Commission to the written question E-0139/08, 13 March 2008.

[37]"The principle of relative stability has, since 1983, provided assurances to the Member States with regard to the share of quotas, thus avoiding annual repetitions of a political debate on the allocation key, which would have made the decision-making on TACs even more complicated..." Green Paper 2001, cit.; section 5.1.4.1.

[38]According to the current regulation: "35. In view of the precarious economic state of the fishing industry and the dependence of certain coastal communities on fishing, it is necessary to ensure the relative stability of fishing activities by allocating fishing opportunities among Member States, based on a predictable share of the stocks for each Member State. 36. Such relative stability of fishing activities (...) should safeguard and take full account of the particular needs of regions where local communities are especially dependent on fisheries and related activities (...)"; Regulation 1380/2013, cit., recitals 35–36.

with their fishing communities. RS itself has promoted a series of practices that has led to RS no longer providing this guarantee. This, along with other factors, has contributed to a current discrepancy between the quotas allocated to MSs and the actual needs and uses of their fleets.[39] However, this amounts to a merely political argument, not a legal one.

This argument would stand as justification for changing the law (through the appropriate legislative procedure). Indeed, in the debate prior to the last reform, the Commission used it to propose amendments.[40] However, as these amendments did not prosper, this argument does not seem to be considered valid legal grounds for bringing an action before the Court of Justice of the EU (CJEU). In this regard, we must also remember that, in a case of a violation of RS, the CJEU does not recognize the right of fishing operators to be compensated for any damage that such a violation may cause to them.[41]

3.3 Allocation Formula Contained in the Derived Legislation

The CJEU has had to rule on RS[42] and in doing so uses the term "principle,"[43] which is also used by many authors. However, when identifying its contents, the CJEU qualifies RS as a "fixed percentage" and as an "allocation formula originally laid down [that] will continue to apply as long as an amending regulation has not been adopted."[44]

While we do not intend here to examine, or reflect upon, the principles of EU law,[45] we must keep in mind the following: like international and national laws, EU law has an unwritten component consisting of general principles that take precedence not over primary law but over secondary law. Some of these principles are rooted in principles of international law or national law (e.g., the presumption of

---

[39]Green Paper 2009, cit., section 5.3.

[40]Ibid.

[41]"88. (...) the principle of relative stability concerns only relations between Member States, it cannot confer individual rights upon private parties, the infringement of which would give rise to a right to compensation (...). 89. (...) The principle of relative stability does not therefore confer on fishermen any guarantee that they can catch a fixed quantity of fish, since the requirement of relative stability must be understood as meaning merely maintenance of a right to a fixed percentage for each Member State in that distribution". Judgment of the Court of First Instance of 19 October 2005, *Cofradía de pescadores de "San Pedro" de Bermeo and Others v Council*, T-415/03; ECLI:EU:T:2005:365; paragraphs 88–89.

[42]Franckx (2012) clearly identifies RS as one of the main of the many fishery issues that the CJEU has had to deal with. For more on this case law, see also: Le Bihan (2003), Sobrido-Prieto (2013), and Sobrino Heredia and Rey Aneiros (1997).

[43]E.g., Judgment in case C-141/05, cit., paragraphs 11, 44, 48, 62, 85, 87.

[44]Ibid., paragraph 86.

[45]See Sobrino Heredia (2009).

innocence),[46] while others are specific principles of EU law (e.g., the principle of institutional balance).[47] These principles have been articulated by the CJEU through an *interpretive* analysis extracted from written EU law and also international law and the national law of the MSs. The CJEU does not create these principles but extracts them from these sources.

Alongside these general principles, each area of the EU also has its peculiarities. In fisheries management, for example, certain principles such as good governance[48] and the precautionary principle[49] are particularly important. RS, however, is not presented as a principle. The legislative acts that give substance to RS qualify it as a "notion"[50] or, in its current regulation, as a "concept."[51]

What is noteworthy is its specific and binding nature and the possibilities for change, flexibility, and even repeal. RS is not a guiding principle to ensure a distribution that tends to provide stability but specific percentages imposed as binding that form part of the *acquis communautaire*. Even for new fishing opportunities, the decision made by the Council is not necessarily based on stability.[52]

RS is a distribution key negotiated between MSs that is laid down in the secondary legislation of the EU.

The implementation of RS as percentages is the result of negotiation between the MSs. Perhaps the best example of this negotiation is the Hague Preferences. Although presented as a safeguard for the benefit of the most vulnerable regions, in practice they are not applied in benefit of any region that may be classified as such (regions where the local populations are especially dependent on fisheries and related activities) but only in the case of Ireland and the UK. This negotiating

[46]E.g. Judgment of the Court of 21 January 2016, *"Eturas" UAB and Others v Lietuvos Respublikos konkurencijos taryba*, C-74/14, ECLI:EU:C:2016:42, paragraph 38.

[47]E.g. Judgment of the Court of 6 October 2015, *Council v Commission*, C-73/14, ECLI:EU:C:2015:663, paragraph 61.

[48]"It is important for the management of the CFP to be guided by principles of good governance. Those principles include decision-making based on best available scientific advice, broad stakeholder involvement and a long-term perspective. The successful management of the CFP also depends on a clear definition of responsibilities at Union, regional, national and local levels and on the mutual compatibility of the measures taken and their consistency with other Union policies", Regulation 1380/2013, cit., recital 14.

[49]"Sustainable exploitation of marine biological resources should be based on the precautionary approach, which derives from the precautionary principle referred to in the first subparagraph of Article 191(2) of the Treaty, taking into account available scientific data", Regulation (EU) No 1380/2013, cit., recital 10. See Proelss and Houghton (2012). And for a general reference about the precautionary principle in the EU see: European Commission, *Communication from the Commission on the precautionary principle*, COM (2000) 0001.

[50]Regulation 170/83, cit., recital 7; Regulation 3760/92 cit., recital 14; Regulation 2371/2002, cit., recital 37.

[51]Regulation 1380/2013, cit., recital 37.

[52]This was expressly stated by the CJEU. Judgment in case C-141/05, cit., paragraph 87.

dimension is not hidden by the benefited parties. To the contrary, both Ireland[53] and the UK[54] appeal to it, and the MSs most strongly opposing the Hague Preferences because they are directly affected do not allege that the beneficiary regions are not dependent on fisheries or that there are other dependent regions that are left out. Rather, they argue that the Hague Preferences have altered the percentages negotiated in 1983.[55]

Although RS is a distribution key negotiated between the MSs of the EU, it does not operate as an agreement between parties imposed as mandatory according to international law but separate from EU law. The binding nature of RS does not derive from the *pacta sunt servanda* principle. Instead, it finds its legal basis in the secondary EU legislation, currently Regulation 1380/2013. So, as stated by the CJEU, RS will continue to apply until an amending regulation is adopted.[56] RS is an allocation that can be modified or repealed by an act of legislation. In this regard, as discussed above, during the last reform that culminated in Regulation 1380/2013, the Commission proposed amendments that finally did not come to fruition.

4 The Discard Ban

One of the negative effects of RS is the discards, which are catches returned to the sea (1). The new regulation imposes a progressive ban on discards but maintains RS as the distribution key for national quotas (2). The coexistence of both RS and the discard ban is a challenge for the first, not in its formal dimension as a quota allocation but in its substantive content on the utilization of the allocated quotas (3).

[53]During the debate prior to the latest reform of the CFP in 2010, Ireland stated that the Hague Preferences in Annex VII of the Hague Resolution were the counterpart to a concession that Ireland had made: access to their exclusive economic zone. It even claimed that it was "not possible to re-open or diminish the principles set out in Annex VII of the Hague Resolution without re-opening the whole issue of access within the 200 mile Exclusive Fisheries Zone". Government of Ireland, *Ireland's response to the Commission's Green Paper on the Reform of the Common Fisheries Policy* (Department of Agriculture, Fisheries and Food, 2010) 16; http://ec.europa.eu/fisheries/reform/docs/ireland_en.pdf.

[54]In the "Transcript of Minister for Europe David Lidington comments to media on the prospects of EU membership for a newly independent Scotland" (extracts from the interviews given to ITV Borders and BBC Scotland on 17 January 2014) published on the British government's website, Lidington warns: "if we look at the UK in the EU, we have got a good deal for Scotland. In terms of fisheries, the Shetland box, the Hague preferences, that wouldn't be guaranteed if Scotland walked away from the UK"; https://www.gov.uk/government/news/prospects-of-eu-membership-for-a-newly-independent-scotland.

[55]These MSs have stated this on various occasions. For example, see the Statements published in the minutes of the Council, January 2008: "Statement by Belgium, Denmark, Germany and the Netherlands concerning the Hague Preferences" (p. 13) and "Statement by the French delegation...2. Implementation of the Hague Preferences" (p. 14); available at http://data.consilium.europa.eu/doc/document/ST-12272-2008-INIT/en/pdf.

[56]E.g., Judgment in case C-141/05, cit., paragraph 86.

4.1 Relative Stability and Discards

The impact of discarding varies by species: some have low survivability when discarded (e.g., cod) whereas others may have higher survival rates (e.g., crustaceans).[57] And discards have positive ecological effects to the extent that discarded fish is food for a range of scavenging species.[58] However, discards are generally a negative practice entailing a massive waste of resources (human and animal food, potential income). In addition, it is probably the single most important reason for the poor quality of fisheries-dependent data that could be used to improve stock assessments.[59]

The nonrecording of discards makes it difficult to know the exact number of discards at a global or regional level, although we know it has reached worrying levels.[60] In the EU, it varies from area to area, but it can be very high.[61]

The practice of discarding occurs for various reasons, which can be divided into two categories.[62] First, discards occur for commercial reasons: wrong sex (where gender is important from a processing and marketing point of view); damaged fish, fish incompatible with the rest of the catch (slime or abrasion could cause damage to target species); inedible fish, rapidly spoiling fish; lack of space on board and high grading (take the best and leave the rest, often related to size). Second, discards also occur due to legal prohibitions: prohibited size, prohibited season, prohibited

---

[57] Andersen et al. (2014), p. 2.

[58] Food subsidies to wildlife as a result of human activity have an important effect on terrestrial and aquatic ecosystems, and intentional discarding at sea is recognized as one of the major global subsidies. Heath et al. (2014).

[59] Wilson and Jacobsen (2009), p. 6.

[60] In 1994, a study by the FAO—Alverson et al. (1994)—estimated that between 17.9 and 39.5 million tonnes (average, 27.0 million tonnes) of fish were discarded each year in commercial fisheries. In 2005 other study by the FAO—Kelleher (2005)—estimated that the weighted discard rate was 8% (proportion of the catch discarded). Based on this discard rate, the average yearly discards for the 1992–2001 period was estimated to be 7.3 million tonnes. The author warns that because of the different method used in this estimate, it was not directly comparable with the previous estimates of 27 million tonnes. In any case, the author states there was evidence to suggest a substantial reduction in discards in recent years. In geographical terms, the Northeast Atlantic (1.4 million tonnes), the Northwest Pacific (1.3 million tonnes) and the Western Central Atlantic (0.8 million tonnes) generated the highest discards.

[61] In 2011, the Commission created a compilation and review of information on the level of discarding in different fisheries within the EU. The Commission distinguished three categories: high discard fisheries (>40%), medium discard fisheries (15–39%), low discard (<15%). In Table 2 (pp. 11–22), data per zone are shown: Region covered/Target Species/Discard rate/Main discarded species/Reason for discarding. The region with the highest percentage of discards was the Southern North Sea, with a discard rate of 71–95%. Although within a region there may be very different discard practices according to species, for example: North Sea IV (English and Welsh fleets) had a general discard rate of 31%, but 89% for Dub. European Commission (2011) Common Fisheries Policy Impact Assessment—EU Discards Annex. See also: Uhlmann et al. (2013).

[62] We take as a reference, although simplifying some reasons and grouping them into two categories, the work of Lucas (1997).

gear (a quota may be given for the capture of a particular species by a particular type of gear), prohibited fishing ground (closed for the capture of one species but open for others), prohibited species (no quota for the particular operator), and quotas reached (often the reason for high grading). This second category of discard occurs because the capture cannot be legally brought to market.

In the EU, prior to the current Regulation 1380/2013, it was not prohibited to discard fish, and discarded fish did not count towards an operator's quota. In this context, the national quota allocation system based on RS contributed to discarding, not because the EU's total TAC was used up but because the share for a particular MS was. While the fleet of one MS may not have used up its quota for a species, another fleet may have done so or may have had no quota at all, in which case this latter fleet would be forced to discard catches of this species.[63] If all quota systems generate discards, the EU system multiplies discards as each national quota generates its own discarding constraints.

This situation is in the process of being eradicated as Regulation 1380/2013 imposes a progressive ban on discards. One of the main reasons for this change was the pressure of public opinion, from both inside and outside of the EU, which had been sparked by striking images in the media.[64]

4.2 The Landing Obligation

In accordance with Regulation 1380/2013, the discard ban is being introduced gradually (between 2015 and 2019) and on a fishery-by-fishery basis.[65] The regulation distinguishes between four categories: small pelagic fisheries (e.g., mackerel, herring, horse mackerel, blue whiting, boarfish, anchovy, argentine, sardine, and sprat), large pelagic fisheries (e.g., bluefin tuna, swordfish, albacore tuna, bigeye tuna, blue and white marlin), fisheries for industrial purposes (e.g., capelin, sandeel, and Norwegian pout), species that define the fisheries (no examples of this category are given in the regulation). In addition to these categories, the regulation also describes certain species-area (*inter alia* fisheries for salmon in the Baltic Sea, fisheries for hake in the North Sea, etc.). It establishes four time frames that include these categories and species-areas and sets four deadlines by which the landing

[63]Green Paper 2009, cit., section 5.3.

[64]Borges (2015), p. 536, highlights an incident that took place in 2008. A UK trawler (The Prolific) was filmed by the Norwegian coastguard throwing five tonnes of fish overboard immediately after leaving Norwegian waters, where discarding is prohibited. This event was widely reported in the press. See, for example, the Guardian newspaper in its edition of 13/8/2008. Its online version even provides a video over 4 min long showing the operation in which the Prolific discarded nearly 80% of its catch. The boat had previously been inspected in Norwegian waters and declared legal, before crossing into UK waters where it dumped its load; http://www.theguardian.com/environment/2008/aug/13/fishing.endangeredspecies.

[65]Art. 15, Regulation 1380/2013, cit.

obligation is to be effective (1 January 2015, 2016, 2017, and 2019). Without going into the details of which species or species-areas are included in each of these time frames, it is important to note that the EU discards affect the demersal species more than the pelagic species,[66] which is why the implementation did not start with them.

So far, the Commission has adopted 5+3 plans. In October 2014, the Commission adopted five discard plans, applicable from 1 January 2015, for certain pelagic and industrial fisheries.[67] One year later, in October 2015, the Commission adopted three discard plans, applicable from 1 January 2016, for certain demersal fisheries.[68]

The discard ban is established as a landing obligation. Catches during fishing activities in Union waters or by Union fishing vessels in international waters (waters not subject to EU or third countries' sovereignty or jurisdiction) must be brought and retained on board the fishing vessels, recorded, landed, and counted against any applicable quotas, except when used as live bait.

There are two possible exceptions to this obligation. First, the regulation allows for fishing operators to continue to discard species that, according to the best available scientific advice, have a high survival rate when released into the sea.[69] For instance, an exemption from the landing obligation exists for Norway lobster caught in pots, traps, or creels in ICES division VIa and subarea VII.[70] Second, to cater for unwanted catches that are unavoidable even when all measures for their reduction are taken, certain *de minimis exemptions* from the landing obligation may be established.[71] For

[66]E.g. the discard rate in the North Sea has been 30–40% by weight for the main demersal fish species (cod, haddock, whiting and plaice) since the 1970s; and around 10% for pelagic fish. Heath et al. (2014).

[67]Commission Delegated Regulation (EU) No 1392/2014 of 20 October 2014 establishing a discard plan for certain small pelagic fisheries in the Mediterranean Sea; Commission Delegated Regulation (EU) No 1393/2014 of 20 October 2014 establishing a discard plan for certain pelagic fisheries in north-western waters; Commission Delegated Regulation (EU) No 1394/2014 of 20 October 2014 establishing a discard plan for certain pelagic fisheries in south-western waters; Commission Delegated Regulation (EU) No 1395/2014 of 20 October 2014 establishing a discard plan for certain small pelagic fisheries and fisheries for industrial purposes in the North Sea; Commission Delegated Regulation (EU) No 1396/2014 of 20 October 2014 establishing a discard plan in the Baltic Sea. OJ 2014 L 370.

[68]Commission Delegated Regulation (EU) 2015/2438 of 12 October 2015 establishing a discard plan for certain demersal fisheries in north-western waters; Commission Delegated Regulation (EU) 2015/2439 of 12 October 2015 establishing a discard plan for certain demersal fisheries in south-western waters; Commission Delegated Regulation (EU) 2015/2440 of 22 October 2015 establishing a discard plan for certain demersal fisheries in the North Sea and in Union waters of ICES Division IIa. OJ 2015 L 336.

[69]Species for which scientific evidence demonstrates high survival rates, taking into account the characteristics of the gear, of the fishing practices and of the ecosystem. Regulation 1380/2013, cit., recital 27, article 15 paragraph 4.b.

[70]Article 2, Commission Delegated Regulation 2015/2438, cit.

[71]The *de minimis exemption* shall apply where scientific evidence indicates that increases in selectivity are very difficult to achieve; or to avoid disproportionate costs of handling unwanted catches, for those fishing gears where unwanted catches per fishing gear do not represent more than a certain percentage, to be established in a plan, of total annual catch of that gear. Regulation 1380/2013, cit., recital 31, article 15 paragraphs 4.c and 5.c.

example, it exists a provision that allows to discard up to a maximum of 7% in 2015 and 2016, and 6% in 2017 of albacore tuna for total annual catches in the albacore tuna directed fisheries using midwater pair trawls (PTM) in ICES sea area VII.[72]

In addition to these two *possible* exemptions, which are to be determined in the corresponding discard plan, fishing operators *must* discard catches of prohibited species (e.g., basking shark)[73] and, since 2015, also predator-damaged fish.[74] These catches of prohibited species and predator-damaged fish cannot be retained on board and must be returned into the sea.[75]

These catches that can (high survivability, *de minimis)* or must (prohibited species, predator-damaged fish) be discarded are not counted against the quota, but they must be documented in the logbook.[76]

4.3 Use of National Quotas

The EU carries out a stock-by-stock management based on TACs (EU) and quotas (MSs). And, as we have seen, allocation keys were basically fixed for each stock-area on an MS basis when they joined the EU. Even if those allocation keys were adequate when fixed—a matter denied by some—the fact is that, as the Commission points out, conditions have changed since then due to different factors as, not intending to be exhaustive, stock development, the evolution of fleets, new fishing strategies on different stocks, changes in demand for given species, or the evolution of imports.[77] One of the effects of this system is the so-called mini-quotas.[78]

[72]Article 3.a, Commission Delegated Regulation 1393/2014, cit.

[73]Species in respect of which fishing is prohibited and which are identified as such in a Union legal act adopted in the area of the CFP. Regulation 1380/2013, cit., recital 27, article 15 paragraph 4.a.

[74]Fish which have been damaged by predators such as fish-eating marine mammals, predatory fish or birds, can constitute a risk to humans, to pets and to other fish by virtue of pathogens and bacteria which might be transmitted by such predators. This fish is a new exception to the landing obligation inserted by Regulation (EU) 2015/812 of the European Parliament and of the Council of 20 may 2015 amending, among others, Regulation (EU) No 1380/2013. JO 2015 L 133, recital 16 and art. 9.a.

[75]European Commission, "1 January 2015: the landing obligation", http://ec.europa.eu/fisheries/cfp/fishing_rules/landing-obligation/index_en.htm.

[76]Ibid.

[77]European Commission (2011) Impact assessment; Commission staff working paper accompanying the document Commission proposal for a Regulation of the European Parliament and of the Council on the Common Fisheries Policy; IA 2011, hereinafter. SEC (2011) 891 final, de 13.7.2011; Section 2.1.4.

[78]During the discussions prior to the last reform of the CFP, Ireland manifested its support for RS but also proposed a relaxation of it. One of the problems that Ireland pointed out was the mini-quotas: "(...) While these allocations may reflect catch history in the 1970's, today they are very small quotas in often distant fisheries. In many cases fleets steam many miles (incurring significant carbon foot prints) to catch small allocations in fisheries where they may or may not also have

There are two ways of easing RS. MSs may exchange all or part of the fishing opportunities allocated to them,[79] or they may make use a year-to-year flexibility mechanism of up to 10% of their permitted landings.[80] Outside of these two options, overfishing is penalized. When the Commission establishes that a MS exceeded the quotas which have been allocated to it, the Commission shall operate deductions in the future quotas of that MS by applying a multiplying factor.[81] For this reason, some exchanges of quotas between MSs are actually regularizations in disguise.[82]

Quota exchanges take place between MSs[83] but not in a fully satisfactory manner, so quotas sometimes have been underutilized.[84] This has been the situation so far, and in the current period it is expected to worsen as a result of the ban on discards.

At this point in time, the traditional mechanisms for providing RS flexibility are still in place and, as we have seen, the landing obligation is in force with some exemptions. In addition are two further elements: discards are taken into account when setting quotas for MSs, and a new flexibility mechanism has been introduced. Indeed, on the one hand, when the landing obligation for a fish stock is introduced, fishing opportunities will be fixed taking into account the change from fixing fishing opportunities that reflect *landings* to fixing fishing opportunities that reflect *catches*, given that, for the first and subsequent years, discarding of that stock will no longer be allowed.[85] In other words, in a scenario with discards, the reference is landings

---

quotas for other stocks caught in mixed fisheries. Given their size it is evident that it is not commercially viable to catch these quotas; indeed many of them are economically unsound and should be redistributed to Member States in a position to utilize them (. . .)." Department of Agriculture, Fisheries and Food of the Government of Ireland (2010), Ireland's Response to the Commission's Green Paper on the Reform of the Common Fisheries Policy, pp. 16–17, section 3.3.1; http://ec.europa.eu/fisheries/reform/docs/ireland_en.pdf.

[79]Regulation 1380/2013, cit., recital 29 and Art. 16.8.

[80]Ibid., Art. 15.9.

[81]Article 105, Council Regulation (EC) No 1224/2009 of 20 November 2009; JO 2009 L 343.

[82]Commission warned some years ago (IA 2011, cit., Section 2.1.4) that the Commission got close to 1000 notifications of swaps per year, 50% of which were nearly permanent, the rest were late year 'regularisations' intended to legitimise excessive catches.

[83]As the Commission informed (ibid.), MSs exchanged more than 10% of their quotas in the period 2005–2008 on an annual basis. The species with the highest swap volumes were redfish, cod and hake and some pelagic species like herring, blue whiting, jack mackerel, mackerel, sprat, anchovy and sandeel.

[84]The Vessels' Owners Cooperative of the Spanish Port of Vigo España (ARVI) conducted a study on the situation. The study analyzes 20 major species subject to quotas in the EU for the period 2008–2014. The report concludes (pp. 52–53) that the remaining portion of the quota that was not eventually utilized by MSs amounted to 1 million tonnes (exactly 1,039,549.78 tonnes) over the period 2008–2014, i.e., an average of 23% (an annual average of roughly 143 thousand tonnes). These underutilized quotas represented a total value of 1833.7 million euros (based on the guide prices). ARVI (2016) Update of the TAC and quota system in face of the ban of discards; Spanish-English bilingual edition, http://www.arvi.org/publicaciones/PuestaTacsCuotasDescartes.pdf.

[85]Regulation 1380/2013, cit., Art. 16.2.

(catches less discards), while when discards are prohibited, the reference is the catches (discounting catches not subject to the landing obligation). On the other hand, flexibility between species is allowed. Catches of species that are subject to the landing obligation and that are caught in excess of quotas of the stocks in question, or catches of species in respect of which the MS has no quota, may be deducted from the quota of the target species provided that they do not exceed 9% of the quota of the target species.[86] This second possibility—flexibility between species—has been seen by some as a breach of RS in favor of the autonomy of MSs, as it allows applying not used-up quotas to other species for which there is no quota, or for which quotas have been exceeded.[87]

In summary, RS maintains its flexibility mechanisms (quota swapping and year-to-year flexibility), the landing obligation is not absolute (with the exemptions of high survivability, *de minimis*, prohibited species, and predator-damaged fish), the fishing opportunities that take into account the effect of discards are expected to increase, and some flexibility between species will be admitted. Will this be enough?

The biggest concern is with the so-called choke species, which may even further reduce quota exchanges. MSs that previously got rid of certain quotas will now need them to cover the discard ban to prevent that other species strangle the catches of its target species. The real challenges lie with the demersal species. Although we will have to wait to see what happens on the fishing grounds, the experience with pelagic species has not been encouraging. In this case, the choke species have not stimulated quota exchanges but rather reduced them.[88]

At this point, it should be remembered that one of the objectives of the CFP is to increase productivity in fisheries by ensuring the rational development of fisheries production and the optimum utilization of the factors of production, in particular labor.[89] A notion that must be understood in the light of other objectives, which include a fair standard of living for the fishing community and supply at reasonable prices.

This management of resources must also take into account the peculiarities resulting from structural and natural disparities between the various fishing regions,

[86]This provision only applies where the stock of the non-target species is within safe biological limits. Regulation 1380/2013, cit., Art. 15.8.

[87]E.g. Spanish Government; interview with Carlos Dominguez, at the moment Secretary General for Fisheries of the Government of Spain; published in March 2013 and reported in various media; see *La Opinion* in its edition of 19.03.2013, http://www.laopinioncoruna.es/mar/2013/03/18/carlos-dominguez-plan-descartes-reducira-pesca-gran-sol-tres-meses-ano/703190.html.

[88]With regard to the issue of choke species and quota swaps, European Commission informs that stakeholders are suggesting that MSs seem to retain quotas rather than increase swapping under the newly introduced landing obligation. Commission staff working document accompanying the document: Communication from the Commission to the European Parliament and the Council—Consultation on the fishing opportunities for 2017 under the Common Fisheries Policy; COM (2016) 396 final, 15.06.2016, p. 10.

[89]And also by promoting technical progress, but it is not what we want to emphasize now.

the impact of the fisheries sector on the whole economy, and the need to effect the appropriate adjustments by degrees.[90]

As the fisheries sector points out, for years discards were socially and politically accepted. When this ceased to be the case, the need arose to ban them. Likewise, there may come a day when the socioeconomic effects of the inability to take advantage of fishing quotas also become socially and politically intolerable. We understand that this is especially relevant to fishing-dependent regions, understanding this category of regions not in the sense of the Hague Preferences formula, which is limited to Ireland and the UK, but in a genuine sense to include all fishing-dependent regions of the EU.[91] Good examples, although not the only ones, of regions in this category are Galicia (Spain), Highlands and Islands (UK), N-E Scotland (UK), Algarve (Portugal), and Peloponnisos (Greece).[92]

5 Final Considerations

Perhaps much of the frustration that RS has generated over the years in some states—mainly in states adversely affected like Spain—derives from the misleading way it has been formulated. The mismatch between the formulation of the RS concept and the concrete allocation of fishing opportunities among MSs, which has been applied annually for more than 30 years, has led some to consider that the concrete allocation constitutes a breach of the RS and hence a breach of EU law. But despite that RS is formulated as a system based on *historical catches* that also takes into account *the needs of regions particularly dependent on fisheries*, the fact is that both considerations serve only as a partial explanation of what RS actually is. As the CJEU has stated, "requirement of relative stability must be understood as meaning that each Member State is to retain a fixed percentage when fishing

[90]See art. 39 TFEU. This article sets out the objectives of the common agricultural policy. However, it should be noted that, as occurs in the art. 38.1 TFEU, references to the common agricultural policy or to agriculture, and the use of the term "agricultural", must be understood as also referring to fisheries.

[91]See Natale et al. (2013). The authors identify and map specific local communities in which, given the conditions of accessibility, employment and size of the fishing fleet, the dependence on fishing activities can be considered particularly relevant, i.e., with ratios above 5%. See also European Commission, *Facts and figures on the Common Fisheries Policy—Basic statistical data* (European Union, Luxembourg, 2014).

[92]European Parliament (2007) Regional dependency on Fisheries, IP/B/PECH/ST/IC/2006-198; study requested by the European Parliament's Committee on Fisheries and carried out by Pavel Salz and Graeme Macfadyen. E.g. combining income dependency on the fisheries sector and the number employed, the top five-ranked NUTS-2 regions were Galicia (Spain), Highlands and Islands (UK), N-E Scotland (UK), Algarve (Portugal) and Peloponnisos (Greece); see table 11, pp. 17–18. About Galicia, see Surís-Regueiro and Santiago (2014); also the latest report published by the Galician Statistics Institute: "Analisis do Sector da Pesca", 2015, http://www.ige.eu/estatico/pdfs/s3/publicaciones/AnaliseSectorPesca.pdf.

opportunities are distributed" and "the distribution formula originally laid down . . . will continue to apply as long as an amending regulation has not been adopted,"[93] which has yet to occur. States have been negotiating the distribution key for years (mainly in 1983 but also in subsequent acts of accession and whenever it has been necessary to allocate quotas for new fishing opportunities). It is a fixed percentage, only altered annually when appropriate by applying the Hague Preferences. Therefore, in our opinion, the arguments that this allocation key identified as RS is not a true reflection of historical catches of MSs, or it does not take into account the needs of all regions particularly dependent on fisheries, probably do not provide a sufficient legal ground for questioning its legality. However, they may be good arguments for negotiation within the framework of a reform process.

One of the effects of RS is discards, i.e., catches returned to the sea. The current legislation, Regulation 1380/2013, keeps RS as a criterion for allocating fishing opportunities among MSs but introduces a gradual ban on discards. Aside from the existing flexibility mechanisms of RS (quota swapping and year-to-year flexibility), Regulation 1380/2013 establishes some exemptions to the landing obligation (high survivability, *de minimis*, prohibited species, and predator-damaged fish), provides an additional mechanism of flexibility (between species), and allows an increase in fishing opportunities to take into account the effect of discards.

The challenge is great, especially in certain mixed fisheries where the discard rate is very high. At this stage at least, it appears that RS could be the loser in the compromise sought in Regulation 1380/2013 between it and the discard ban. Exactly to what extent the RS is affected will depend on how insufficient the regulation's mechanisms turn out to be and whether any further corrective action is taken. RS will continue to be the distribution key, but the MSs could not be able to satisfactorily use their quotas. While national quotas were not being fully used prior to the ban on discards, mainly due to deficiencies in quota swapping, the discard ban might not improve this. Although we are still at the implementation phase, and the real challenge is with the demersal species, the experience with pelagic species seems to indicate no increase in quota swapping and so far is having exactly the opposite effect.

From a legal point of view, to what extent is it relevant that MSs cannot fully exploit their fishing quotas? To answer this question, we need to keep in mind the objectives of the CFP. The EU has to increase productivity in fisheries by ensuring the rational development of fisheries production and the optimum utilization of the factors of production, in particular labor. The EU has to exploit fisheries resources in such a way that ensures the sustainability of marine ecosystems and also provides reasonable income to those who depend on fishing activities while taking into account the interests of consumers. The inability of a state to fully use their fishing quotas clearly has a negative effect on those who make a living from fishing, and also on the price and supply. This effect is logically more damaging in fishing-dependent regions, a category that should be freed from the shackles of the Hague

[93]Judgments in cases C-87/03 and C-100/03, cit., paragraph 27.

Preferences, which strips it of its substance by restricting it to regions in two MSs (Ireland and the UK) while, in its true sense, this category includes many other regions.

References

Alverson DL, Freeberg MH, Pope JG, Murawski SA (1994) A global assessment of fisheries bycatch and discards. FAO fisheries technical paper no. 339

Andersen P, Andersen J, Mardle S (2014) What's going to happen with the CFP Reform discard policy? In: Conference papers and presentations of the 17th IIFET conference, International Institute of Fisheries Economics and Trade (IIFET), Brisbane, 7–11 July 2014

Borges L (2015) The evolution of a discard policy in Europe. Fish Fisheries 16(3):534–540

Churchill R, Owen D (2010) The EC Common Fisheries Policy. Oxford University Press, Oxford

Del Vecchio A (1982) Sull'incidenza della normativa comunitaria sui tratati in materia di pesca fra Stati membri della CEE e Stati terzi. Rivista di diritto internazionale 65:571–582

Franckx E (2012) Sea fisheries cases before the European Court of Justice. In: Wolfrum R (ed) Max Planck Encyclopedia of Public International Law. Oxford University Press, Oxford, pp 42–47

Heath MR, Cook RM, Cameron AI, Morris DJ, Speirs DC (2014) Cascading ecological effects of eliminating fishery discards. Nat Commun 5:1–8. doi:10.1038/ncomms4893

Holden MJ (1985) The procedures followed and the problems met by the European Economic Community in implementing the scientific recommendations of the International Council for the exploration of the sea on total allowable catches. In: Papers presented at the expert consultation on the regulation of fishing effort (fishing mortality), Rome, 17–26 January 1983. FAO Fisheries Rep 289(3):215–470

Kelleher K (2005) Discards in the World's Marine Fisheries. An update. In: FAO fisheries technical paper no. 470. FAO, Rome

Le Bihan D (2003) La jurisprudence communautaire et le principe de stabilite relative dans la politique commune de la pêche. In: Cudennec A, Gueguen-Hallouet G (eds) Le juge communautaire et la mer. Bruylant, Brussels, pp 143–160

Lostado i Bojo R (1985) La Política Común de la Pesca en la CEE y España. Revista de Estudios Agro-Sociales 131:39–69

Lucas I (1997) A study of the options for utilization of by catch and discards from marine capture fisheries. In: FAO fisheries circular no. 928 FIIU/C928. FAO, Rome

Meseguer Sánchez JL (1981) El Derecho comunitario y los derechos de pesca de los Terceros Países. Revista de Instituciones Europeas 8(1):53–65

Natale F, Carvalho N, Harrop M, Guillen J, Frangoudes K (2013) Identifying fisheries dependent communities in EU coastal areas. Mar Policy 42:245–252

Penas Lado E (2016) The Common Fisheries Policy: the quest for sustainability. Wiley-Blackwell

Proelss A, Houghton K (2012) The EU Common Fisheries Policy in light of the precautionary principle. Ocean Coastal Manag 70:22–30

Rey Aneiros A (2001) La Unión Europea frente a las transformaciones del Derecho internacional de la pesca. Tirant lo Blanch, Valencia

Sobrido-Prieto M (2013) Los jueces españoles en su papel de jueces comunitarios: el intercambio de cuotas pesqueras de anchoa en el Golfo de Vizcaya y el principio de estabilidad relativa en las sentencias del tribunal de Luxemburgo, de la Audiencia Nacional y del Tribunal Supremo. Revista de Derecho comunitario europeo 45:691–718

Sobrino Heredia JM (1990) Les Accords de pêche conclus par l'Espagne et les pays en voie de développement. In: Velas P (ed) Aspects du droit international économique. Pedone, Paris, pp 203–237

Sobrino Heredia JM (2003) La participación de la Unión Europea en las transformaciones del Derecho de Pesca. Cuadernos de derecho pesquero 2:77–98

Sobrino Heredia JM (2009) La politique communautaire de conservation et gestion des ressources halieutiques et quelques problèmes concernant l'application du droit de l'Union européenne. In: Casado Raigón R, Cataldi G (eds) L'évolution et l'état actuel du droit international de la mer. Mélanges de droit de la mer offerts à Daniel Vignes. Bruylant, Bruxelles, pp 865–885

Sobrino Heredia JM, Rey Aneiros A (1997) La Jurisprudencia del Tribunal de Justicia de las Comunidades Europeas sobre la Política Común de la Pesca. Xunta de Galicia, Santiago de Compostela

Surís-Regueiro JC, Santiago JL (2014) Characterization of fisheries dependence in Galicia (Spain). Mar Policy 47:99–109

Treves T (1976) La Communauté européenne et la zone économique exclusive. Annuaire français de droit international 22(1):653–677

Uhlmann SS, van Helmond ATM, Stefánsdóttir EK, Sigurðardóttir S, Haralabous J, Maria Bellido J, Carbonell A, Catchpole T, Damalas D, Fauconnet L, Feekings J, Garcia T, Madsen N, Mallold S, Margeirsson S, Palialexis A, Readdy L, Valeiras J, Vassilopoulou V, Rochet M-J (2013) Discarded fish in European waters: general patterns and contrasts. ICES J Mar Sci: 1–11. doi:10.1093/icesjms/fst030

Wilson DC, Jacobsen RB (2009) Governance issues in mixed-fisheries management: an analysis of stakeholder views. IFM—Innovative Fisheries Management, Aalborg University, Aalborg

Some Recent Questions Regarding the European Union's Public Access Fisheries Agreements

Gabriela A. Oanta

1 Introduction

Public access fisheries agreements, also called 'international fisheries agreements' or 'sustainable fisheries partnership agreements' in the most recent Common Fisheries Policy (CFP) reform, are one of the main manifestations of the EU's external fisheries activity at the international level. They are therefore one of the elements that best define the EU's international legal personality, which is explicitly provided for under Art. 47 TEU.

According to Art. 4(37) of Regulation (EU) No 1380/2013 on the current CFP, these agreements are concluded 'with a third state for the purpose of obtaining access to waters and resources in order to sustainably exploit a share of the surplus of marine biological resources, in exchange for financial compensation from the Union, which may include sectoral support'.[1] These public access fisheries

[1]Regulation (EU) No 1380/2013 of the European Parliament and of the Council *on the Common Fisheries Policy, amending Council Regulations (EC) No 1954/2003 and (EC) No 1224/2009 and repealing Council Regulations (EC) No 2371/2002 and (EC) No 639/2004 and Council Decision 2004/585/EC*, O.J. L 354/22 (2013). Art. 4(37) Regulation 1380/2013 is completed by Arts. 31 and 32 of the same normative act in relation to the principles and objectives of the sustainable fisheries partnership agreements as well as to the financial aid that will be given by the EU on the basis of these fishing agreements. For an overview of the current regulation of these agreements, see Sobrino Heredia and Oanta (2015), pp. 71–80.

G.A. Oanta (✉)
University Institute of European Studies "Salvador de Madariaga", University of A Coruña, Campus de Elviña, s/n, A Coruña 15071, Spain

Jean Monnet Module "European Union's Integrated Maritime Policy", University of A Coruña, Casa do Lagar, Campus de Elviña s/n, 15071 A Coruña, Spain
e-mail: gabriela.oanta@udc.es

G. Andreone (ed.), *The Future of the Law of the Sea*,
DOI 10.1007/978-3-319-51274-7_3

agreements are divided into tuna agreements, on the one hand,[2] and mixed or multi-species agreements, on the other.[3] They must also be distinguished from private access agreements, which are concluded between private companies based in EU Member States and third countries.[4]

In the last 30 years, the EU has concluded more than 30 such agreements, affording its fishing fleet access to very diverse stocks in the respective partner country's economic exclusive zone (EEZ).[5] Undoubtedly, this fisheries treaty activity has been possible due to the EU's exclusive international competence in this field. As is well known, in 1998 the EU concluded both the United Nations Convention of 10 December 1982 on the Law of the Sea (UNCLOS) and the Agreement of 28 July 1994 relating to the implementation of Part XI thereof[6] and also made a Declaration concerning its exclusive competences with regard to matters governed by the UNCLOS and that Agreement.[7] Moreover, this exclusive competence is not restricted only to the maritime waters under the sovereignty or jurisdiction of EU Member States as, according to Art. 1(2) of Regulation (EU) No 1380/2013, it also extends to activities carried out by EU fishing vessels in the waters of third countries or on the high seas, as well as by European citizens 'without prejudice to the primary responsibility of the flag State'.

[2]The EU has concluded twelve such tuna agreements so far, by virtue of which EU fishing vessels have been able to fish tunas stocks in the Indian Ocean and the Pacific Ocean. Specifically, it has concluded them with Cape Verde, Comoros, Côte d'Ivoire, Gabon, Equatorial Guinea, Kiribati, Liberia, Madagascar, Mauritius, Mozambique, São Tomé and Príncipe, the Seychelles, and the Salomon Islands. See Le Manach et al. (2013), pp. 257–266.

[3]Seven mixed or multispecies agreements have been signed so far, affording the EU's fishing vessels access to very diverse fish stocks in the EEZs of the following countries: Greenland, Guinea, Guinea-Bissau, Morocco, Mauritania, Micronesia and Senegal. See *Fishing for Coherence in West Africa. Policy Coherence in the Fisheries Sector in Seven West African Countries* (2008).

[4]Molenaar (2002), pp. 137–138.

[5]For an overview of the international fisheries agreements concluded by the EU in the last 30 years, see, e.g., Andreone (2007), pp. 326–347; Ould Ahmed Salem (2009); Ruiloba García (2005), pp. 333–345; Sobrino and Oanta (2015), pp. 61–85; Van der Burgt (2013) and Witbooi (2012).

[6]Council Decision 98/392/EC *concerning the conclusion by the European Community of the United Nations Convention of 10 December 1982 on the Law of the Sea and the Agreement of 28 July 1994 relating to the implementation of Part XI thereof*, O.J. L 179/1 (1998).

[7]For the first time, an international treaty contains an EU's declaration of such characteristics. Concretely, the EU affirmed that its Member States had transferred it competences in this field and, therefore, 'in this field it is for the Community to adopt the relevant rules and regulations (which are enforced by the member States) and, within its competence, to enter into external undertakings with third States or competent international organizations'. See Division for Ocean Affairs and the Law of the Sea, Office of Legal Affairs, 37 *Law of the Sea Bulletin* (1998), pp. 7–14. On the declaration of competences made by the EU in the field of fisheries, see, e.g., Lijnzaad (2014), pp. 187–207.

By virtue of this competence, the EU has been able to take part in different international fora and defend its fisheries interests at the global level.[8] As a result, for the last 30 years it has been an active actor on the international fisheries scene.[9] Moreover, the EU fish market is the largest in the world and, at the same time, depends on both fishing imports and fishing captures in waters not under the sovereignty or jurisdiction of its Member States.[10] Currently, approximately 25% of EU fishing captures are made in such waters, approximately 8% are enabled by fisheries agreements with third countries and approximately 20% are carried out in the high seas, basically in areas under the jurisdiction of regional fisheries management organisations (RFMOs).[11]

The aim of this chapter is to present the EU's treaty activity in the field of fisheries in light of the most relevant case law of the CJEU published in 2014 and 2015. Recent practice by EU fishing vessels has highlighted the need to look to the CJEU's position to clarify certain pending aspects of the EU's fisheries treaty activity on the international stage, such as, *first*, the issue of the European Commission's competence to represent the EU before the International Tribunal for the Law of the Sea (ITLOS), in a case in which the fisheries treaty activity of an international organisation, amongst other issues, was analysed—*Council v European Commission* (C-73/14) (Sect. 2); *second*, the necessary legal basis for the adoption of a normative act by virtue of which a third-country fishing vessel could fish in the waters of an EU Member State, as well as the scope of the international fisheries agreements—*European Parliament and European Commission v Council* (joined cases C-103/12 and C-165/12) (Sect. 3); and, *third*, certain aspects resulting from the application of the fishing agreement signed between the EU and Morocco and its successive Protocols—*Ahlström and Others* (C-) and

[8]See, e.g., De Yturriaga Barberán (2009), pp. 269–297, Sobrino Heredia (2002), pp. 53–82 and Treves (2008), pp. 1–20.

[9]As is well known, the EU is one of the most important coastal entities, has one of the largest long-distance fleet, being the third fishing power globally, having an important transformer sector of fishing.

[10]It is estimated that 90% of the fishing resources globally are in the developing countries EEZs. As is known, according to Art. 56(1)(a) of UNCLOS, '[i]n the exclusive economic zone, the coastal State has (a) sovereign rights for the purpose of exploring and exploiting, conserving and managing the natural resources, whether living or non-living, of the waters superjacent to the seabed and of the seabed and its subsoil'. In addition, Art. 61 of UNCLOS provides that '1. The coastal State shall determine the allowable catch of the living resources in its exclusive economic zone' and Art. 62 of UNCLOS regarding the utilization of the living resources stipulates that '1. The coastal State shall promote the objective of optimum utilization of the living resources in the exclusive economic zone without prejudice to article 61. 2. The coastal State shall determine its capacity to harvest the living resources of the exclusive economic zone. Where the coastal State does not have the capacity to harvest the entire allowable catch, it shall, through agreements or other arrangements and pursuant to the terms, conditions, laws and regulations referred to in paragraph 4, give other States access to the surplus of the allowable catch'.

[11]Currently, the EU is a member of fourteen RFMOs. Regarding the EU's participation in the framework of these organizations, see, e.g., Antonova (2015), pp. 125–143, Franckx and Van den Bossche (2010), pp. 419–448 and Vázquez Gómez (2002).

Front Polisario v Council (T-512/12) (Sect. 4). In the author's view, all of these judgments will be very useful for the General Court in Luxembourg in case T-180/14 regarding the action for annulment brought on 14 March 2014 by Front Polisario against the Council in relation to Decision 2013/785/EU of 16 December 2013 on the conclusion of the Protocol between the EU and Morocco setting out the fishing opportunities and financial contribution provided for in the fisheries partnership agreement between the EU and that country. The Chapter will conclude with some final remarks.

2 The European Commission's Competence to Represent the EU Before the ITLOS

The judgment of the Court (Grand Chamber) of 6 October 2015 in the case *Council v European Commission* (C-73/14)[12] was the first decision of the Luxembourg court related to the European Commission's capacity to present allegations before an international court without prior authorisation from the Council. In this sense, it put on the table the question of who is responsible for a breach by European fishing vessels of a fishing agreement concluded by the EU: the EU Member State acting as the flag State or the EU itself? This judgment is a doubtless part of a larger procedural action of the Council that has been encouraged by the entry into force of the Lisbon Treaty,[13] which contains, amongst other things, a new system for the EU's international representation through a new division of competences in the field of external action.[14]

Through this judgment, the Council questioned the legality of the European Commission's Decision of 29 November 2013 regarding the submission of written comments on behalf of the EU to the ITLOS in the framework of Case No 21 on the request for an advisory opinion submitted by the Sub-Regional Fisheries Commission (SRFC).[15] In that case, the ITLOS was asked, *inter alia*, whether the flag State

[12]EU:C:2015:663. For an extensive analysis of this judgment, see Oanta (2016), pp. 208–216.

[13]Sánchez-Tabernero (2015), pp. 1057–1073.

[14]With regard to the changes in the division of competences in the field of external action resulting from the entry into force of the Lisbon Treaty, it should be mentioned that the EU's international representation is currently provided, depending on the field in question, by the European Council's President (Art. 15(5) and (6) TEU), the High Representative of the Union for Foreign Affairs and Security Policy (Art. 18 TEU), or the European Commission (Art. 17(1) TEU).

[15]The SRFC is a RFMO created on 29 March 1985. It is formed by seven African countries, namely: Cape Verde, Gambia, Guinea, Guinea-Bissau, Mauritania, Senegal and Sierra Leona; it is based in Dakar (Senegal). For more details, see the SRFC official website http://www.spcsrp.org. Accessed 29 April 2016. This request for an advisory opinion was prepared in February 2013 in the framework of 'Atelier sur la lutte contre les pêches illicites, non déclarées et non règlementées (PINN)', which took place in Dakar on 25–26 February 2013. See http://www.spcsrp.org/medias/csrp/comm/at_PINN_publication_web.pdf. Accessed 29 April 2016.

or an international agency—such as the EU—would be responsible for the violation of the fisheries legislation of a coastal State by a fishing vessel with a fishing licence granted under an international fisheries agreement signed with that coastal State.[16]

The Council claimed, first, infringement of the principle of conferral of powers laid down in Art. 13(2) TEU, as well as of the principle of institutional balance, and, second, infringement of the principle of sincere cooperation. With regard to the principle of conferral of powers and the principle of institutional balance, the Council stated that Art. 218(9) TFEU—which provides that it may adopt a decision establishing the EU's positions in a body set up by an international agreement that could adopt acts having legal effects for the EU—and Art. 16(1) TEU had been infringed. As to the infringement of the principle of sincere cooperation, the Council claimed that the Commission had not submitted a proposal for a decision on the position to be adopted on behalf of the EU before the ITLOS, as required under Art. 218(9) TFEU, and also that it had not cooperated with it in good faith in the preparation of the written statement submitted to this international court in Case No 21.[17]

The EU submitted written statements on two occasions—on 29 November 2013 and on 13 March 2014—in the proceedings opened by the ITLOS.[18] The first time the European Commission acted before the ITLOS on the basis of the decision adopted on 5 August 2013,[19] without previously submitting its comments to the Council for approval, 'despite the latter's request',[20] notifying the Council the same day (namely on 29 November 2013).[21] As a result of this action by the Commission

[16]The questions made by the SRFC to the ITLOS, which are regarding the phenomena of illegal, unreported and unregulated fishing, were the following: (1) What are the obligations of the flag State in cases where illegal, unreported and unregulated fishing activities are conducted within the Exclusive Economic Zones of third party States? (2) To what extent shall the flag State be held liable for illegal, unreported and unregulated fishing activities conducted by vessels sailing under its flag? (3) Where a fishing license is issued to a vessel within the framework of an international agreement with the flag State or with an international agency, shall the State or international agency be held liable for the violation of the fisheries legislation of the coastal State by the vessel in question? (4) What are the rights and obligations of the coastal State in ensuring the sustainable management of shared stocks and stocks of common interest, especially the small pelagic species and tuna? See http://www.itlos.org/index.php?id=252&L=0%20and%207%253D2. Accessed 29 April 2016.

[17]The European Commission considered that, under Art. 335 TFEU, it had the capacity to represent the EU in the judicial proceedings. Hence it decided to submit a written statement to the ITLOS on behalf of the EU, as well as to take part in the oral proceedings before the international court. Nevertheless, a few interveners in the case denied that this provision allowed the European Commission to represent the EU before the ITLOS.

[18]*Request for an advisory opinion submitted by the Sub-Regional Fisheries Commission (SRFC), Order of 24 May 2013, ITLOS*. See also *Request for an advisory opinion submitted by the Sub-Regional Fisheries Commission (SRFC), Order of 20 December 2013, ITLOS*. For an overview of these issues, see Becker (2013) and Oanta (2014a), pp. 301–304.

[19]Decision C (2013) 4989 final.

[20]EU:C:2015:663, para 37.

[21]Ibid, paras 20–32.

in late 2013, on 10 February 2014, the Council, acting under Art. 263 TFEU, filed the action for annulment addressed by the Court in this judgment.

In relation to ITLOS Case No 21, it has to be mentioned that the European Commission argued before the ITLOS that the responsibility of the flag State or 'international agency'—as would be its case—for infringement of the national fisheries legislation of a coastal State depended on the content of the applicable international agreement and that, in the absence of such a conventional act, the general rules concerning the international responsibility of the State would apply. Specifically, it argued that the flag State of a fishing vessel operating in the EEZ of a third State would be responsible for any violations by it of the coastal State's national legislation.[22]

The Court considered, on the one hand, that Art. 335 TFEU 'provided a basis for the Commission to represent the European Union before the ITLOS in Case No 21'[23] since, as it had ruled in its judgment *Reynolds Tobacco and Others v Commission*,[24] Art. 335 TFEU 'is the expression of a general principle that the European Union has legal capacity and is to be represented, to that end, by the Commission'.[25] On the other hand, it found that the Commission had fulfilled the obligation to consult the Council before acting before the ITLOS and, therefore, had not infringed the principle of sincere cooperation, as a working paper on the allegations it wished to present to the ITLOS had been referred to the Council and revised twice.[26]

Underlying this power struggle between the Commission and the Council is an important pronouncement by the ITLOS with profound consequences for the EU's treaty practice in the field of fisheries, namely, the international court ultimately attributed the international responsibility for infringement of a coastal State's legislation by a vessel flying the flag of an EU Member State and fishing under a fisheries agreement to the EU. In other words, the EU can no longer hide behind the shield of the vessel, shifting the responsibility to the flag State; instead, the EU itself must deal with the consequences of such infringements.

In the author's view, in its Advisory Opinion in Case No 21, the ITLOS seems, first, to accept the Commission's assessment in considering (para 170) that the responsibility of an international organisation, as a result of the infringement of a coastal State's fisheries legislation by a vessel flying the flag of a Member State in possession of a fishing licence obtained under a fisheries agreement depends on the existence in the agreement of specific provisions relating to liability in the case of

---

[22] *Request for an advisory opinion submitted by the Sub-Regional Fisheries Commission (SRFC), Written Statement by the European Commission on Behalf of the European Union, 29 November 2013, ITLOS*, paras 83 and 92.

[23] EU:C:2015:663, para 59.

[24] C-131/03P, EU:C:2006:541, para 94.

[25] EU:C:2015:663, para 58.

[26] Ibid, paras 86–88.

such infringements; it stated that, in the absence of such provisions, the general rules of international law would apply, namely that the responsibility would correspond to the Member State that was the flag State.

However, the ITLOS qualified this consideration in the following paragraphs (paras 171 and 172) on the basis of the due diligence obligation applicable to the international organisation.[27] In this case, the Court considered that the international organisation, as the only contracting party of the fisheries agreement with the coastal State, must ensure that vessels flying the flag of a Member State respect the fishing regulations of the coastal State and do not engage in illegal, unreported and unregulated fishing (IUU fishing) in that State's EEZ; the EU must fulfil its due diligence obligation. Otherwise, the ITLOS considered (para 173) that only the international organisation, not its Member States, will be responsible under the fisheries agreement. That is, if the international organisation fails to comply with its obligation of due diligence, the coastal State (SBFC's Member States) may hold it liable for the infringement of fishing regulations by a fishing vessel flying the flag of one of its Member States when that vessel fishes in its EEZ within the framework of a fisheries agreement concluded between that organisation and the coastal State.[28]

In the author's view, this should lead to a change in the EU's fisheries treaty activity with a view to including 'competence clauses' in the fisheries agreements. These clauses are intended for mixed agreements involving shared competences between the EU and its Member States, which is not the case with fisheries agreements, which are agreements affecting the EU's exclusive competences. However, such an inclusion would result in greater security for third countries and for the EU itself. Likewise, it has to be mentioned that many of these fisheries agreements affect marine areas that are the site of abundant IUU fishing, activities that constitute internationally wrongful acts and that go beyond the scope of the pure conservation of living marine resources for which the EU has exclusive competence.

[27]On the issue of due diligence, see, e.g., Barnidge (2006), pp. 81–121 and Ouedraogo (2011), pp. 307–346.

[28]*Request for an advisory opinion submitted by the Sub-Regional Fisheries Commission (SRFC), Advisory Opinion, 2 April 2015, ITLOS*, pp. 62–63. Moreover, the ITLOS considered that SRFC Members could ask an international organisation or its Members, provided they were Parties to UNCLOS, to inform them of who would be responsible for each specific issue. Both the international organisation and those States should facilitate the concerned information. Otherwise, it would result 'in joint and several liability of the international organization and the member States concerned' (para 174). For a general overview of the EU's international legal responsibility and the shared responsibility between the EU and its Member States for an internationally wrongful act committed, see, e.g., Cortés Martín (2013), pp. 189–199, Gaja (2013) and Palkokefalos (2013), pp. 385–405.

3 The Scope of the Public Access Fisheries Agreements

The judgment of the Court (Grand Chamber) of 26 November 2014 in the joined cases *European Parliament and European Commission v Council of the European Union* (joint cases C-103/12 and C-165/12)[29] addressed the legal basis for Council Decision 2012/19/EU of 16 December 2011.[30] This Decision had been used to adopt the Declaration on the granting of fishing opportunities in EU waters to fishing vessels flying the Venezuelan flag in the EEZ off the coast of French Guiana. Both the European Parliament (EP) and the Commission considered that the legal basis chosen by the Council was wrong. Moreover, the EP claimed that the Decision had been adopted on the basis of an incorrect procedural provision. For its part, the Commission alleged, amongst other things, that the Council had failed to respect the EP's institutional powers when it adopted the Decision.

With the adoption of Decision 2012/19/EU, the Council sought to fill a gap in the legislation regarding the access of fishing vessels flying the Venezuelan flag to the EEZ of an EU Member State—in this case, France's EEZ off the coast of French Guiana.[31] The aim was to circumvent the process of negotiating and concluding an international agreement in order to respond rapidly to the need to provide an international title for access to the French Guiana waters, which had no impact for fisheries in the EU as a whole.[32] Hence, the Council considered that it would be more appropriate to issue a unilateral declaration in the above terms that would fulfil the same function as a fisheries agreement, generating international rights and obligations for the affected parties.

In the author's view, the cornerstone of this action for annulment is the legal basis used by the Council to adopt Decision 2012/19/EU, which, as noted, sought to offer a quick legal response to an activity that had existed for decades. Thus, the Council invoked Art. 43(3) TFEU in conjunction with Art. 218(6)(b) TFEU, whilst the EP and the Commission held that the contested Decision should have been adopted according to Art. 43(2) TFEU in conjunction with Art. 218(6)(a)(v) TFEU since it amounted to an international agreement for a third country—in this case Venezuela—to access and engage in fishing activities in EU waters and, therefore, required the EP's prior approval.

[29]EU:C:2014:2400. For a larger study of this judgment, see Oanta (2016), pp. 200–208.

[30]O.J. L 6/8 (2012).

[31]French Guiana is one of six French overseas departments (Guadalupe, French Guiana, Martinique, Mayotte, Reunion and Saint Martin) and one of the EU's nine outermost regions (together with: the Azores, the Canary Islands, Guadalupe, Madeira, Martinique, Mayotte, Reunion and Saint Martin). See 'The Outermost Regions: European Regions of Assets and Opportunities' (Luxembourg, 2012). With regard to this case, it should be noted that the fishing vessels flying the Venezuelan flag had been fishing in that EEZ for several decades and, moreover, that the French Guiana processing industry has begun to rely on those fish landings, which are of great economic and social importance for the region's population.

[32]Opinion of Advocate General Sharpston in the joined cases European Parliament and Commission v Council, C-103/12 and C-165/12, EU:C:2014:334, para 108.

Regarding the legal basis for the adoption of an international fisheries agreement, this case is thought to reflect the tension of recent years, following the entry into force of the Lisbon Treaty, between the EP and the Commission, on the one hand, and the Council, on the other, in relation to the legislative procedure for the adoption of fisheries legislation. As is well known, prior to 1 December 2009, the EP had played only a marginal role in the legislative process in the field of the CFP. However, today, it has recognised legislative powers under Art. 43 TFEU.[33] Thus, Art. 43(2) TFEU provides for the ordinary legislative procedure for the adoption of provisions that are 'necessary for the pursuit of the objectives' of the CFP, whilst Art. 43(3) includes a reserved executive procedure for the 'fixing and allocation of fishing opportunities'. This situation has been interpreted by part of the doctrine[34] as a *sui generis* procedure and an exception to the legislative procedure under Art. 43(2) TFEU.

At the same time, Art. 218 TFEU also reflects the significant increase in the EP's influence in the adoption by the EU of fisheries treaties.[35] Art. 218(6)(b) TFEU provides that the Council, 'on a proposal by the negotiator', may conclude an agreement between the EU and a third country or international organisation 'after consulting the European Parliament', which must issue an opinion 'within a time-limit which the Council may set depending on the urgency of the matter. In the absence of an opinion within that time-limit, the Council may act'. On the other hand, in accordance with Art. 218(6)(a)(v) TFEU, unless it falls within the scope of the Common Foreign and Security Policy, the Council shall adopt the decision concluding an agreement between the EU and a third country or international organisation subject to the approval of the EP with respect to those agreements related to fields to which the ordinary legislative procedure applies or, where the EP's consent is required, the special legislative procedure. Moreover, in an urgent situation or emergency, the EP and the Council 'may [. . .] agree upon a time-limit for consent'.

The choice of the legal basis for such a legislative act is of extraordinary importance since, if it is wrong, the concluding act could be invalidated, thereby vitiating the EU's consent to be bound by the agreement signed.[36] In addition, as stated in the CJEU case law,[37] the choice of legal basis for an EU act must be based on objective factors amenable to judicial review, such as, in particular, the aim and

[33]See De Sadeleer (2014), p. 801.

[34]Ibid.

[35]The same position has been expressed by professor Yves Petit. See Petit (2015), p. 64.

[36]Opinion 2/00, of 6 December 2001, EU:C:2001:664, para 5; Opinion 1/08, of 30 November 2009, EU:C:2009:739, paras 108–110.

[37]Judgment of the Court Parliament v Council, C-130/10, EU:C:2012:472, para 42; Judgment of the Court United Kingdom v Council, C-431/11, EU:C:2013:589, para 44.

content of the act.[38] The Court held that the purpose of the statement concerning the allocation of fishing opportunities to vessels flying the Venezuelan flag in the EEZ off the coast of French Guiana was not to ensure 'the fixing and allocation of fishing opportunities' in the sense of Art. 43(3) TFEU but rather to offer the Latin American country the opportunity to participate in the exploitation of fisheries resources in the EEZ of French Guiana, under the conditions set by the EU, and to ensure compliance with the requirement that the CFP provisions regarding conservation and control and other CFP regulations be met.[39]

As for the question of the issues raised by the notion of an international agreement concluded in the field of fisheries, in the present case the EU had offered to allow a limited number of fishing vessels flying the Venezuelan flag to operate in relation to part of the surplus allowable catches in French Guiana's EEZ.[40] The Court considered that the offer made to Venezuela was not a technical implementing measure but rather a measure involving the adoption of an autonomous decision, which should be made in the light of the EU policy interests pursued through its common policies, particularly its CFP.

In this judgment, the Court once again decided on a very broad concept of agreement. Indeed, as noted in its case law, on the one hand, it is irrelevant whether a treaty consists of a single document or several related legal instruments and, on the other hand, the term 'agreement' must be understood in a general sense to designate any kind of binding commitment expressed by a subject of international law regardless of its formal designation.[41]

In this case, Advocate General Sharpston expressed another position, moving away from this notion of 'agreement' and considering that the EU's international legal personality allowed it to issue a unilaterally binding declaration. However, he noted that having the capacity to adopt a treaty 'does not suffice to conclude that, in accordance with the principle of conferral, the EU is competent to do so'.[42] The Advocate General further considered that there were two possibilities regarding the legal nature of the declaration made by the EU in the contested decision: either it was a unilaterally binding instrument for the EU or it was a unilateral declaration that would 'produce legal effects only when subsequently accepted by the third State in whose favour it was made (in which case it is only one side of an international agreement)'.[43] Finally, the Advocate General concluded that

---

[38]EU:C:2014:2400, para 51.

[39]Ibid, paras 75, 77 and 78. Concretely, it was about the paras 1 and 3 of the EU's Declaration, which was an Annex of the Decision 2012/19/EU. For a presentation of conservation and management measures adopted by the EU, see Oanta (2015), pp. 247–251. See also Sobrino Heredia et al. (2010), pp. 193–256.

[40]Later, on 26 March 2012, the European Commission adopted Decision C (2012) 2162, which authorised 38 fishing vessels flying the Venezuelan flag to fish in French Guiana's EEZ.

[41]The first position of the Luxembourg Count in this regard was with the occasion of the Opinion 1/75 (EU:C:1975:145). This Opinion has been repeatedly recalled in different occasions. See, amongst others, Opinion 2/92 (EU:C:1995:83), para 8; EU:C:2014:334, para 83.

[42]Opinion of Advocate General Sharpston, EU:C:2014:334, para 64.

[43]Ibid, para 72.

Venezuela had not agreed to be bound by the Declaration 'as an agreement concluded between it and the EU'.[44]

Based on its interpretation of the act as an international agreement, however, the Court decided to annul the contested Decision as it considered that it should have been adopted by virtue of Art. 43(2) TFEU in conjunction with Art. 218(6)(a)(v) TFEU. Furthermore, the Court decided to maintain the effects of the Decision, as requested by both the Commission and the Council,[45] until the adoption of a new decision in this field, under the TFEU provisions. This happened on 14 September 2015, with the adoption of Council Decision (EU) 2015/1565.[46]

Although the Court agreed with the Advocate General in declaring the contested Decision null and void and in maintaining its effects until a new decision could be adopted in accordance with the provisions of Art. 43(2) TFEU in conjunction with Art. 218(6)(a)(v) TFEU, it differed in its qualification of the act, which, in the author's view, can hardly be considered an international agreement. In so doing, the Court missed an excellent opportunity to rule, for the first time, on the EU's capacity to issue binding unilateral acts[47] in the field of international maritime affairs.

Thus, the Court strengthened its broad and, undoubtedly, conservative interpretation of the notion of 'international agreement'. On the one hand, this position will facilitate the conclusion of fisheries agreements; on the other, it is so permissive that it blurs the legal scope of such agreements. In keeping with the Opinion of the Advocate General, in the author's view, the CJEU could have addressed the nature of a unilateral declaration and its applicability to the EU's international fisheries activity. However, rather than embarking down the unexplored path of unilateral declarations, the CJEU opted to take a more prudent position, that is, to benefit from the broad notion of 'international agreement' that it defends in its case law and that perhaps best fits the factual context in which the case unfolded.

---

[44]Ibid, para 81.

[45]For its part, the EP had stated that it would not take a negative position toward such a solution.

[46]This new Decision differs from the annulled Decision only with regard to the legal basis used for its adoption, namely Art. 43(2) TFEU in conjunction with Art. 218(6)(a)(v) TFEU. See Council Decision 2015/1565/EU *on the approval, on behalf of the European Union, of the Declaration on the granting of fishing opportunities in EU waters to fishing vessels flying the flag of the Bolivarian Republic of Venezuela in the exclusive economic zone off the coast of French Guiana*, O.J. L 244/55 (2015). See also Proposal for a Council Decision on the approval, on behalf of the European Union, of the Declaration on the granting of fishing opportunities in EU waters to fishing vessels flying the flag of the Bolivarian Republic of Venezuela in the exclusive economic zone off the coast of French Guiana, COM (2015) 1 final, 12.01.2015.

[47]De Pietri (2015), pp. 22–32. On the issue of unilateral acts under the international law, see, e.g., Degan (1994), pp. 149–266 and Tomuschat (2008), pp. 1487–1507.

4 Some Aspects Arising from the Application of the Fishing Agreement Signed by the EU and Morocco and Its Successive Protocols

The activity of European fishing vessels carried out under the successive fisheries agreements concluded by the EU and Morocco[48] has sparked numerous controversies due to both possible imbalances in the allocation of rights and obligations to the parties—which have given rise to recurring criticism in the socio-economic sector of European fisheries (particularly in Spain)—and, more recently, fishing in waters off the Western Sahara coasts.[49] The CJEU examined some of the issues raised by this fisheries treaty activity on two occasions in 2014 and 2015, focusing mainly on the European fishing activity conducted in Western Sahara waters.

First, the judgment of the Court of 9 October 2014 in *Ahlström and Others* (C-565/13)[50] refers to a question submitted under the provisions of Art. 267 TFEU in the context of criminal proceedings before a Swedish court, in which the defendants had been accused of engaging in illegal fishing practices in Western Sahara waters between April 2007 and May 2008;[51] it addresses the interpretation of the most recent fisheries agreement concluded between the EU and Morocco, which entered into force on 1 April 2007.[52]

Second, the judgment of the EU's General Court (GC) of 10 December 2015 in *Front Polisario v Council* (T-512/12)[53] refers to an action for annulment brought by the Front Polisario in relation to Council Decision 2012/497/EU of 8 March 2012 on the conclusion of an Agreement in the form of an Exchange of Letters between the EU and Morocco concerning reciprocal liberalisation measures on agricultural products, processed agricultural products, fish and fishery products, the replacement of Protocols 1, 2 and 3 and their Annexes and amendments to the Euro-Mediterranean Agreement establishing an association between the EU and its

[48]For a detailed analysis of the different fishing agreements signed between the EU and Morocco, see, e.g., Lahlou (2005), pp. 39–46, Milano (2006), pp. 1–33 and Sobrino Heredia (2012), pp. 235–261.

[49]See Chapaux (2012), pp. 217–237 and Dawidowicz (2013), pp. 250–276.

[50]EU:C:2014:2273. For a larger study of this judgment, see Andreone (2014), pp. 680–686; Oanta (2016), pp. 216–219.

[51]For an overview of the illegal fisheries phenomena see Oanta (2014b), pp. 149–197.

[52]It has been the fourth fishing agreement signed between these two subjects under international law: the first one was signed on 25 May 1988, the second one on 1 May 1992, the third on 1 December 1995. The Protocol in force of the 2006 Agreement was published through the Council Decision 2013/720/EU (O.J. L 328 (2013)).

[53]Judgment of the General Court of 10 December 2015 Front populaire pour la libération de la saguia-el-hamra et du rio de oro (Front Polisario) v Council, T-512/12, EU:T:2015:953. See Gosalbo Bono (2016), pp. 21–77, King (2014), pp. 71–89 and Soroeta Liceras (2016), pp. 202–238.

Member States, on the one hand, and Morocco, on the other hand.[54] Although this second judgment raises multiple international legal issues of great significance for the Western Sahara,[55] it really addresses only a few issues related to the notion of public access fisheries agreements concluded by the EU with Morocco. The GC's judgment in the case *Front Polisario v Council* (T-180/14) is expected to be more relevant for the field of public access fisheries agreements concluded by the EU with third countries, as the action for annulment brought by the Front Polisario will refer specifically to fisheries activities off the coast of the Western Sahara under the provisions of the Protocol signed between the EU and Morocco setting out the fishing opportunities and financial contribution provided for in the fisheries partnership agreement between them.[56]

Regarding the judgment of 9 October 2014 in *Ahlström and Others*, in the author's view the scope of the so-called 'exclusivity clause' that accompanies the public access fisheries agreements is perhaps the most interesting point for the purposes of this chapter. It should be noted that this clause excludes any type of fishing done by EU Member States' vessels outside the framework of the said agreements. Consequently, the European fleet's fishing activities in the waters covered by the agreements must be carried out solely and exclusively within the framework of the agreements.

In this regard, it has to be mentioned that the exclusivity clause is the most important issue to be addressed by the Court in this case. The Court was essentially asked whether Art. 6 of the fisheries agreement concluded by the EU with Morocco in 2006 must be interpreted as precluding any possibility for EU vessels to carry out fishing activities in waters under the sovereignty or jurisdiction of Morocco on the basis of a licence issued by Moroccan authorities without the intervention of the EU's competent authorities.

It should also be stressed that the aforementioned Art. 6 stipulates that EU fishing vessels 'may fish in Moroccan fishing zones only if they are in possession of a fishing licence issued under this Agreement. The exercise of fishing activities by Community vessels shall be subject to the holding of a licence issued by the competent Moroccan authorities at the request of the competent Community authorities'. Moreover, '[f]or fishing categories not covered by the Protocol [setting out the fishing opportunities and financial contribution provided for in the Fisheries Agreement ('the Protocol")], licences may be granted to Community vessels by the Moroccan authorities'. However, the granting of such licences is dependent on the

[54] O.J. L 241/2 (2012). The Association Agreement EU-Morocco was concluded in Brussels on 26 February 1996 (O.J. L 70/2 (2000)).

[55] Indeed, this judgment addresses issues unrelated to the purpose of this Chapter, namely: the legitimacy of Front Polisario when submitting applications to the CJEU, the legal status of the Western Sahara and the holder of sovereignty over Western Sahara resources, and the existence of an absolute prohibition on concluding an international agreement that could be applied to a territory controlled *de facto* by a State whose sovereignty is not recognised over that territory under international law. For an overview of these issues, see Soroeta Liceras (2016).

[56] Action brought on 14 March 2014 by Front Polisario against the Council (T-180/14).

receipt of a favourable opinion from the European Commission. Furthermore, the potential access by EU vessels to a third country's waters will be determined under a bilateral public access fisheries agreement concluded by the EU with the said third country, and only then, on the one hand, will the Council be responsible for granting fishing opportunities according to the provisions of the agreement and, on the other, will the Commission be able to grant fishing licences to EU Member States for them to grant to vessels flying their flag.

With regard to fisheries, amongst other things, the exclusivity clause seeks to prevent EU vessels from fishing outside the framework of a public access fisheries agreement[57] or failing to contribute to the long-term conservation of fisheries resources, as stated in the second written statement it submitted to the ITLOS on behalf of the EU on 13 March 2014 in Case No 21.[58]

The Court found that 'it cannot be accepted that Community vessels should be able to access Moroccan fishing zones in order to carry out fishing activities' through the conclusion of a specific contract 'with a Moroccan company holding a licence issued by the Moroccan authorities to Moroccan owners [...] or by using any other legal instrument in order to access those fishing zones for the purpose of carrying out such activities there outside the scope of the Fisheries Agreement and, consequently, without the intervention of the competent European Union authorities'.[59] Therefore, the aforementioned Art. 6 excludes any possibility for EU vessels to carry out fishing activities in the fishing areas of a third country with which the Union has concluded a public access fisheries agreement on the basis of a licence issued by the authorities of that country without the intervention of the competent EU authorities.

In the author's view, this judgment reinforces the value of the public access fisheries agreements concluded by the EU with more than 30 countries as necessary instruments for responsible and sustainable fishing; they contain the same obligations for the EU vessels fishing in third-country waters as those imposed on all EU fishing vessels fishing in EU waters. In so doing, the EU is seeking to prevent European vessels from changing their flag or fishing under a private access fisheries agreement. In such a context, the EU would not be able to hold fishing vessels that infringe the international, EU and national legislation regarding fish stocks conservation accountable.

Regarding the GC's judgment of 10 December 2015 in *Front Polisario v Council*, it is considered that the ninth plea in law used by Front Polisario in this case is the most relevant for the field of fisheries. It has to be underlined that Front Polisario relied, on the one hand, on the Association Agreement between the EU

---

[57]This clause is also provided for in Art. 31(6)(b) Regulation (EU) 1380/2013.

[58]*Request for an advisory opinion submitted by the Sub-Regional Fisheries Commission (SRFC). Second Written Statement by the European Commission on Behalf of the European Union, 13 March 2014, ITLOS*, para 27.

[59]EU:C:2014:2273, paras 33–35.

and Morocco and, on the other hand, on the principles of UNCLOS; the fisheries Agreement or its most recent Protocol did not receive special attention in this case.

In relation to the Association Agreement, Front Polisario claimed that this treaty infringed 'the right to self-determination and the rights which derive from that, in particular, sovereignty over natural resources and the primacy of the interests of the inhabitants of Western Sahara'.[60] With respect to the UNCLOS, Front Polisario argued that, according to the provisions of this Convention, the people of Western Sahara had sovereignty over, firstly, the waters adjacent to the coast of Western Sahara and, secondly, the infringement of the basic criterion resulting from the UNCLOS, the Association Agreement, the Protocol 4 of the fisheries partnership agreement concluded by the EU with Morocco in 2006 and the Agreement of an exchange of letters concerning the provisional application of the Agreement on cooperation in the sea fisheries sector between the European Community and Morocco initialled in Brussels on 13 November 1995.[61]

This judgement has various positive contributions in this field. Thus, the General Court remembers that in the case *Intertanko and Others*, the Court of Justice held that 'the nature and the broad logic of that convention prevent the Courts of the European Union from being able to assess the validity of an EU measure in the light of that convention'[62] and also reiterates 'that the EU institutions enjoy a wide discretion as regards whether it is appropriate to conclude an agreement with a non-member State which will be applied on a disputed territory'.[63] Although the General Court considers correct the Council's argument of not being liable for any actions committed by a country, which has an agreement concluded with the EU,[64] the EU underlines the special situation of the Western Sahara, 'which is in fact administered by a non-member State, in this case the Kingdom of Morocco, although it is not included in the recognised international frontiers of that non-member State', and also the fact that Morocco neither has any mandate granted by the United Nations or by another international body for the administration of the Western Sahara territory nor transmits to the United Nations information relating to that territory, according to Art. 73(e) of the United Nations Charter.[65] Finally, the General Court has decided to annul the provisions of Council Decision 2012/497/ EU referring to Western Sahara.

In the author's view, this judgment is only the first judicial step regarding the EU–Morocco relations that affect Western Sahara in the field of fisheries. This decision has already been the subject of an appeal before the Court. We are referring to the case *Council v Front Polisario* (C-104/16 P), which has been

[60]EU:T:2015:953, para 189.

[61]O.J. L 306/1 (1995). See EU:T:2015:953, paras 190–191.

[62]Judgment of 3 June 2008 *Intertanko and Others*, C-308/06, EU:C:2008:312, para 65. See also EU:T:2015:953, para 195.

[63]EU:T:2015:953, para 223.

[64]Ibid, paras 230–231.

[65]Ibid, paras 232–233.

brought before the Court on 11 March 2016. It has to be mentioned that on 7 April 2016, the President of the Court has ordered this case to be judged through the accelerated procedure in accordance with Art. 133 of the Court's Rules of Procedure and on 13 September 2016 the Advocate General in this case has published his Opinion. Finally, on 21 December 2016 the Court of Justice has published its judgment in this case, deciding to set aside the judgment of the General Court of 10 December 2015 as well as to dismiss the action brought by the Front Polisario as inadmissible.[66]

In addition, the General Court will have to publish its judgment in the case *Front Polisario v Council* (T-180/14). In this case, Front Polisario relies on 12 pleas in law in support of its action for annulment of Council Decision 2013/785/EU of 16 December 2013 on the conclusion, on behalf of the European Union, of the Protocol between the European Union and the Kingdom of Morocco setting out the fishing opportunities and financial contribution provided for in the Fisheries Partnership Agreement between the European Union and the Kingdom of Morocco.[67] It claims that the contested Decision, amongst other things, is contrary to the objectives of the CFP and also represents an infringement of the UNCLOS provisions as Morocco sets fishing quotas for waters not under its sovereignty, as well as authorises EU vessels to exploit fisheries resources that are under the sole sovereignty of the Sahrawi people.

5 Final Remarks

Despite the EU's extensive experience with fisheries treaty activity at the global level, recently the CJEU's intervention has been needed to shed light on several relevant issues in this field.

Thus, the judgment in the case *Council v Commission* (C-73/14) solved a jurisdictional problem, reasserting the exclusive nature of the EU's competences with regard to the conservation and management of living marine resources and, therefore, the European Commission's right to explain the EU's position on issues affecting fisheries resources before an international court. It likewise highlighted the issue of the EU's possible international legal responsibility for infringements of the provisions of public access fisheries agreements.

In the joined cases *Parliament and Commission v Council* (C-103/12 and C-165/12), the Court considered, first, that public access fisheries agreements should be adopted according to Art. 43(3) TFEU in conjunction with Art. 218(6)(a)(v) TFEU, thereby strengthening the EP's role in the field of fisheries. Second, the judgment found that a unilateral declaration made by the EU regarding part of the surplus allowable catches in the EEZ of one of its Member States that is later accepted by a

[66]EU:C:2016:232. EC:C:2016:677. EC:C:2016:973.

[67]O.J. L 349/1 (2013).

third country should be considered part of an agreement concluded by the EU and the said country on the authorisation of exploitation under the conditions set out in the declaration. This legal solution is consistent with the CJEU's classic case law, which has interpreted public access fisheries agreements broadly; however, it also represents a missed opportunity to address the international scope of a unilateral declaration made by the EU in relation to these issues.

Finally, in *Ahlström and Others* (C-565/13) and *Front Polisario v Council* (T-512/12), the CJEU handed down two interesting judgments on various extremely important aspects of the public access fisheries agreement (including the corresponding Protocol) concluded between the EU and Morocco. In the first one, the Court showed the EU's clear position, regarding the exclusivity clause as a tool to reinforce the role of the public access fisheries agreement in achieving responsible and sustainable fisheries in the waters under the jurisdiction of a third country and also for fighting the reprehensible practice of fishing under private fisheries agreements that encourages overexploitation of resources for profit motives. And, in the second one, the General Court has made the first step forward in the international recognition of the special situation that Western Sahara is living under the *de facto* control of Morocco, although this African country does not any legitimacy on Western Sahara under international law.

References

Andreone G (2007) La nouvelle génération des accords de pêche conclus par la Communauté européenne avec les pays tiers. Annuaire du Droit de la Mer 12:326–347

Andreone G (2014) Jurisprudence. Annuaire du Droit de la Mer 19:680–686

Antonova A (2015) Reforming European Union participation in fisheries management and conservation on the high seas. Ocean Yearb 29:125–143

Barnidge R-P (2006) The due diligence principle under international law. Int Community Law Rev 8(1):81–121

Becker M-A (2013) Sustainable fisheries and the obligations of flag and coastal states: the request by the sub-regional fisheries commission for an ITLOS Advisory Opinion. ASIL Insights 17

Chapaux V (2012) La cuestión del Acuerdo de Pesca entre las Comunidades Europeas y Marruecos. In: Ponce de León M, Arts K, Pinto Leite P (eds) El derecho internacional y la cuestión del Sahara Occidental. International Platform of Jurists for East Timor, Leiden, pp 217–237

Cortés Martín J-M (2013) European exceptionalism in international law? The European Union and the system of international responsibility. In: Ragazzi M (ed) Responsibility of international organizations: essays in memory of Sir Ian Brownlie. Brill, Leiden, pp 189–199

Dawidowicz M (2013) Trading fish or human rights in Western Sahara? Self-determination, non-recognition and the EC-Morocco Fisheries Agreement. In: French D (ed) Statehood and self-determination. Reconciling tradition and modernity in international law. Cambridge University Press, Cambridge, pp 250–276

De Pietri D (2015) Comentario a la sentencia de la Gran Sala del Tribunal de Justicia de 26 de noviembre de 2014, Asuntos acumulados C-103/2012 y C-165/2012, Parlamento Europeo y Comisión/Consejo. La Ley Unión Europea 24:22–32

De Sadeleer N (2014) Base juridique d'une déclaration du Conseil de l'Union européenne adressée à un État tiers. Revue des affaires européennes 4:801–806

De Yturriaga Barberán J-A (2009) The European community and some problems of the law of the sea concerning fisheries. In: Casado Raigón R, Cataldi G (dirs) L'évolution et l'état actuel du droit international de la mer. Mélanges de droit de la mer offerts à Daniel Vignes. Bruylant, Bruxelles, pp 269–297

Degan V-D (1994) Unilateral act as a source of particular international law. Finnish Yearb Int Law 5:149–266

Franckx E and Van den Bossche K (2010) The role of the European Union in Regional Fisheries Organizations. In: Russell D-A, VanderZwaag D-L (eds) Recasting transboundary fisheries management arrangements in light of sustainability principles: Canadian and international perspectives. Nijhoff, Leiden/Boston, pp 419–448

Gaja G (2013) The Relations Between the European Union and Its Member States from the Perspective of the ILC Articles on Responsibility of International Organizations. SHARES Research Paper 25

Gosalbo Bono R (2016) El Frente Polisario, las normas del derecho internacional y el derecho de la Unión Europea. Apuntes en torno a la sentencia del Tribunal General, de 10 de diciembre de 2015, T-512/12, Frente Polisario/Consejo. Revista de Derecho Comunitario Europeo 53:21–77

King S-J (2014) The emergence and Policits of the Polisario front. In: Boukhars A, Roussellie J (eds) Perspectives on Western Sahara: myths, nationalisms and geopolitics. Rowmand and Littlefield, Lanham, pp 71–89

Lahlou A (2005) Une nouvelle approche de la relation halieutique Union européenne – Maroc: d'un accord de dimension politique à une convention à vocation économique. In: Casado Raigón R (dir) L'Europe et la mer. Pêche, navigation et environnement marin/Europe and the Sea. Fisheries, Navigation and Marine Environment. Bruylant, Bruxelles, pp 39–46

Le Manach F, Andriamahefazafy M, Harper S, Harris A, Hosch G, Lange G-M, Zeller D, Sumaila U-R (2013) Who gets what? Developing a more equitable framework for EU fishing agreements. Marine Policy 38:257–266

Lijnzaad L (2014) Declarations of competence in the law of the sea, a very European affair. In: Lodge M-W, Nordquist M-H (eds) Peaceful order in the World's oceans. Essays in Honor of Satya N. Nandan. Brill/Nijhoff, Leiden/Boston, pp 187–207

Milano E (2006) The new fisheries partnership agreements between the European Community and the Kingdom of Morocco: fishing too south? Leiden J Int Law 1:1–33

Molenaar E-J (2002) Netherlands fisheries in a European and international legal context. Neth Yearb Int Law XXXIII:137–138

Oanta G-A (2014a) The international tribunal for the law of the sea and the polar regions. Law Pract Int Courts Tribunals 13(3):286–305

Oanta G-A (2014b) Illegal fishing as a criminal act at sea. In: Trapp K-N, Papastavridis E (eds) Criminalité en Mer/Crimes at Sea. Martinus Nijhoff Publishers, Leiden/Boston, pp 149–197

Oanta G-A (2015) New steps in the control of illegal, unreported and unregulated fishing. In: Koch H-J, König D, Sanden J, Verheyen R (eds) Legal regimes for environmental protection: governance for climate change and ocean resources. Brill/Nijhoff, Leiden/Boston, pp 229–257

Oanta G-A (2016) Tres sentencias claves para la delimitación del contorno jurídico de las competencias convencionales de la Unión Europea en el ámbito pesquero. Revista de Derecho Comunitario Europeo 53:201–231

Ouedraogo A (2011) La neutralité et l'émergence du concept de due diligence en droit international. L'affair de l'Alabama revisitée. J Hist Int Law 13:307–346

Ould Ahmed Salem Z (2009) Les écueils du "partenariat": l'Union européenne et les accords de pêche avec l'Afrique. Karthala, Nouakchott

Palkokefalos I (2013) Shared responsibility aspects of the dispute settlement procedures in the law of the sea convention. J Int Dispute Settlement 4(2):385–405

Petit Y (2015) Politique commune de la pêche. Annulation de la décision 2012/19/UE du Conseil pour défaut de bases juridique et procédurale adéquates. Revue de droit rural 435:64

Ruiloba García E (2005) Les accords de pêche entre la Communauté européenne et des pays tiers. In: Casado Raigón R (dir) L'Europe et la mer. Pêche, navigation et environnement marin/ Europe and the Sea. Fisheries, Navigation and Marine Environment. Bruyland, Bruxelles, pp 333–345

Sánchez-Tabernero S-R (2015) La ilegalidad de las decisiones híbridas en el marco de la celebración de acuerdos mixtos. Revista de Derecho Comunitario Europeo 52:1057–1073

Sobrino Heredia J-M (2002) L'Union européenne face aux transformations du droit de la mer. In: Cataldi G (dir) La Méditerranée et le droit de la mer à l'aube du 21e siècle/The Mediterranean and the Law of the Sea at the dawn of the 21st Century. Bruylant, Bruxelles, pp 53–82

Sobrino Heredia J-M (2012) Les relations euro-marocaines concernant la pêche. In: Del Valle Gálvez A (dir) Inmigración, seguridad y fronteras. Dykinson, Madrid, pp 235–261

Sobrino Heredia J-M, López Veiga E-C, Rey Aneiros A (2010) La integración del enfoque ecosistémico en la política pesquera común de la Unión Europea. Tirant lo Blanch, Valencia

Sobrino Heredia J-M, Oanta G-A (2015) The sustainable fisheries partnership agreements of the European Union and the objectives of the common fisheries policy: fisheries and/or development? Span Yearb Int Law 19:61–85

Soroeta Liceras J (2016) La sentencia de 10 de diciembre de 2015 del Tribunal General de la UE (T-512/12), primer reconocimiento en vía judicial europea del estatuto del Sahara Occidental y de la subjetividad internacional del Frente Polisario. Revista General de Derecho Europeo 38:202–238

Tomuschat C (2008) Unilateral acts under international law. In: Meziou K (dir) Droit et culture: mélanges en l'honneur du Doyen Yadh Ben Achour. Centre de Publications Universitaire, Tunis, pp 1487–1507

Treves T (2008) The European community and the European Union and the law of the sea: recent developments. Indian J Int Law 48(1):1–20

Van der Burgt N (2013) The contribution of international fisheries law to human development. An analysis of multilateral and ACP-EU fisheries instruments. Martinus Nijhoff, Leiden/Boston

Vázquez Gómez E (2002) Las organizaciones internacionales de ordenación pesquera: la cooperación para la conservación y la gestión de los recursos vivos del alta mar. Consejería de Agricultura y Pesca, Sevilla

Witbooi E (2012) Fisheries and sustainability. A legal analysis of EU and West African agreements. Martinus Nijhoff, Leiden/Boston

The Protection of Biodiversity in the Framework of the Common Fisheries Policy: What Room for the Shared Competence?

Marta Chantal Ribeiro

1 Introduction

The reform of the Common Fisheries Policy—especially anchored in Regulation (EC) No. 2371/2002, of 20 December, on the conservation and sustainable exploitation of fisheries resources—undertaken by Regulation (EU) No. 1380/2013, of 11 December,[1] clearly aims to enhance the protection of marine biodiversity.[2] Based on the basic regulation of 2002, the concern with species and ecosystems, that is, beyond the immediate sustainability of the targeted stocks, was a recurrent and priority aspect in the decision-making process of the EU regulations applicable to the management of fisheries. In the reform of 2013, this desideratum was reinforced, as stated notably in Art. 2 of Regulation No. 1380/2013: application of the ecosystem-based approach, fostering the collection of scientific data, decisions taken under the best available scientific advice, new strategy for discards, and coherence with the European Union environmental legislation. It is precisely the latter objective that will be in the core of the discussions: "The CFP shall, in

This chapter is based on a version published in Portuguese, entitled "A protecção da biodiversidade marinha no quadro do Regulamento (UE) n.° 1380/2013. A perspectiva do Estado-Membro costeiro", in Pueyo Losa & Jorge Urbina (Coord.), La Gobernanza Marítima Europea. Retos planteados por la reforma de la política pesquera común, Navarra, Thomson Reuters - Aranzadi, November 2016, pp. 73–105 [research project: The Reform of the International and European Fisheries Governance. Challenges for the Spanish Fishing Industry (DER2013-45923-R), Spain].

[1]O.J. L 354/22 (2013).

[2]In general see Churchill and Owen (2010).

M.C. Ribeiro (✉)
Faculty of Law, University of Porto, Porto, Portugal

Interdisciplinary Centre of Marine and Environmental Research, Porto, Portugal
e-mail: mchantal@direito.up.pt

G. Andreone (ed.), *The Future of the Law of the Sea*,
DOI 10.1007/978-3-319-51274-7_4

particular: (. . .) be coherent with the Union environmental legislation, in particular with the objective of achieving a good environmental status by 2020 as set out in Article 1(1) of Directive 2008/56/EC."[3] The reinforcement of the environmental dimension in the management of fisheries pursued by Regulation No. 1380/2013 has, in fact, the consequence of amplifying opaque and overlapping legal solutions inherited from the previous regulation. This is noticeable, especially, in the interaction with the goals set forth by the *Marine Strategy Framework Directive* (2008/56/EC) and the directives under which the Natura 2000 network was being developed (92/43/EEC and 2009/147/EC). Furthermore, the balance established by Regulation No. 1380/2013 between the exclusive competence of the European Union for the conservation of fisheries resources[4] and, in the domain of the shared competences,[5] the competence of the coastal Member States for the protection of marine biodiversity is highly controversial. These two aspects are the axes of the analysis developed in the following pages. Attention will be focused on the interpretation of Arts. 11, 19, and 20—Art. 11 because of its direct connection with marine protected areas (MPAs) and Arts. 19 and 20 because these articles favor the protection of the ecosystems in general.

A prior clarification must be made concerning the terminology "conservation of marine ecosystems" used by Regulation No. 1380/2013. Our understanding is that the word "conservation" is used in a wide sense, including, on the one hand, the conservation of ecosystems from which the fish stocks and the continuity of fisheries are directly dependent (this is the obvious example of Art. 8) and, on the other hand, the protection of ecosystems in a strict environmental sense, that is, ecosystems negatively affected by fisheries but with no direct relation to the sustainability of fish stocks.

2 Interaction with the Marine Strategy Framework Directive and the Natura 2000 Network: Interpretation of Art. 11 of Regulation No. 1380/2013

One of the novelties introduced by Regulation No. 1380/2013 is Art. 11 on the conservation measures that are necessary for the purpose of complying with Member States' obligations under Art. 13(4) of Directive 2008/56/EC, of 17 June (*Marine Strategy Framework Directive*—MSFD)[6]; Art. 4 of Directive 2009/147/

[3]See paras. 11 and 25 of the Preamble, Art. 2(5)(j) and Art. 11 of Regulation No. 1380/2013.

[4]Art. 3(1)(d) TFEU. For the history of the exclusive competence of the European Union for the conservation of marine biological resources under the common fisheries policy, see Churchill and Owen (2010) p. 3 et seq. and p. 302 et seq. See also the case law initiated by the *AETR/ERTA* case (31.03.1971, Case 22/70), the *Kramer* case (14.07.1976, Case 3, 4 e 6/76) and the *Commission v. United Kingdom* case (05.05.1981, Case 804/79).

[5]Art. 4(2)(e), TFEU.

[6]O.J. L 164/19 (2008).

EC, of 30 November (*Birds Directive*);[7] or Art. 6 of Directive 92/43/EEC, of 21 May (*Habitats Directive*).[8]

The *Birds* and *Habitats* Directives are the legal bases for the implementation of the Natura 2000 network, that is, a coherent European ecological network of protected areas called "special protection areas" (*Birds*) and "special areas of conservation"[9] (*Habitats*). On the other hand, Natura 2000 network is an important axis of the MSFD. In fact, marine protected areas are crucial[10] for the achievement or maintenance of a good environmental status in the marine environment. Therefore, for this purpose, besides the Natura 2000 network, national and international networks of MPAs are also relevant.[11]

According to scientific data, repeatedly across the years, fishing activities are one of the major threats for the marine biodiversity, namely due to the overexploitation of stocks, by-catch, and damage caused by the fishing nets. Hence, the restriction and/or prohibition of fishing activities are very common measures of the management plans of MPAs. In this scenario, it is important to ascertain whether Art. 11 facilitates or hinders the accomplishment by the coastal Member States of the "2020" goals, that is:

> On the one hand, by 2020, at least, 10 per cent of coastal and marine areas, especially areas of particular importance for biodiversity and ecosystem services, are conserved through effectively and equitably managed, ecologically representative and well connected systems of protected areas. This goal was launched by the Contracting Parties to the Convention on Biological Diversity (CBD) in 2004.[12]
>
> On the other hand, the achievement or maintenance of a good environmental status in the marine environment by the year 2020 at the latest, as laid down in Art. 1 of the MSFD.

---

[7]O.J. L 20/7 (2010).

[8]O.J. L 206/7 (1992).

[9]Under Art. 4 of *Habitats* Directive, the designation of a special area of conservation implies three steps. First step: Member States propose a list of sites; second step: European Commission adopts a list of sites of Community importance; third step: Member States designate the special areas of conservation. The sites benefit from a preventive protection since the first step. See Art. 4(5) and Art. 6(2)(3)(4) of the *Habitats* Directive as interpreted by the European Union Court of Justice (ECJ) in the *Dragaggi* case (13.01.2005, Case C-117/03), in the *Bund Naturschutz in Bayern eV and others v. Freistaat Bayern* case (14.09.2006, Case C-244/05, paras. 41, 44 and 46) and in the *Commission v. Spain case* (Iberian lynx, 20.05.2010, Case C-308/08, para. 21).

[10]See paras. 6, 7 e 18 of the Preamble and Art. 13(4) of the Directive No. 2008/56/CE. See also, European Commission (2012).

[11]See Annex I and Annex III, Table 1 (Habitat types) of the Directive No. 2008/56/CE.

[12]See Conference of the Parties to the CBD: COP 7-2004 (Decision VII/30, Annex II, Target 1.1) and COP 10-2010 (Nagoya), *The Strategic Plan for Biodiversity 2011–2020 and the Aichi Biodiversity Targets* (Decision X/2, Annex, IV, 13, Target 11): "by 2020, at least (...) 10 per cent of coastal and marine areas, especially areas of particular importance for biodiversity and ecosystem services, are conserved through effectively and equitably managed, ecologically representative and well connected systems of protected areas and other effective area-based conservation measures, and integrated into the wider (...) seascapes". The European Union is Contracting Party of the CBD since 21 March 1994.

The analysis of Art. 11 is divided into three sections: the geographical scope, the substantive scope, and the prescriptive competence of the coastal Member States.

2.1 The Geographical Scope

For the purpose of this study, at present, the issue of the geographical scope of the legal instruments under analysis is uncontroversial. The MSFD has the clearest wording by stating its application to internal waters, territorial sea, exclusive economic zone, and the continental shelf, including the areas beyond 200 nautical miles (Art. 2 and Art. 3(1)(a)(b)). In the areas of the continental shelf beyond 200 nautical miles (or "outer continental shelf"), the achievement or maintenance of a good environmental status is faced with major legal barriers, given the high seas regime of the water column overlying the seabed (Arts. 78 and 87, UNCLOS).[13] Articles 13(4)(5) and 15 reveal this concern and establish a procedure to mitigate the effects of the high sea regime. The *Birds* and *Habitats* Directives share the same wide geographical scope, as confirmed by the European Union Court of Justice (ECJ). Thus, the *Habitats* Directive is also applicable to the natural habitats, habitats of species, and sedentary species of the continental shelf beyond 200 nautical miles.[14]

Turning the attention to Regulation No. 1380/2013, according to Art. 1(2) the Common Fisheries Policy has a more ambitious geographical scope embracing all the fishing activities carried out in the maritime areas under jurisdiction of the Member States (*Union waters*) and the fishing activities carried out outside the Union waters by fishing vessels flying the flag of Member States and registered in the European Union (*Union fishing vessels*).[15] The expression "Union waters" must be widely interpreted, including the sedentary species of the seabed (Art. 77(4), UNCLOS). It is noteworthy to clarify, however, that the geographical scope of Art. 11 is confined to the Union waters.

2.2 The Substantive Scope: The MPAs. The Complementary Effect of the MSFD with Regard to the Habitats Directive

The wording of Art. 11 raises no doubts when it comes to the delimitation of its substantive scope. Article 11 applies only to conservation measures that are deemed necessary for the effectiveness of MPAs. Differently, the *Habitats* Directive raises

[13]United Nations Convention on the Law of the Sea, of 10 December 1982.

[14]*Commission v. United Kingdom case*, 20.10.2005, Case C-6/04, Col. I- 9017, paras. 115, 117 and 120. See also Churchill and Owen (2010), p. 65; and European Commission (2012), p. 20, paras. 3 and 4.

[15]E.g., Regulation (EC) No. 734/2008, of 15 July, O.J. L 201/8 (2008).

several hermeneutical issues that must be addressed, taking into account that this directive is a milestone for the designation of MPAs.

Unlike the *Birds* Directive, which is applicable to all wild bird species, the *Habitats* Directive establishes a selective protection restricted to the natural habitats listed in Annex I, the habitats of species listed in Annex II, and the species listed in Annexes IV and V. After reading these annexes, the prevailing conclusion is that coastal biodiversity is clearly privileged. The protection given to the biodiversity of the open and deep seas is fragmented and exhibits serious gaps. Benefiting from the practice followed by some Member States,[16] the European Commission was sensible to these weaknesses of the *Habitats* Directive and, after a process started in 2003, came up with a solution developed in the document entitled "Guidelines for the establishment of the Natura 2000 network in the marine environment. Application of the Habitats and Birds Directives," of May 2007.[17] By means of revision of the *Interpretation Manual of European Union Habitats*, concluded in 2007, it is possible to extend the application of Annex I of the *Habitats* Directive to important deep sea ecosystems, that is, hydrothermal vent fields, cold coral reefs, and seamounts.[18] This approach is simple and quick; nevertheless, it raises many questions unsolved so far, namely:[19]

> First, does not address the gaps concerning deep sea and open sea species. Measures adopted in the framework of conservation of fisheries resources mitigate only part of the problem.[20]
>
> Second, does not identify which deep sea and open sea habitats should be classified as 'priority natural habitat' (e.g., Art. 4, *Habitats* Directive).
>
> Third, there is no clear timetable for the implementation of *Habitats* Directive in the marine environment.
>
> Fourth, neither the *Guidelines* nor the *Interpretation Manual* have binding force. Thus, it depends on the willingness of the Member States to comply with the extension operated by the revision of the *Interpretation Manual* in 2007.

It must be highlighted that some of these problems, such as gaps, the lack of a timetable, and the *soft law* nature of the *Guidelines* and of the *Interpretation Manual*, are indirectly mitigated by the timetable and framework established by

[16]See the case of Portugal. In 2002, a seamount located in the Portuguese exclusive economic zone (*Banco D. João de Castro*) was listed as site of Community importance, under the code 'Reefs', following the Portuguese proposition. See Decision 2002/11/EC, of 28 December 2001, O.J. L 5/16 (2002).

[17]For a more detailed insight see Ribeiro (2013), p. 585 et seq.

[18]At present, *Interpretation Manual of European Union Habitats*, European Commission (DG Environment), EUR 28, April 2013. See, for instance, code 1170 (reefs).

[19]The European Commission did an evaluation of the *Birds* and *Habitats* Directives to ensure that they are *'fit for purpose'*. See http://ec.europa.eu/environment/nature/legislation/fitness_check/index_en.htm. Accessed 31 January 2017.

[20]E.g., as regards deep-sea sharks and the orange roughy (*hoplostethus atlanticus*) see the 'zero' tolerance established by the Regulation (EU) No. 1367/2014, of 15 December, O.J. L/1 366 (2014).

the MSFD.[21] In respect of the gaps, for instance, the MSFD extends the protection potentially to all species and ecosystems[22] and, in Art. 13(4), takes into account other regimes, domestic or international, applicable to the designation of MPAs. The MSFD and these regimes have, consequently, the virtue of complementing the protection given by the Natura 2000 network. This is the case, notably, of the OSPAR Convention[23] and the Barcelona Convention systems.[24] The *OSPAR List of Threatened and/or Declining Species and Habitats*,[25] for instance, gives a wider protection to species (e.g., orange roughy, sharks, cod) and ecosystems (e.g., seamounts, hydrothermal vents fields).[26] Furthermore, Contracting Parties may give protection to other species and habitats types and also to areas of ecological relevance.

In conclusion, in the framework of Regulation No. 1380/2013, networks of MPAs benefit from a particular attention. Presumably, the designation of an MPA facilitates the adoption by the European Commission/Union of conservation measures (restrictions or prohibitions of fishing) when there is involvement of fishing vessels flying the flag of a Member State other than the Member State that has designated the MPA. It should be highlighted that Art. 8 of Regulation No. 1380/2013 applies to a different situation. While Art. 11 concerns to MPAs in the sense of the Convention on Biological Diversity ("holistic" MPAs[27]), Art. 8 applies to "biologically sensitive areas" intrinsically related with fish stock recovery (sectoral

[21]In respect of the timetable, see European Commission (2012), para. 17: "The timetable is also different. The MSFD requires that measures are taken to achieve or maintain GES [good environmental status] by 2020. There is no formal timetable set for achieving FCS [favourable conservation status] according to the HD [Habitats Directive]. The MSFD could therefore provide an additional stimulus for the implementation of conservation measures under the Habitats and Birds Directives, if measures to achieve FCS for species and habitats protected by HD and equivalent measures for wild birds are incorporated into or cross-referenced under the programme of measures within the respective marine strategies".

[22]See European Commission (2012), paras. 38 and 45, and p. 20, para. 1. See also the MSFD, Annex III, Table 1, Habitats types and Biological features.

[23]Convention for the Protection of the Marine Environment of the North-East Atlantic (OSPAR Convention), of 22 September 1992. In the OSPAR legal framework it must be highlighted the Annex V on the Protection and Conservation of the Ecosystems and Biological Diversity of the Maritime Area, of 23 July 1998, and the Recommendation 2003/3 on a Network of Marine Protected Areas, as revised by the *Recommendation 2010/2 on amending Recommendation 2003/3 on a network of Marine Protected Areas*, OSPAR 10/23/1, Annex 7.

[24]Convention for the Protection of the Marine Environment and the Coastal Region of the Mediterranean, as revised in 10 June 1995. In the Barcelona Convention framework it must be highlighted the Protocol Concerning Specially Protected Areas and Biological Diversity in the Mediterranean, as revised in 10 June 1995.

[25]Reference Number: 2008-6, OSPAR Commission. See also *Descriptions of habitats on the OSPAR list of threatened and/or declining species and habitats (Reference Number: 2008-07)*, OSPAR Commission.

[26]For the Mediterranean see: http://www.unepmap.org/index.php?module=content2&catid=001001001. Accessed 31 January 2017.

[27]See Molenaar and Elferink (2009), pp. 6–7; and Ribeiro (2014), pp. 185–191.

MPAs). It might be argued, therefore, that Art. 8 consubstantiates the European Union's answer to the international call for the protection of vulnerable marine ecosystems in the fisheries context.[28]

Finally, taking into account the three-step process for the designation of "special areas of conservation" under *Habitats* Directive, it must be stressed that Art. 11 shall be applicable since the inclusion of a natural habitat in the National List of Sites (first step). Only this interpretation complies with the ECJ case law and with the duty of the Member States to give preventive protection to the sites.[29]

2.3 *The Prescriptive Competence of the Coastal Member State. Grounds and Solutions for an Interpretation of Art. 11 Consistent with the Shared Nature of the Environmental Competence*

Article 11, as previously described, clearly deals with pure protection of the marine environment. This area is conceived as shared competence according to Art. 4(2)(e) of the Treaty on the Functioning of the European Union (TFEU), to which applies the important principles of subsidiarity and proportionality developed by Art. 5(3)(4)[30] of the Treaty on European Union (TEU).[31] We should

[28]See in particular the UNGA (United Nations General Assembly) Resolution no. 61/105, 08.12.2006, A/RES/61/105, paras. 80 83; Resolution no. 64/72, 04.12.2009, A/RES/64/72, paras. 119, 122–123; and Resolution no. 66/68, 06.12.2011, A/RES/66/68, paras. 121–126, 128–129, 131–132 and 135.

[29]See note 9.

[30]The compliance with the principles of subsidiarity and proportionality precedes the effects described by Art. 2(2) TFEU.

[31]Prior to the Regulation No. 1380/2013 came into force, there were discussions about the competent level for regulating fisheries inside MPAs: the coastal Member State or the European Union? At that time several Authors argued in favor of the coastal Member State competence. See Schwarz (2004), Owen (2004) and Ribeiro (2013), pp. 694–703. In the latter, we did a short analysis of the *pros and cons* of giving prevalence to the European Union level. *Pros*: coordination, consistency and coherence of the measures adopted; publicity; wider acceptance of the measures (less conflicts); increased facility in the adoption of measures applicable to large geographical areas (with or without MPAs in place). *Cons*: long decision-making processes; the prevailing economic rationale of the decision makers in the framework of fisheries; the prevailing power of the European Commission and the Council concerning the timing and content of the measures, in other words, it is easy to predict the adoption of measures—or the absence of measures—quite different from the ones proposed by the coastal Member State, which has a closer knowledge of the right balance of interests for ensuring an effective protection of the marine biodiversity (p. 698). It might also be argued that a coastal Member State may manipulate the measures so that its fishing fleet might get a competitive advantage. This scenario is real; however, the European Commission has several ways of controlling the measures without the need of emptying the regulatory powers of the coastal Member State.

expect, therefore, a more coherent and proportionate regime regarding the prescriptive—notably legislative—competence of the coastal Member State, taking into account that the *Birds* and *Habitats* Directives, as well as the MSFD, rely completely on the coastal Member State for the adoption of the required conservation measures.[32] The system established by Art. 11, however, follows the same approach found in other provisions inherited from Regulation No. 2371/2002, notably Arts. 13, 19, and 20. More clearly:

> First, according to Art. 5(1) of Regulation No. 1380/2013, "Union fishing vessels shall have equal access to waters and resources in all Union waters". This principle of equal access to waters and resources derogates the exclusivity for the fishing vessels flying the flag of the coastal State, set forth by Art. 19,[33] Arts. 56 and 62,[34] and Art. 77[35] of UNCLOS.
>
> Second, paragraphs 2, 3 and 4 of Art. 5 establish two main derogations (12 nautical miles; outermost regions: 100 nautical miles)[36] to the equal access to EU waters and resources, notwithstanding, the ordinary situation is the access to the EU waters – including frequently the 12 nautical mile and the 100 nautical mile zones – of fishing vessels flying the flag of diverse Member States.
>
> Third, the range of the regulatory powers of the coastal Member State depends on the flag of the fishing vessel. Actually, the decision-making process and the intensity of the control made by the European Commission differ depending whether the fishing vessels fly the flag of the coastal Member State or the flag of other Member States. It should be highlighted that both the decision-making process and the type of control made by the European Commission differ also from one provision to another. In this study the comparison will be focused on Arts. 11, 19 and 20.
>
> Fourth, in the context of fisheries, a wide or restrictive interpretation of the prescriptive competence of the European Union concerning the pure protection of the marine environment will affect, in the same extent, the international competence of the coastal Member State (Art. 3(2) TFEU).

2.3.1 The Prescriptive Competence of the Coastal Member State: Fishing Vessels Flying Its Flag

Article 11 does not raise any relevant criticism in the case of the need for the adoption of conservation measures applicable to fishing vessels flying the flag of the coastal Member State. Within the boundaries of MPAs, in any maritime zone under national jurisdiction (territorial sea, exclusive economic zone, continental shelf), the coastal Member State is empowered to unilaterally regulate fishing activities carried out by fishing vessels flying its flag (Art. 11(1)). The conservation measures

---

[32]E.g., Art. 6 of *Habitats* Directive and Arts. 13, 15 and 18 of MSFD.

[33]Territorial sea.

[34]Exclusive economic zone. In this maritime zone the exclusivity for the fishing vessels flying the flag of the coastal State is not absolute given the regime set forth by Art. 62(2) of UNCLOS.

[35]Continental shelf.

[36]These derogations were inherited from the legal framework applicable to the Common Fisheries Policy before the Regulation No. 1380/2013. The question is whether they will be extended after 31 December 2022 (Art. 5(4)).

adopted by the coastal Member State must, nevertheless, comply with three cumulative requirements:

> First, the measures must be compatible with the objectives set out in Art. 2 of Regulation No. 1380/2013. Article 2 sets forth a large list of objectives. With respect to the objectives of socio-economic nature, our understanding is that its assessment must take into consideration the regime set out in Art. 6(3)(4) and Art. 7 of *Habitats* Directive (assessment of the implications for the site[37] and exceptions) and in Art. 14 of MSFD (exceptions). These commands prescribe strict criteria for an inversion of the hierarchy between environmental objectives and socio-economic objectives. Inclusively, when doing the analysis of the articulation between Art. 6 of *Habitats* Directive and Art. 14 of MSFD, the European Commission itself concluded in the sense of the prevalence of the most favourable solution to the protection of the environmental objectives, as follows: "the MSFD exceptions cannot take precedence over Article 6 of the Habitats Directive as the Treaty requires that stricter provisions take precedence when more than one applies to the same issue".[38]
>
> Second, the measures must meet the objectives of the relevant Union legislation that they intend to implement.
>
> Third, the measures must be at least as stringent as measures prescribed by European Union law. In other words, the coastal Member State must respect this minimum standard of protection (*"measures under Union law"*), nonetheless, he can go further in the intensity of protection based on Art. 193 TFEU. Under this article the coastal Member State can maintain or introduce "more stringent protective measures", provided that such measures are compatible with the Treaties and notified to the European Commission.

What if the coastal Member State does not comply with these requirements? In this event, considering the general control for which is competent the European Commission, this institution (and also other Member States) might bring the Member State before the ECJ in the context of an action for infringement of European Union law (Arts. 258 to 260 TFEU).

2.3.2 The Prescriptive Competence of the Coastal Member State: Fishing Vessels Flying the Flag of Other Member States or Third States

The legal scenario changes completely if the measures adopted by the coastal Member States, under the same circumstances, are liable to affect fishing vessels flying the flag of other Member States (Art. 11(2–6)). In this case, the decision-making process might be long and with an unpredictable outcome, involving Member States with direct management interest in the fishery to be affected by such measures, the Advisory Councils, the European Commission, and, when this institution makes use of Art. 43 TFEU, the European Parliament and the Council, as well as the Economic and Social Committee. More clearly:

---

[37] In the *Landelijke Vereniging* case (07.09.2004, Case C-127/02), on the mechanical fishing of cockles, the ECJ included the fishing activities in the concept of 'project' for the purpose of assessment of its implications for the site.

[38] See European Commission (2012), para. 61 et seq., notably para. 67.

First, the coastal Member State – called "the initiating Member State" – must request the adoption of the relevant measures by initiating a procedure near the European Commission. The initiating Member State shall provide the European Commission and the other Member States having a direct management interest with relevant information on the measures required, including their rationale, scientific evidence in support and details on their practical implementation and enforcement (Art. 11(3)).

Second, the initiating Member State and the other Member States having a direct management interest may submit a joint recommendation, as referred to in Art. 18(1)(2), within six months from the provision of sufficient information. The Commission shall adopt the measures, by means of delegated acts (Art. 46)[39], taking into account any available scientific advice, within three months from receipt of a complete request (Art. 11(2)(3)).

When comparing the wording of Art. 11(3) and the wording of Art. 18(3), we came into the conclusion that the European Commission *must* adopt – not a *mere empowerment* or *option* – the required conservation measures, provided that the requirements set out by Art. 11(1) are met. This conclusion is supported by the purpose of Art. 11(1) and also by the fact that the *Birds* and *Habitats* Directives and the MSFD rely completely on the *original*[40] regulatory powers of the coastal Member State for the adoption of the required conservation measures.[41] This reasoning also explains the exclusive power of initiative of the coastal Member State in the context of Art. 11.[42]

It is not clear, however, whether the European Commission, in cooperation with the Member States[43], can influence the shape of the measures, taking into account the available scientific advice, with the aim of avoiding the rejection of the measure. A positive answer seems to be more in line with the spirit of the legislator (*mens legislatoris*).

It should be highlighted that occasionally the conservation measures to be adopted might affect fishing vessels flying the flag of third States. This might occur namely in the fishing grounds overlying the continental shelves of Member States beyond 200 nautical miles. This situation is expressly addressed in Art. 13(5) and Art. 15 of MSFD, as well as in Art. 18(4)[44] of Regulation No. 1380/2013. In this event, only the European Union can propose the restriction or prohibition of a fishing activity to the relevant regional fisheries management organisation or, when direct negotiation is adequate, to third States. The intermediation of the European Union is the consequence of its exclusive competence at the international level for adopting decisions concerning, strictly, the conservation of fishing

[39]See also Art. 290 TFEU.

[40]It is important to remember that the limits of European Union competences are governed by the principle of conferral (Art. 3(6) and Art. 5(1)(2) TEU; Art. 7 TFEU) and the use of shared competences is governed by the principles of subsidiarity and proportionality (Art. 5(3)(4) TEU followed by Art. 2(2) TFEU).

[41]E.g., Art. 6(1)(2) and Art. 4(5) of *Habitats* Directive, and Arts. 13, 15 and 18 of MSFD. It is of the coastal Member State the power and duty to adopt preventive measures and conservation measures.

[42]See the difference of Art. 12 of Regulation No. 1380/2013. Emergency measures can be adopted by the European Commission at the reasoned request of a Member State or on its own initiative.

[43]See Art. 11(6) and Art. 18(2) of Regulation No. 1380/2013.

[44]"Where the conservation measure applies to a specific fish stock shared with third countries and managed by multilateral fisheries organisations or under bilateral or multilateral agreements, the Union shall endeavour to agree with the relevant partners the measures that are necessary to achieve the objectives set out in Article 2".

resources (Art. 3(2) TFEU).[45] This exclusive competence of the European Union in the fisheries domain must not threaten the internal competence of Member States when acting in the context of shared competences, such as the *Birds* and *Habitats* Directives or the MSFD and, likewise, must not endanger the international competence of Members States when acting in the context of shared competences, such as the protection of marine environment (e.g., OSPAR Convention, Barcelona Convention, CCAMLR[46]).[47] Both competences-exclusive and shared-must be articulated, giving high relevance to the principle of sincere cooperation,[48] and a clear border must be established between the mere conservation of fishing resources (competence of the European Union) and the protection of marine environment (competence of the Member States and of the European Union, the latter exclusively in the area of pre-emption by common rules).[49] In order to keep the balance established by the Member States when ratifying the TEU and TFEU, the abusive appropriation of competence by the European Union must be refrained, taking into account the supreme principle of conferral of competences (Art. 5(1)(2) and Art. 48(2),[50] TEU). In other words, the system laid down by the Members States in the TEU and TFEU requires that the scope of the exclusive competences, both at internal and external levels, must be subject to a restrictive interpretation.

Third, if the joint recommendation is deemed not to be compatible with the requirements referred to in Art. 11(1), the European Commission may submit a proposal in accordance with the Treaty, that is, Art. 43(1)(2) and Art. 289(1) TFEU. According with these provisions, the conservation measures will be jointly adopted by the European Parliament and the Council. In our understanding, before the referred submission of the proposal under Art. 43, the Member States may amend the joint recommendation and restart the procedure before the European Commission.

Fourth, if not all Member States succeed in agreeing on a joint recommendation to be submitted to the European Commission (absence of a joint recommendation), two things might happen: this institution may submit a proposal in accordance with the Treaty (Art. 11 (3))[51] or, in the case of urgency, the European Commission shall adopt temporary conservation measures (Art. 11(4)(5)). These measures shall be limited to those in the absence of which the achievement of the objectives associated with the establishment of the

---

[45]"The Union shall also have exclusive competence for the conclusion of an international agreement when its conclusion is provided for in a legislative act of the Union or is necessary to enable the Union to exercise its internal competence, or in so far as its conclusion may affect common rules or alter their scope". Emphasis added. See the case law cited in note 4.

[46]An interesting case was brought before the ECJ in November 2015 concerning the *Convention on the Conservation of Antarctic Marine Living Resources*, of 20 May 1980 (CCAMLR). See Case C-626/15: http://data.consilium.europa.eu/doc/document/ST-15523-2015-INIT/en/pdf. Accessed 31 January 2017. This Convention goes beyond the pure conservation of fishery resources. See, for instance, the wide concept of "marine living resources", which embraces all marine species (Art. I (2)).

[47]See, in general, Wouters et al. (2009).

[48]See Art. 4(3) TEU: "Pursuant to the principle of sincere cooperation, the Union and the Member States shall, in full mutual respect, assist each other in carrying out tasks which flow from the Treaties (…)".

[49]See again Art. 3(2) TFEU: "The Union shall also have exclusive competence for the conclusion of an international agreement when (…) in so far as its conclusion may affect common rules or alter their scope". In general, for a deep analysis of the complex ECJ case law, see Rosas (2015).

[50]The proposals for the amendment of the Treaties may serve "either to increase or to reduce the competences conferred on the Union in the Treaties". Emphasis added.

[51]See Arts. 43 and 289(1) TFEU.

conservation measures, in accordance with the *Birds*, *Habitats* and MSFD Directives and the Member State's intentions, is in jeopardy. According with paragraph 4, hence, the conservation measures adopted by the European Commission must be consistent with the *initiating* Member State's intentions. The use of paragraphs 4 and 5 will occur possibly when the debate between the initiating (coastal) Member State and the other Member States is extreme, based in conflicting objectives: urgent environmental protection, on the one hand, and socio-economic reasons, on the other hand.

It is noteworthy to mention that Art. 11(4)(5), when compared with Art. 12, set out a more generous procedure and time limit. Under Art. 11(5), the urgency measures shall apply for a maximum period of 12 months, which may be extended for the same period provided the conditions that justified the measures continue to exist.

The articulation between paragraphs 3 and 4 of Art. 11 raises some doubts. In our understanding, flexibility must be given to the European Commission. Instead of a mandatory choice between submitting a proposal under Art. 43 TFEU or adopting emergency measures, the European Commission may combine both options, taking into account that the ordinary legislative procedure may be long.

Fifth, the Commission shall facilitate cooperation between the Member State concerned and the other Member States having a direct management interest in the fishery in the process of implementation and enforcement of the measures adopted under Art. 11(2)(3)(4).

Article 11 itself raises some doubts of interpretation, as previously explained. The main obstacles, however, emerge when we compare Art. 11 with other provisions, notably Arts. 19 and 20. The most relevant obstacles are the overlapping of regimes and the different balance between exclusive and shared competences (developed in Sect. 3, *infra*). The case of the 12 nautical mile zone, for instance, is obvious. In fact, when the conservation measures to be adopted by the coastal Member State are liable to affect fishing vessels flying the flag of other Member States, Art. 20(2)(3)(4) sets out a more respectful legal solution regarding the environmental competence of the coastal Member State. In other words, the decision-making process laid down in Art. 20(2)(3)(4) is centered in the coastal Member State, and the European Commission can only exercise an external—but important[52] and necessary[53]—control. Therefore, when fishing vessels flying the flag of other Member States are affected, the adoption of conservation measures by the coastal Member State is easier under such article. This leads to the absurd conclusion that adopting measures for the protection of ecosystems is easier when there is no designation of MPAs. Art. 11 thus, in the case of fishing vessels flying the flag of other Member States, establishes a disproportionate solution favoring the exclusive competence of the European Union (conservation of fishery resources) rather than the shared competence of the coastal Member State (environment: designation and regulation of MPAs). In order to ensure compliance with the system of competences established in TFEU and directives, and coherence in the application of Arts. 11 and 20, we propose that Regulation No. 1380/2013 must be interpreted as follows:

[52]See Art. 20(4).

[53]The control is required, for instance, to avoid disproportion, distortion and manipulation of objectives by the coastal Member State.

First, in the 12 nautical mile zone, when the conservation measures to be adopted by the coastal Member State are liable to affect fishing vessels flying the flag of other Member States, the application of Art. 20(2)(3)(4)[54] should prevail over Art. 11.

Second, in a future revision of Regulation No. 1380/2013 (*de iure condendo*), the legal solution set out in Art. 20(2)(3)(4) should be extended to the exclusive economic zone and the continental shelf when the conservation measures applicable in the MPAs are liable to affect fishing vessels flying the flag of other Member States.

3 Beyond Art. 11: The Contribution of Other Provisions for the Protection of Marine Biodiversity

The express concern of Regulation No. 1380/2013 with MPAs (Art. 11) does not diminish the importance of other provisions that also contribute, directly, to the protection of ecosystems: Art. 8 concerning the protection of fish stocks recovery areas, described as biologically sensitive areas (equivalent to vulnerable marine ecosystems); Arts. 12 and 13 concerning emergency measures; and Arts. 19 and 20 concerning national measures. With the exception of Art. 8, these articles derive from the former basic regulation (No. 2371/2002); nonetheless, some important changes have been inserted, namely, in the decision-making process towards a lighter procedure and the strengthening of the coastal Member State's regulatory powers.

Under this new legal framework, the coastal Member State can adopt general measures with the aim to protect ecosystems and species outside the MPAs while complementing their effects. Let us see Arts. 19 and 20 more closely and the interaction between them and with Art. 11.

3.1 Scope of Art. 20: Conservation Measures Adopted by the Coastal Member State in the 12 Nautical Mile Zone

Within 12 nautical miles of its baselines, the coastal Member State may, on the one hand, take nondiscriminatory measures for the conservation and management of fish stocks and, on the other hand, take nondiscriminatory measures for the *"maintenance or improvement of the conservation status of marine ecosystems."* We will give relevance to the goals evidenced in italics, given their importance for the achievement of the objectives laid down in the *Birds* and *Habitats* Directives, and

[54]We reject the application of Art. 20(1) in the context of MPAs as regards the possibility of the European Union calling back its regulatory competence. This possibility must be subject to a restrictive interpretation in the sense that European Union cannot unilaterally replace the coastal Member State regarding the initiative and legislative competence for the protection of marine environment, as enshrined in the *Birds* and *Habitats* Directives and MSFD.

MSFD. It must be underlined that all these legal instruments set out objectives beyond the MPAs, addressing both the protection of ecosystems and species.[55] Under the former basic Regulation (No. 2371/2002), the practice confirmed the inclusion of species in the scope of Art. 9 (now Art. 20).[56]

The prescriptive competence of the coastal Member State is subject to three cumulative requirements (Art. 20(1)):

> First, that the European Union has not adopted, namely under Art. 43 TFEU, measures addressing *conservation* specifically for that area or specifically addressing the problem identified by the coastal Member State concerned. This requirement must be clarified. In the context of the pure conservation of marine ecosystems and species – a domain of elusive borders, that is not always easy to establish, in relation to the conservation of fish stocks and associated ecosystems![57] – the European Union cannot arbitrarily replace the coastal Member State regarding its initiative and legislative competence, as enshrined in the *Birds* and *Habitats* Directives and MSFD. Actually, as mentioned before, also the European Union must comply with the principle of sincere cooperation (Art. 4(3), TEU). If the coastal Member State does not take adequate measures or initiatives the right option for the European Commission is making use of the action for infringement of, notably, these directives (Arts. 258 and 260 TFEU).
>
> Second, the conservation measures adopted by the coastal Member State must be compatible with the objectives set out in Art. 2.
>
> Third, the conservation measures must be at least as stringent as measures prescribed by European Union law. We recall here the reasoning developed in Art. 11 about this *minimum standard of protection*.

3.1.1 Twelve Nautical Mile Zone: Fishing Vessels Flying the Flag of the Coastal Member State

In the 12 nautical mile zone, provided that the three requirements described above are met, the prescriptive competence of the coastal Member State is absolute with respect to fishing vessels flying its flag. The only additional requirement is the duty of that State to make publicly available appropriate information concerning the measures adopted (Art. 20(3)). This command in paragraph 3 is a novelty, and so is paragraph 4 of Art. 20, both introduced in 2013 due to the amendment of the decision-making processes.

According to paragraph 4, if the European Commission considers that a measure adopted under Art. 20 does not comply with the conditions set out in paragraph 1, it may, subject to providing relevant reasons, request that the coastal Member State concerned amends or repeals the relevant measure. In our understanding, however,

[55] In the case of species, see Art. 12 et seq., and Annexes IV-VI of *Habitats* Directive. In general, see Annex I of MSFD.

[56] See Churchill and Owen (2010), pp. 192–193. See the following Decisions of the European Commission: C (2004) 3229, of 24 August, and 2005/322/EC, of 26 February.

[57] It is worthy of analysis the Council Regulation (EC) No. 1967/2006, of 21 December, concerning management measures for the sustainable exploitation of fishery resources in the Mediterranean Sea.

in the case of fishing vessels flying the flag of the coastal Member State, the more favorable regime set out in Art. 19 (*all Union waters*) must prevail over Art. 20 (1 and 4), provided that, by means of an extended interpretation, the application of Art. 19 to the conservation of ecosystems is accepted, as the practice indicates (see Sect. 3.2., *infra*). Therefore, under these circumstances, the European Commission can only make use of the general powers of control inherent to the action for infringement of European Union law (Arts. 258 and 260 TFEU).

3.1.2 Twelve Nautical Mile Zone: Fishing Vessels Flying the Flag of Other Member States

Where conservation measures to be adopted by the coastal Member State are liable to affect fishing vessels of other Member States, the decision-making process is substantially different involving duties of coordination (Art. 6(4)), consultation, and motivation. More clearly, the conservation measures shall be adopted by the coastal Member State only after consulting the European Commission, the relevant Member States, and the relevant Advisory Councils[58] on a draft of the measures, which shall be accompanied by an explanatory memorandum that demonstrates, inter alia, that those measures are nondiscriminatory. For the purpose of such consultation, the coastal (consulting) Member State may set a reasonable deadline, which shall, however, not be shorter than two months (Art. 20(2)). Still, Regulation No. 1380/2013, when compared with Regulation No. 2371/2002, clearly strengthens the regulatory powers of the coastal Member State, taking into account that it is (always) up to this State to make the final decision.

Paragraph 3 (publicity of the measures by the coastal Member State) and paragraph 4 (control of the measures by the European Commission) gain increased relevance when conservation measures are liable to affect fishing vessels of other Member States. These requirements are very important in performing an adequate counterbalance of the devolution of prescriptive competence to the coastal Member State. The external control of the measures by the European Commission, notably, may prevent distortion and manipulation of objectives by the coastal Member State (e.g., disguised competitive gain in a particular fishery). What if the coastal Member State does not amend or repeal the measure as requested by the European Commission? In our view, in the context of conservation of ecosystems and species, the answer is making use of the action for infringement of European Union law (Arts. 258 and 260 TFEU), being the final decision of the ECJ.

We fully agree with the devolution of regulatory powers operated by Art. 20 (2) of Regulation No. 1380/2013. The solutions enshrined in Art. 20 convey a fair balance of interests and are more respectful of the environmental competence of the coastal Member State. It is surprising, in our point of view, that Art. 11 does not

[58]The powers of the Advisory Councils are not irrelevant. See Art. 44(3)(4) of Regulation No. 1380/2013.

follow the same approach precisely in a situation (MPAs) where the environmental competence of the coastal Member State—and, consequently, the inherent regulatory powers—should obviously prevail. The fact that Art. 11 also applies to the exclusive economic zone and continental shelf is not a convincing argument. We recall, consequently, our interpretation developed in Sect. 2.3.2., in fine: in the 12 nautical mile zone, when the conservation measures to be adopted by the coastal Member State are liable to affect fishing vessels flying the flag of other Member States, the application of Art. 20(2)(3)(4) should prevail over Art. 11.

A final remark concerning the pragmatic solution adopted by Portugal and Spain, by which the solutions laid down in Art. 20 were circumvented: these two Member States signed a bilateral fisheries agreement in Brussels, on 24 March 2014, establishing a regime, based on principles of reciprocity and national treatment, for the adjacent areas (Minho and Guadiana) of their respective territorial seas in the Atlantic Ocean.[59]

3.2 Scope of Art. 19: Conservation Measures Adopted by the Coastal Member State Applicable to Fishing Vessels Flying Its Flag or to Persons Established in Its Territory. Grounds for the Inclusion of Marine Ecosystems by Means of an Extended Interpretation

According to paragraph 1 of Art. 19, Member States may adopt "measures for the conservation of fish stocks in Union waters" provided that those measures fulfill three cumulative requirements:

> First, measures must apply solely to fishing vessels flying the flag of that Member State or, in the case of fishing activities which are not conducted by a fishing vessel, to persons established in that part of its territory to which the Treaty applies (Art. 355 TFEU).
>
> The second and third requirements are identical to those set out in Art. 20(1): compatibility of measures with the objectives set out in Art. 2 and those measures must be at least as stringent as measures under European Union law.

Furthermore, paragraphs 2 and 3 of Art. 19 stipulate some duties of information:

> The Member State shall, for control purposes, inform the other Member States concerned of provisions adopted, and the Member States shall make publicly available appropriate information concerning the measures adopted.

The noncompliance with these requirements and obligations may end in an action for infringement of European Union law initiated by the European Commission or other Member States (Arts. 258 to 260 TFEU).

Besides the duties of information, the main innovation introduced by Art. 19, when compared with the former Art. 10 of Regulation No. 2371/2002, is the

---

[59]See, namely, Art. 4(3) and Art. 5(4) of Decreto No. 21/2014, of 8 August, DR I/152, p. 4139, available at https://dre.pt/application/file/55236009. Accessed 31 January 2017.

geographical scope, that is, the measures adopted by a Member State may be applicable in the territorial sea, exclusive economic zone, and continental shelf under its jurisdiction or under the jurisdiction of other Member States (*Union waters*).[60] Consequently, within European Union waters, a coastal Member State may establish a unified regime for the fishing vessels flying its flag. In these circumstances, in the 12 nautical mile zone, the requirement whereby measures must be "at least as stringent as measures under Union law" should be extended to measures adopted by the relevant Member States in accordance with Art. 20(2).

The big question[61] about Art. 19 is the scope of the conservation measures. In fact, for the purpose of protection of marine biodiversity, the wording of Art. 19 may generate controversy, taking into account that it refers only to "measures for the conservation of fish stocks in Union waters." Focusing particularly on the exclusive economic zone and the continental shelf,[62] does this mean that the coastal Member State is restrained from adopting measures with the aim to protect ecosystems or species envisaged by the European Union environmental legislation, such as the MSFD?[63]

CHURCHILL and OWEN writing about Art. 10 of Regulation No. 2371/2002, the predecessor of Art. 19, acknowledge the following: "The authors have been unable to ascertain whether the failure of Article 10 expressly to apply to marine ecosystems was an oversight on the part of those drafting the Regulation or was intentional."[64] In our understanding, only an oversight on the part of those drafting the regulations is admissible. In fact, there are several irrefutable arguments in favor of an extended interpretation of Art. 19,[65] in order to include the protection of ecosystems in the exclusive economic zone and continental shelf:

> First, the predominant link of nationality, that is, between the coastal Member State that adopt the conservation measures and the fishing vessels to which the measures apply. The link of nationality is fully respected by Arts. 11 and 20. Why Art. 19 would be different in the case of conservation of marine ecosystems?
>
> Second, the shared nature of the environmental competence and the powers and duties of the coastal Member State set out in the *Birds* and *Habitats* Directives and MSFD. The

[60]See Art. 4(1)(1): "Union waters' means the waters under the sovereignty or jurisdiction of the Member States, with the exception of...".

[61]Another question is the articulation of Art. 19 with Art. 20. In the 12 nautical mile zone, outside the boundaries of MPAs but taking into account the limits described previously (see note 54 and Sect. 3.1 of this chapter), Art. 19 might be important when the European Union call back the regulatory powers according with Art. 20(1). Art. 19 provides legal basis for the adoption of other measures by the coastal Member State applicable to his fishing vessels, as long as the requirement of minimum standard of protection is met. Another issue of articulation was addressed in Sect. 3.1.1 of this chapter.

[62]Art. 20, for the 12 nautical mile zone, expressly embraces the conservation of ecosystems.

[63]E.g., ecosystems characterized by dispersion, namely, cold-water coral reefs and sponge aggregations.

[64](2010), p. 191.

[65]We expressed our point of view for the first time in *A protecção da biodiversidade marinha...* (2013), pp. 715–717.

interpretation by which the wording of Art. 19 expresses the exclusivity of the European Union for adopting measures for the conservation of marine ecosystems would contravene the system of competences established by the TFEU and the directives referred above.

Third, the global system of the Regulation No. 1380/2013. The use of Art. 13 (emergency measures) by the coastal Member State is subject to strict conditions (e.g., *"[o]n the basis of evidence of a serious threat (...) to the marine ecosystem"*) and is limited in time (*"measures shall apply for a maximum period of three months"*). Art. 13, hence, does not provide legal basis for a sufficient and enduring protection of ecosystems in the exclusive economic zone and continental shelf. In our point of view, Art. 13 combined with Arts. 11, 19 (extended interpretation) and 20, only shows real usefulness with regard to its possible application to fishing vessels flying the flag of other Member States, in particular for fishing activities carried out in the exclusive economic zone and continental shelf.[66]

This extended interpretation was followed by Portugal in Portaria No. 114/2014, of 28 May,[67] and was accepted by the European Commission. The facts are quite easy to describe:

Following the requests of Portugal and Spain, in 2005, the European Union adopted the Regulation No. 1568/2005, of 20 September,[68] regarding the protection of deep-water coral reefs and other vulnerable deep-sea ecosystems from the effects of fishing in large areas of the Macaronesian region, that is, waters around the Azores and Madeira Archipelagos and Canary Islands (Fig. 1). This Regulation prohibits the use, by the European Union fleet, of any gillnet, entangling net or trammel net at depths greater than 200 meters and any bottom trawl or similar towed nets operating in contact with the bottom of the sea, including in areas of the exclusive economic zone and continental shelf beyond 200 nautical miles,[69] notably, where hydrothermal vent fields are located.

In 2014 Portugal extended the geographical scope to other parts of the exclusive economic zone and a larger area of the continental shelf beyond 200 nautical miles in order to protect diverse deep-sea ecosystems, such as seamounts and hydrothermal vent fields (Fig. 1). Therefore, according with Portaria No. 114/2014, in those larger areas, fishing vessels flying the flag of Portugal are prohibited from using several nets operating in contact with the bottom of the sea. The Portaria is clearly anchored in Art. 19 of Regulation No. 1380/2013.

In 8 July 2015 Portugal requested to the European Commission the extension, to the rest of the European Union fleet, of the prohibition contained in the Portaria No. 114/2014. The request was based in Art. 15 of the MSFD. The European Commission acknowledged the legitimacy of the Portuguese request, but no measures have been taken so far.

A final remark concerning the pragmatic solution adopted by Portugal and Spain, circumventing the limits set out by Art. 19, for some areas of the exclusive economic zones adjacent to Madeira and Canary Islands and the mainland: the bilateral fisheries agreements signed by these two Member States respectively in

---

[66]In the 12 nautical mile zone the main benefit seems to be the shorter deadline for consultation.

[67]DR I/102, p. 2977. Available at https://dre.pt/application/file/25346153. Accessed 31 January 2017.

[68]O.J. L 252/2 (2005). The Regulation No. 1568/2005 amends Regulation (EC) No. 850/98, of 30 March.

[69]The Portuguese submission to the Commission on the Limits of the Continental Shelf (CLCS) was formally deposited in 11 May 2009, with the No. 44. See the official website of the CLCS: http://www.un.org/depts/los/clcs_new/commission_submissions.htm. Accessed 31 January 2017.

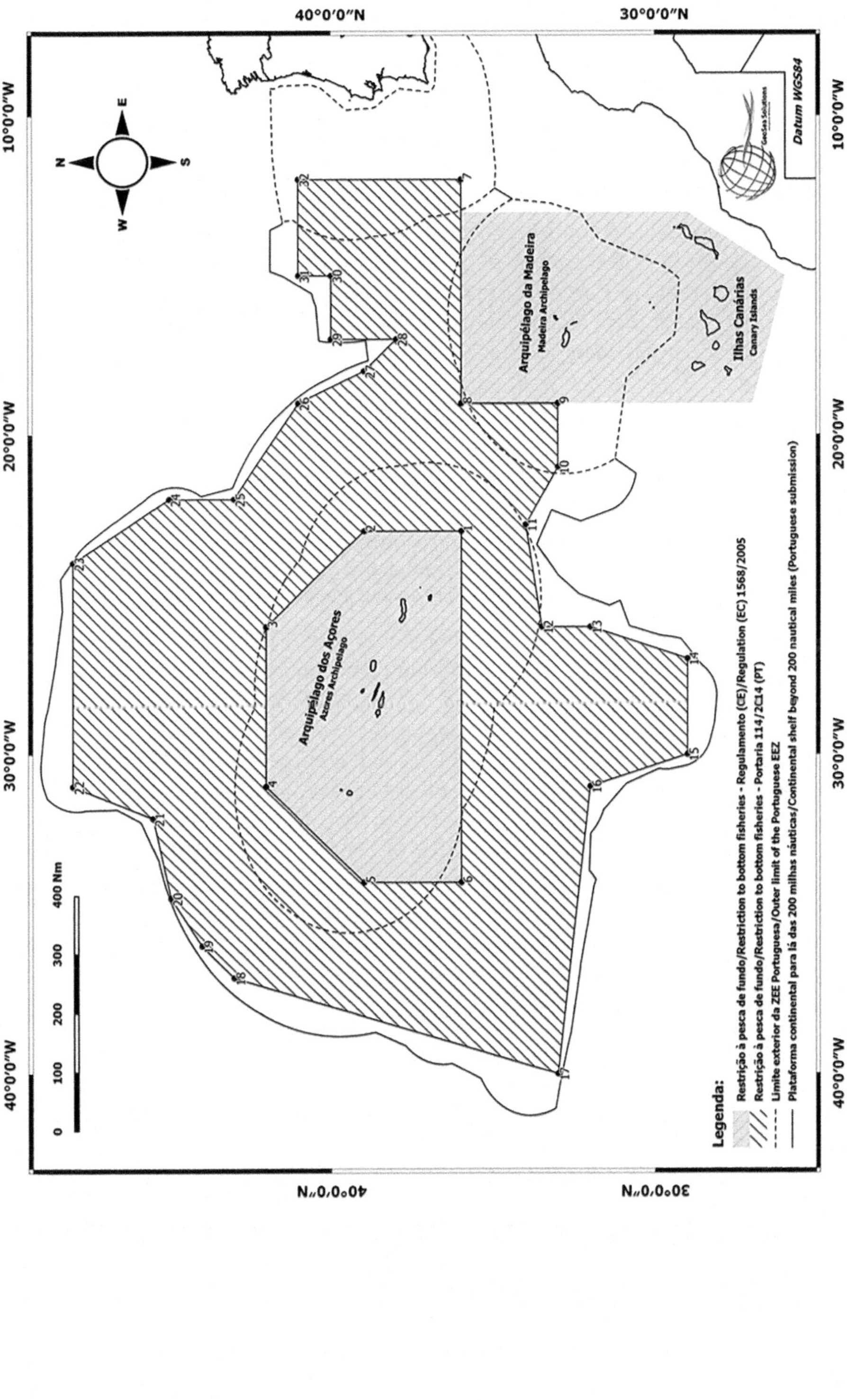

Fig. 1 Restriction to bottom fisheries. Courtesy of *GeoSea Solutions, Lda*

Porto, on 9 May 2012,[70] and in Brussels, on 24 March 2014,[71] are based on principles of reciprocity and national treatment.

4 Concluding Remarks: Balance and Tension Between Exclusive and Shared Competences

Article 11 of Regulation No. 1380/2013 is applicable to MPAs only. The provisions do not raise any relevant critic in the case of the need for the adoption of conservation measures applicable to fishing vessels flying the flag of the coastal Member State. Presumably the designation of an MPA would facilitate the adoption of restrictive or prohibitive measures applicable to fishing vessels flying the flag of other Member States; nonetheless, in the 12 nautical mile zone, the adoption of measures by the coastal Member State seems to be easier when there is no MPA (Art. 20).

Therefore, when the conservation measures to be adopted by the coastal Member State are liable to affect fishing vessels flying the flag of other Member States, within the 12 nautical mile zone, the decision-making process set out by Art. 20(2)(3)(4) should prevail over Art. 11, that is:

> First, the conservation measures should be adopted by the coastal Member State, with obligations of prior coordination (Art. 6(4)), consultation, motivation and publicity;
>
> Second, the European Commission can play an important external control.

In a future revision of Art. 11 of Regulation No. 1380/2013, this decision-making process, which favors the prescriptive competence of the coastal Member State with respect to fishing vessels flying the flag of other Member States, should be extended to the exclusive economic zone and the continental shelf.

Only these interpretations ensure coherence and compatibility with the environmental (and prescriptive) competence of the coastal Member State as fully respected by the *Birds* and *Habitats* Directives (Art. 6) and by the MSFD (e.g., Arts. 13 and 15), according to the principle of conferral of competences and the principles of subsidiarity and proportionality. Otherwise, the other Member States will benefit from a significant power to influence the degree of environmental protection in maritime zones that are not under their jurisdiction, and the European Union will be legitimized to exercise regulatory powers that genuinely belong to the coastal Member States:

---

[70] Related to tuna (traditional pole-and-line fishing gear) and black scabbard fish (longline fishing gear). Exclusive economic zones adjacent to Madeira and Canary Islands. See Art. 9(1) of Decreto No. 8/2013, of 9 May, DR I/89, p. 2756, available at https://dre.pt/application/file/260696. Accessed 31 January 2017.

[71] Exclusive economic zones adjacent to mainland (Atlantic Ocean only). See Decreto No. 21/2014, cit., Art. 3.

Within the 12 nautical mile zone, Art. 19 must be interpreted together with Art. 20 when the conservation measures are applicable only to fishing vessels flying the flag of the coastal Member State, that is:

> First, the tacit system of control set out by Art. 19 – based on Arts. 258 and 260 TFEU – should prevail over the regime established by Art. 20(4).
>
> Second, measures adopted by the coastal Member State under Art. 19 must eventually take into account the measures adopted by other coastal Member States under Art. 20(2).

The conservation of ecosystems, especially in the exclusive economic zone and continental shelf, must be included in the scope of Art. 19 by means of an extended interpretation. Only this interpretation complies with the link of nationality (flag), the shared nature of the environmental competence, and the global system enshrined in Regulation No. 1380/2013. This understanding was followed by Portugal in Portaria No. 114/2014 in articulation with the duties set out by the MSFD.

In the framework of the exclusive competence of the European Union—conservation and management of fishery resources—there is an important devolution of regulatory powers to the coastal Member State (e.g., Arts. 19 and 20), inclusively when fishing vessels flying the flag of other Member States are liable to be affected by the measures (Art. 20). In the framework of the shared competences—MPAs and conservation of ecosystems (Arts. 11 and 19; exception of Art. 20)—under a literal interpretation, there is a controversial appropriation of regulatory powers by the European Union, with possibly important consequences at the external level. All in all, the pretension of the fisheries framework to dominate the environmental protection is clearly evidenced in Regulation No. 1380/2013. Besides issues of conflicting competences and proportionality, acknowledging the importance of an effective control by the European Union to avoid distortion and manipulation, the fundamental question is whether that dominance is the best option for the oceans' health and the consequent sustainability of fisheries.

References

Churchill R, Owen D (2010) The EC Common Fisheries Policy. OUP, Oxford

European Commission (2012) Links between the Marine Strategy Framework Directive (MSFD 2008/56/EC) and the Nature Directives (Birds Directive 2009/147/EEC (BD) and Habitats Directive 92/43/EEC (HD)). Interactions, overlaps and potential areas for closer coordination. http://ec.europa.eu/environment/nature/natura2000/marine/docs/FAQ%20final%202012-07-27.pdf. Accessed 31 January 2017

Molenaar EJ, Elferink AGO (2009) Marine protected areas in areas beyond national jurisdiction. The pioneering efforts under the OSPAR convention. Utrecht Law Rev 5(1):5–20

Owen D (2004) Protecting marine SACs and SPAs from fishing activities: who has the power to impose restrictions. In: Jürgen Ritterhoff J, Gubbay S, Zucco C (eds) Marine protected areas and fisheries. Proceedings of the International Expert Workshop held at the International Academy for Nature Conservation, Isle of Vilm, Germany. Bundesam für Naturschutz, Bonn, pp 61–65

Ribeiro MC (2013) A protecção da biodiversidade marinha através de áreas protegidas nos espaços marítimos sob soberania ou jurisdição do Estado: discussões e soluções jurídicas contemporâneas. O caso português. Coimbra Editora, Coimbra

Ribeiro MC (2014) Marine protected areas: the case of the extended continental shelf. In: Ribeiro MC (ed) 30 years after the signature of the United Nations Convention on the Law of the Sea: the protection of the environment and the future of the Law of the Sea. Coimbra Editora, Coimbra, pp 179–207

Rosas A (2015) EU External Relations: Exclusive Competence Revisited. Fordham Int Law J 38 (4):1073–1096

Schwarz K-A (2004) Legal aspects of marine protected areas and fisheries. In: Jürgen Ritterhoff J, Gubbay S, Zucco C (eds) Marine protected areas and fisheries. Proceedings of the International Expert Workshop held at the International Academy for Nature Conservation, Isle of Vilm, Germany. Bundesam für Naturschutz, Bonn, pp 45–56

Wouters J, Jong S, Marx A, De Man P (2009) Study for the Assessment of the EU's Role in International Maritime Organisations. Final Report, Leuven Centre for Global Governance Studies, Katholieke Universiteit Leuven

Marine Scientific Research: Taking Stock and Looking Ahead

Emmanuella Doussis

1 Introduction

Since the Challenger expedition in the 1870s, which is considered as the advent of modern oceanography,[1] marine scientific research (hereafter MSR) has considerably evolved. New methods of research covering a wide area of scientific interest such as biology, chemistry, geology, and geophysics, as well as advanced technology stemming from simple techniques (dredging, sediment coring, towing of platforms carrying video recorders, and echo sounding traverses) to very sophisticated and extremely expensive ones (such as remotely operated vehicles, known as ROVs, capable of diving to great depths to carry out research and retrieve samples from the deep sea) have been put forward in order to enhance our knowledge on the marine environment.[2] This scientific (r)evolution has inevitably increased the interest of the coastal States in the potential economic exploitation of their offshore resources and has consequently grown their appetite for further expanding their jurisdiction in the oceans.

While scientific understanding of the role of the oceans has considerably progressed since the nineteenth century, we still know very little of this huge, abyssal, and often inaccessible, natural asset. Although oceans represent a very

[1]The Challenger expedition, led by British naturalist John Murray and Scottish naturalist Charles Wyville Thompson between 1872 and 1876, is considered to be the first true oceanographic expedition organized to gather data on a wide range of ocean features, including ocean temperatures, currents, marine life and geology of the seafloor.

[2]For brief general background information on the nature of MSR conducted in the oceans see Leary (2007), pp. 183–188.

E. Doussis (✉)
Institute of European Integration and Policy, National and Kapodistrian University of Athens, Athens, Greece
e-mail: edoussis@gmail.com

G. Andreone (ed.), *The Future of the Law of the Sea*,
DOI 10.1007/978-3-319-51274-7_5

essential part of our planet, paradoxically they are the least known and thus the least understood geographical and geomorphological areas. As one commentator has, quite eloquently, noticed: "until quite recently we did not know what was at the bottom of the oceans. Nor did we know what the bottom of the ocean was made of. In most areas, we did not even know where the bottom of the ocean actually was."[3] This is actually the case not only for the deep sea, where only 8% has been explored and mapped to this date, but also for smaller and more crowded marine areas such as the Mediterranean. For instance, general information on deep-sea resources and issues of biosecurity in this marine region is still missing. Furthermore, there is lack of marine habitat maps and information on small-scale fisheries, as well as a complete inventory of the biodiversity.

Consequently, there is a strong need to develop further knowledge of the marine environment. The interest, however, does not only lie in knowing and better understanding what actually occupies their hidden realm. A better knowledge of the marine environment could also have important practical applications. It could, for instance, grow the capacity of coastal States to combat climate change and respond to increasing anthropogenic pollution or promote sustainable policies and management of their resources, mineral or biological, not to mention the role that some potentially valuable biological resources of the seabed, yet unexplored, may play in the future.

However, this need to develop further knowledge of the marine environment is being restricted by rules of law. The seas and oceans of the world consist of a complex mosaic of different maritime zones, where different legal regimes apply. To enter these waters, researchers, being a State, an international organization, or a private institution, should—in some cases—request and obtain the authorization to do so by following several procedures from different administrative services. So the first question that arises is what potential controls could be held on research projects. In other words, how is MSR regulated? Is the applicable legal framework suitable for the current emergent needs? Does it encourage or not the conduct and promotion of marine scientific research?

This chapter critically explores the international legal regime, which operates to regulate marine scientific research. The first part outlines the general characteristics of this regime. It begins with a brief legislative history to illustrate the factors that have influenced the shape of the current legal framework. It then gives a brief overview of the current regime (Sect. 2). The second part then goes on to consider implementation concerns, as well as some unsettled questions that could lead to potential confusion when the regime is being interpreted and applied in practice (Sect. 3). It concludes with some general remarks regarding how marine scientific research can be more effective, a factor of great importance in combatting global ocean threats (Sect. 4).

[3]See Leary (2007), p. 8.

2 The MSR Legal Regime

2.1 *From Geneva to Montego Bay: A Brief Legislative History*

The regulation of MSR is a relative newcomer to the law of the sea. Until the 1950s, it was not perceived as necessary. MSR has been conducted more or less freely on the high seas.[4] However, the gradual expansion of national jurisdictions on the continental shelf and the recognition of the increasing importance of its resources led to calls for the development of the legal framework in this area. Several coastal States wanted to protect their freshly accorded rights from potential unwanted researchers.

The first attempt to develop MSR regulation arose during the first UN Conference on the Law of the Sea in 1958. However, among the four Conventions adopted,[5] only the Convention on the Continental Shelf contained a few provisions on MSR. In its article 5, it recognized to the coastal State sovereign and exclusive rights for the purpose of exploring its continental shelf and exploiting its natural resources. Any research concerning the continental shelf was subject to limited control by the coastal State, especially where MSR might infringe upon these rights.[6] Therefore, a distinction concerning the nature of the research activities between fundamental (undertaken only for scientific purposes carried out with the intention of open publication) and applied (resource-related) research was embodied in the relevant provisions.[7] Research activities qualified as fundamental would normally be conducted without restrictions, while those qualified as applied research were subject to the coastal States' consent.

MSR was specifically addressed neither in the case of the territorial sea nor in the case of the high seas. Regulation within the territorial sea was considered to be an act of sovereignty and, thus, under exclusive control of the coastal State. In other words, any MSR conducted by foreign States should be subject to a coastal State's

[4]See Treves (2012), para. 5.

[5]The Convention on the High Seas, the Convention on the Territorial Sea and Contiguous Zone, the Convention on the Continental Shelf and the Convention on Fishing and Conservation of the Living Resources of the High Seas.

[6]According to article 5 (8): "*the consent of the coastal State shall be obtained in respect of any research concerning the continental shelf and undertaken there. Nevertheless the coastal State shall not normally withhold its consent if the request is submitted by a qualified institution with a view to pure scientific research into the physical or biological characteristics of the continental shelf, subject to the proviso that the coastal State shall have the right, if it so desires, to participate or to be represented in the research, and that in any event the results shall be published*".

[7]For a general discussion see Caflisch and Piccard (1978), pp. 848–852.

consent.[8] Within the high seas, although MSR was not expressly listed as a freedom, it was generally accepted as such.[9]

Thus, the legal framework set forth in Geneva would result in a simultaneous application of a different regime in the same maritime space. Whereas MSR on continental shelf was subject to the consent of the coastal State, it was nevertheless free when conducted on the superjacent waters (waters above), belonging to the high seas.

All these elements would form the basis of a more detailed MSR regime, adopted a few years later in Montego Bay within the framework of the United Nations Convention on the Law of the Sea (hereafter UNCLOS). However, the way was not paved with nenuphars. During the negotiations, held from 1973 to 1982, MSR regulation proved to be one of the most delicate and difficult issues to resolve.[10] The major researching (and, of course, mostly developed and having the necessary funding) States crossed swords with the newly independent and developing coastal States on a number of conflicting issues: the distinction between fundamental or pure and applied research; the extent of the coastal States' control over MSR, especially in the emerging exclusive economic zone (hereafter EEZ); and dispute settlement.[11] Both sides put forward claims and arguments. Researching States claimed a liberal regime for MSR, without restrictions, and open publication of the results of benefit to all. On the other hand, coastal States had a special interest in research activities conducted within waters under their jurisdiction. Several (mostly developing) States strongly believed (rather understandably) that an unlimited right to conduct MSR would lead to abuses on the part of the researching States because it would inevitably have some direct or indirect bearing on their natural resources or might serve as a disguise for other operations related to the exploration and exploitation of natural resources or even intelligence gathering activities.[12] Some countries called for the establishment of an international body responsible for regulating MSR in all marine areas.[13]

While these arguments and proposals were not entirely convincing, it was nevertheless clear that some balance should be found between conflicting interests: the interest of researchers in facilitating the conduct and promotion of MSR and the interests of the coastal States in protecting their rights within the waters under their jurisdiction. Thus, the final result incorporated in UNCLOS, signed in Montego Bay in 1982, was a product of compromise trying to accommodate concerns stemming from both sides.

---

[8]For further analysis see Stephens and Rothwell (2015), p. 563.

[9]Leary (2007), p. 191.

[10]For a brief description see de Marffy (1985).

[11]UN, DOALOS (hereafter: DOALOS Guide) (2010), p. 3.

[12]See Caflisch and Piccard (1978), p. 850.

[13]For a brief description of these proposals see Leary (2007), pp. 191–193.

2.2 *Current Regime Under UNCLOS: Consent v. Freedom*

The 1982 UNCLOS compensated the prior indigence by devoting an entire part, consisting of 28 articles, to the subject of marine scientific research. Part XIII (articles 238–265) describes in detail the legal framework within which all research activities must be carried out in order to "promote the study of the marine environment," proclaimed in the preamble of the Convention.

A simple lecture on the first articles gives the impression of a rather liberal regime. The general rule is that all States, coastal or not, have the right to conduct MSR subject to rights and duties of other States.[14] The same right to conduct MSR is recognized in competent international organizations, i.e. organizations with competence in marine science, such as the International Seabed Authority or the UNESCO International Oceanographic Commission. The right to conduct MSR is directly associated with the obligation to promote and facilitate MSR,[15] which has been convincingly described as a "principle of positive engagement" for the purpose of increasing knowledge for the benefit of all mankind on what is its major natural environment: the ocean.[16]

Nevertheless, the general right to conduct MSR is not an absolute one as it is restrained by subsequent principles and rules. Some of them are justified by the due respect to other international rules or legitimate uses of the sea. Thus, marine scientific research shall be conducted exclusively for peaceful purposes, with appropriate scientific methods and means compatible with the Convention and in conformity with regulations under the Convention,[17] including those for the protection and preservation of the marine environment.[18] The issue of liability is also addressed in these general provisions, providing that researching States or international organizations shall be responsible and liable for damage resulting from measures taken in contravention to the UNCLOS[19] regime and for pollution arising from MSR.[20]

Other principles and rules, though not unjustified, seem to complicate the applicable regime, and their implementation in practice might create great confusion to researchers when preparing, planning, and conducting a research project. The need to balance the interests of the researching States and the interests of the coastal States resulted in an area-by-area approach to rights in connection with MSR. Thus, the rules vary in accordance to the legal status of the marine areas in which the research is being conducted. The general idea concerning MSR is that the closer to the shore of a coastal State, the greater its consent powers to control the research activities.

---

[14] Article 238.

[15] Article 239.

[16] See Pancracio (2010), p. 377.

[17] Article 240.

[18] For further analysis see Kirk (2015).

[19] Article 263 (2).

[20] Article 263 (3).

Therefore, within the territorial sea, the coastal State, being a full sovereign, has complete control over marine scientific research activities.[21] It has the exclusive right to regulate, authorize, and conduct MSR. This jurisdiction is not even limited by the right of innocent passage as it is expressly provided that conducting MSR during passage through territorial waters renders a passage noninnocent.[22] Consequently, all research activities within the territorial sea require the coastal State's express consent through diplomatic channels.

UNCLOS extended the MSR regulation to the emerging EEZ. However, the regime governing MSR both in the EEZ and on the continental shelf is more complicated than the one governing the territorial sea because the coastal State's consent is subject to conditions.[23] Within these maritime zones, the coastal State has both jurisdiction over MSR and the right to regulate, authorize, and conduct research activities. Its consent for MSR activities conducted by third States or international organizations is also required. However, in this case, the coastal State does not have an unlimited discretion to withhold such consent. It can do so only in four cases, expressly enumerated in the Convention, that concern projects (a) of direct significance for the exploration and exploitation of natural resources, whether living or nonliving; (b) that involve drilling into the continental shelf; (c) that involve construction, operation, or use of artificial islands; and (d) that contain incorrect information provided to the coastal State or if the researching State or competent international organization has outstanding obligations to the coastal State from a prior research project.[24] The coastal State is given further guarantees as it has the right to require the suspension of cessation of any MSR activities if they are not conducted under the conditions set forth in Part XIII of UNCLOS.[25]

However, the consent has to be granted in normal circumstances,[26] provided that the research activities are carried out for peaceful purposes and undertaken in order to increase the knowledge of the marine environment for the benefit of mankind. The consent must be explicit, except for two cases in which the Convention provides the possibility of a presumed[27] and an implied[28] consent, under specific conditions. However, these two possibilities have been ignored by State practice.

---

[21] Article 245.

[22] Article 19 (2).

[23] Article 246.

[24] Article 246 (5).

[25] Article 253.

[26] Article 246 (3).

[27] According to article 247, the consent of the coastal State is presumed if that state is a member of or has a bilateral agreement with an international organization that aims at conducting MSR, by itself or under its auspices, in the EEZ or on the continental shelf of the coastal State, and further provided that the coastal State either explicitly approved the project when the decision was initially made or the coastal State did not object to the decision within a period of 4 months after notification.

[28] According to article 252 the consent of the coastal State is implied provided that it has not reacted within a period of 4 months after the required information has been provided by the researching State or the competent international organization.

This constant give and take of guarantees between researching and coastal States attests the difficulties in balancing the conflicting interests of both sides. Coastal developing States feared that freedom of scientific research would increase inequalities between the rich and the poor. Thus, marine scientific activities should be controlled as much as possible. Consequently, researchers also have procedural obligations to follow not only before undertaking a research activity (to provide the coastal State with all necessary information at least 6 months before the starting date of the research activities)[29] but also after having been granted consent to conduct MSR. This is to ensure the right of the coastal State to participate, if it so desires, in the research project and to give the coastal State access to data and information about any major changes in the project.[30]

There is also a provision concerning the continental shelf beyond 200 miles, according to which coastal States may not withhold consent to foreign researchers to conduct MSR, unless it is for specific areas publicly designated by those States as areas in which exploitation or exploration operations are occurring or will occur within a reasonable period of time.[31] It should be noted that the water column above the outer continental shelf belongs to the high seas, where MSR is freely conducted.

In the maritime zones beyond national jurisdiction—in the deep seabed, that is the area beyond the continental shelf called "the Area," as well as in the high seas—MSR may be conducted by all States with due regard for other rules under the Convention, such as the duty to protect the marine environment.[32] In the high seas, MSR has been expressly accorded the status of a high sea freedom.[33] Thus, in this case, only the flag State of the ship conducting research activities has jurisdiction.

These provisions raise some remarks that are worth noting. The first is that the balance seems to weigh more on the side of the coastal States, whose sovereign rights have undoubtedly been reinforced. The extension of the MSR regime to EEZs and the upgrading of the coastal State's consent powers have restrained freedom of scientific activities in larger areas of the sea at the expense of scientific research. However, and this is the second remark, the consent regime applicable to the EEZ and on the continental shelf is not absolutely clear. For instance, the provisions related to the procedural obligations of the researchers are subject to different interpretations or even controversy.[34] What are the limits in the coastal State's right to participate, if it desires so, in the research project? Which are the appropriate official channels for the communication of MSR projects? Who assesses the data required prior or during the research activities? Which decisions

[29]Article 248.

[30]Article 249.

[31]Article 246 (6).

[32]Articles 256 and 257.

[33]Articles 87 and 257.

[34]See Jarmache (2003).

of the coastal State are justiciable?[35] Arguably, the rights of the researchers are not well defined, and this ambiguity may delay or even discourage potential research projects.[36]

3 From Theory to Practice: Implementing the MSR Regime

MSR is regulated by the relevant provisions of the 1982 UNCLOS, which actually counts 167 contracting parties, including the EU.[37] It is worth noting that only few coastal States have enacted special national legislation to prescribe procedures necessary for conducting MSR, but overall it seems that their practice is more or less consistent with the UNCLOS requirements.[38] Moreover, the almost universal acceptance of the Convention and the influence of its Part XIII on State practice indicate that many of the MSR provisions reflect customary international law and, thus, are applicable to all users of the oceans.[39] Other legal instruments, either universal or regional, complement the general framework by encouraging State parties to cooperate for the promotion of MSR.[40]

Obviously, international law offers a general framework for conducting and promoting MSR. The question concerns how this regime is applied in practice and if it is effective. There are three components related to the practical implementation of the MSR legal framework. The first concerns its spatial dimension, while the second refers to its functional application. The third component relates to who is involved.

3.1 Where? The Spatial Dimension

In many parts of the world, maritime zone maps are not yet completely drawn as there are still pending disputes, open issues, or even "unfinished business"[41]

[35]Under article 297 (2), the coastal State denying consent or ordering the suspension or cessation of MSR in its EEZ or on the continental shelf is not obliged to subject itself to the dispute resolution settlement. For further analysis see Roach (1996).

[36]For further discussion concerning the difficulties for foreign researchers to obtain an approval permit see Xue (2009), p. 215.

[37]http://www.un.org/Depts/los/reference_files/chronological_lists_of_ratifications.htm. Accessed: 9 Mar 2016.

[38]For a review of the State practice see the site of the Intergovernmental Oceanographic Commission of UNESCO, http://www.ioc-unesco.org. Accessed: 9 Mar 2016.

[39]However, this is not the case for some provisions, such as the one referring to the possibility of implied consent, which is ignored in State practice, see Treves (2012), par. 16 and 17.

[40]Article 13.

[41]This expression is mentioned by Gavouneli referring to the Agreement concluded in 2009 between Greece and Albania, which was declared as unconstitutional by the Albanian Constitutional Court, Gavouneli (2015), p. 276.

(e.g., agreements concluded but not yet in force). Obviously, this situation affects the conduct and promotion of MSR activities and is not so encouraging for potential researchers. From which coastal State are they going to request permission to undertake a research in disputed areas?

A very characteristic example is the Adriatic and the contiguous Ionian seas. This maritime region links seven countries: Italy, Slovenia, Croatia, Bosnia and Herzegovina, Montenegro, Albania, and Greece. A particular feature of this marine region is that many coastal States have not claimed all maritime zones that they are entitled to establish under international law.[42] The result is that large areas of the Adriatic and Ionian marine regions remain beyond the jurisdiction of coastal States and under the regime of the high seas.

In fact, the current jurisdictional picture is rather complex.[43] All coastal States have established a 12 nm territorial sea, with the exception of Greece, which maintains a 6 nm territorial sea, and Bosnia and Herzegovina, a special case due to its particular geographic situation.[44] Within this zone, coastal States have exclusive control over MSR activities, and their express consent is required.

The coastal states also have jurisdiction on the continental shelf, where they exercise substantial control over MSR activities. This zone does not need to be proclaimed as it exists *ab initio* and *ipso facto*, but the narrow sea space does not permit them to enjoy the maximum jurisdictional rights permitted under international law. However, the relative maritime boundaries have not been yet fully established.[45] With the exception of three delimitation agreements in force (the 1968 agreement between Italy and former Yugoslavia, the 1977 agreement between Italy and Greece, and the 1992 agreement between Italy and Albania), the rest of the maritime boundaries remain to be agreed upon, including some territorial sea boundaries. This includes, for example, the southern boundary of the Slovenian territorial sea with Croatia, as the dispute is currently being subjected

[42]For an explanation see Vidas (2008), pp. 9–10.

[43]A list of the relevant national legislation is provided in the website of DOALOS, http://www.un.org/Depts/los/LEGISLATIONANDTREATIES/europe.htm. Accessed: 2 Oct 2015.

[44]Bosnia and Herzegovina has actually a very limited coastline on the Adriatic Sea, the Neum corridor, which is enclosed between two parts of the Croatian coastline. It could be said that it is an almost landlocked country.

[45]It is worth noting that some States, including Greece and Italy, provide in national legislations that in the absence of delimitation agreements the medial line will apply provisionally. For Greece, see art. 156, Law No. 4001/2011 for the Operation of Electricity and Gas Energy Markets, for Exploration, Production and transmission of networks of Hydrocarbons and other provisions, published in the Government Gazette No. 179, Part One, 22 August 2011, text available at: http://www.ypeka.gr/LinkClick.aspx?fileticket=l3TNzx1rKsM%3D&tabid=765&language=en-US. Accessed: 9 Mar 2016. For Italy, see art. 1 (3), Legge No. 61 di 8 febbraio 2006, *Istituzione di Zone di Protezione Ecologica Oltre il Limite Esterno del Mare Territorial*, Gazzetta Ufficiale No. 52 del 3 marzo 2006, text available at: http://www.camera.it/parlam/leggi/06061l.htm. Accessed: 9 Mar 2016.

to arbitration,[46] not to mention the maritime boundaries between Greece and Albania. In 2009, after lengthy negotiations, the two States signed a continental shelf delimitation agreement with a built-in mechanism for automatic extension to any future maritime zones that might be proclaimed. However, a year later, the Albanian Constitutional Court declared—rather unconvincingly[47]—the agreement as unconstitutional.

Undoubtedly, the list of problems is endless. In 2003, Croatia proclaimed an ecological and fisheries protection zone[48] on the water column above its continental shelf. Although this zone is not mentioned in UNCLOS, its establishment derives from the rights of coastal States to claim an EEZ, and thus the legal regime may be identical to the regime of an EEZ. Thus, MSR activities in this zone are subject to the coastal State's consent. Nevertheless, the Croatian act raised strong protests on the part of the neighboring countries, especially Slovenia, which also declared an ecological protection zone with overlapping jurisdiction with the Croatian one.[49] The dispute has taken not only legal but also political proportions as it was linked to the accession of Croatia to the European Union, and the two countries agreed to follow the route of arbitration. Italy has also declared an ecological protection zone, but it does not apply to the Adriatic and Ionian seas.[50]

There is also another issue of concern. As EEZs have not been proclaimed (with the exception of the derivative zones of Croatia and Slovenia already mentioned), MSR activities on the continental shelf are subject to the consent of the coastal State, whereas they are free when conducted on the superjacent waters, belonging to the high seas. That is why, in practice, several coastal States (including Greece) require either notification or permission on research activities undertaken in the high seas in order to ensure that these activities do not infringe upon their sovereign rights on the seabed.[51] The real question for these States is if there is anything else they can do to ensure that the resources lying on the seabed are treated appropriately.

This jurisdictional picture could change with the establishment of EEZs or even derivative zones, which will reinforce the coastal States' rights to control and

[46]In 2009 the two States signed an agreement to submit their dispute to arbitration. For further information see Territorial and Maritime Arbitration between Croatia and Slovenia, www.pca-cpa.org. Accessed: 9 Mar 2016. For a brief commentary see Cataldi (2013).

[47]According to international law, a State cannot invoke its domestic deficiencies to contest the validity of a duly signed international agreement. For further analysis see Noussia (2010).

[48]Decision on the extension of jurisdiction of the Republic of Croatia in the Adriatic Sea, 53 *Law of the Sea Bulletin*, 2004, pp. 68–69.

[49]Act on the proclamation of the ecological protection zone and on the continental shelf, 60 *Law of the Sea Bulletin*, 2006, pp. 56–57.

[50]Legge No. 61 di 8 febbraio 2006, *Istituzione di Zone di Protezione Ecologica Oltre il Limite Esterno del Mare Territorial*, Gazzetta Ufficiale No. 52 del 3 marzo 2006, text available at: http://www.camera.it/parlam/leggi/060611.htm. Accessed: 5 Nov 2015. For further analysis see Scovazzi (2005).

[51]See Strati (2012), p. 50.

benefit from MSR conducted in areas currently belonging to the high seas.[52] Undoubtedly, the next necessary step should be the delimitation of the maritime boundaries. Although tempting, this scenario is not so desirable. Some coastal States (being also researchers) would rather maintain the current *status quo* because otherwise their rights to conduct free MSR, as well as other activities, up to the limits of the territorial sea of their neighbors will be restricted. Others, although flirting with the idea of proclaiming an EEZ, hesitate to do so; their act could open a Pandora's box, as the example of the dispute between Croatia and Slovenia reveals. Thus, if no exclusive economic zones are proclaimed in these parts of the high seas over the continental shelf under national jurisdictions, problems and concerns relating to the conduct and promotion of MSR will still remain to the detriment of marine scientific activities.

3.2 *Which Activities Fall Under MSR? The Functional Dimension*

Although many proposals have been discussed during the negotiations,[53] UNCLOS does not provide a definition for MSR. Looking back at the *travaux préparatoires*, it seems that the most controversial issue was the difficulty of clearly distinguishing between fundamental and applied research.[54] Many developing States strongly believed that the acceptance of such a distinction would inevitably lead to abuses. However, the simple rejection of the difference and the submission of both activities to discretionary coastal State consent do not eliminate potential abuses, as several incidents especially in the South China Sea reveal.[55]

A careful reading of the UNCLOS provisions, especially those concerning the conduct of MSR in the EEZ and on the continental shelf, sheds light on an implicit distinction between fundamental and applied research, affecting the discretionary powers of the coastal State to uphold its consent. Even if the precise terms are not explicitly used, it is obvious that the activities where the coastal State should *normally grant* its consent refer to fundamental research (*projects undertaken exclusively for peaceful purposes and in order to increase scientific knowledge of*

[52]See the Report prepared for the DG MARE of the European Commission, *Cost and benefits arising from the establishment of maritime zones in the Mediterranean Sea*, June 2013, p. 165, text accessible at: http://ec.europa.eu/maritimeaffairs/documentation/studies/documents/maritime-zones-mediterranean-report_en.pdf. Accessed: 9 Mar 2016.

[53]DOALOS Guide (2010), pp. 4–5.

[54]DOALOS Guide (2010), p. 5.

[55]For ex. the *Impeccable* incident in the South China Sea, where a USA surveillance ship was conducting undersea passive sonar operations and acoustic data gathering, provoking the reactions of China. For further discussion, see Agnihotri and Agarwal (2009).

the marine environment for the benefit of all mankind). On the other hand, those where consent may be withheld concern applied research (*projects of direct significance for the exploration and exploitation of natural resources, that involve drilling into the continental shelf*, etc.).[56]

However, in practice, it is very difficult to distinguish the two types of activities as no objective criteria have been set forth. The Geneva regime was more effective in that respect as it provided the criterion for open publication of the results in order to make a distinction between the two. Thus, fundamental research is conducted with the intention of open publication of the results, while applied research is undertaken with the intention of producing certain practical results. Certainly, all fundamental research may acquire some practical relevance, but, as Lucius Caflisch suggested, "this does not mean that such research is undistinguishable from applied research."[57] As the same author argues, "even in borderline cases where the planned research is partly fundamental in nature and partly aimed at obtaining practical results," the requirement of open publication will not be necessarily detrimental to the coastal State's interests as "it will in fact be the coastal State which will mainly benefit from these results."[58] This is because it enjoys exclusive resource jurisdiction over the area in which the research is carried out. Nevertheless, even if MSR is conducted under the watchful eye of the coastal State, the latter might be unwilling to publish the results and UNCLOS gives full discretion in that respect. Coastal States would not be willing to share any information concerning resources lying in maritime zones under national jurisdiction. They will even try to protect from any abusive appropriation those lying in waters outside national jurisdictions.

Yet it can be argued that even if the Convention had incorporated a definition for MSR, it might have been outdated as science and technology evolve quicker than legal regimes. Regardless of how persuasive this argument may be and in line with the position of the negotiators that concluded that a definition would be superfluous,[59] the lack of a clear definition of marine scientific activities and their means of execution may lead to different interpretations as practice proves.[60] Therefore, it creates great uncertainty about the activities covered by the MSR regime and those that are not.

There is, indeed, a legal grey zone concerning jurisdiction. For instance, it is not certain if all forms of data collection, routine operational activities, or hydrographical surveys (collection of information for the making of navigational charts and safety of navigation)[61] can be subject to the MSR regime. Some authors

---

[56] Article 246 (3) and (5).

[57] Caflisch and Piccard (1978), p. 850.

[58] Caflisch and Piccard (1978), p. 851.

[59] Bork et al. (2008), p. 303.

[60] Some States limit or enlarge the meaning of the term, according to their own interests. For the American practice, for ex., see Roach (2001), p. 9.

[61] See Bateman (2009).

even suggest that activities directed at shipwrecks and other forms of underwater cultural heritage[62] come within the scope of MSR regime and are, thus, subject to the coastal State consent.[63]

There is also great controversy as to whether military surveys, which relate to data collection for military purposes, are subject to the MSR regime.[64] The equipment used for this type of activities is often the same as that used in marine scientific research. Some authors suggest, however, that the intended use of the information collected from such activities by the military would exempt this category from the MSR regime.[65] On the contrary, others argue that military surveys cannot be distinguished from MSR as the respective motives cannot be easily determined.[66] They seem to suggest that all marine data collection activities should be covered by the MSR regime; otherwise, they could be carried out in the coastal States' EEZ without any restrictions.

Another important activity, of which the inclusion in the MSR regime is hotly debated, is bioprospecting.[67] It relates to the access to genetic resources and involves collection and analysis of information, data, or samples aimed at increasing humankind's knowledge of the valuable compounds and genetic materials. The possible commercialization of the results would have "the practical effect of transforming the activity into one that is of direct significance to the exploitation of a natural resource."[68] It seems that, as in the case of military surveys, it is the intended use of the data collected from such activities, rather than the practical nature of the activities themselves, that distinguishes them from pure scientific research.

Unfortunately, there is no clear answer for all these concerns as there is no definition of MSR activities and the means of their execution.

3.3 Who Is Involved? The Unexplored Duty of Cooperation

MSR is open to all States and their research institutions, as well as competent international organizations. Certainly, the coastal States are the most interested not only in conducting and promoting scientific research but also in ensuring protection of their natural resources and economic interests.

---

[62]See Dromgoole (2010).

[63]*Contra* Roach (1996), p. 60.

[64]See Xue (2009) p. 222 and Bork et al. (2008), p. 305.

[65]See Roach (1996), p. 61.

[66]See Xue (2009), pp. 218–219.

[67]See Jorem and Tvedt (2014).

[68]See Stephens and Rothwell (2015), p. 568.

But the real question is: do the coastal States have adequate means to study and understand by themselves their adjacent marine environment? It seems that capacities in terms of institutions and equipment are very unevenly distributed.[69] For instance, in the Mediterranean, only a few States have large research vessels able to undertake research in the high seas.[70] In order to reinforce their research capacity, they might conclude agreements with foreign researchers States.[71] UNCLOS encourages international cooperation in MSR between States and competent international organizations.[72] These actors are even invited to conclude bilateral or multilateral agreements to create favorable conditions for the conduct of MSR and integrate the efforts of scientists in studying the marine environment.[73]

Indeed, cooperation is very much needed in a domain such as MSR, which requests considerable investments in human and financial resources. Advantages could be gained from networking and better cooperation between research institutions. In fact, several international research projects do exist. A characteristic example is the "Argo floats" project; launched in 2000, the project boasts an impressive network of data collection *in situ*, covering economic exclusive zones in the Atlantic, the Pacific, and the Indian oceans.[74] The objective of this project is the continuous monitoring of the temperature, salinity, and velocity of the upper ocean with all data being relayed and made publicly available within hours after collection. However, even these routine operational activities may raise several legal questions that do not receive unanimous answers.[75]

It should also be noted that international cooperation is not always a given. Jurisdictional uncertainty and legal ambiguities may impact the conduct of these

[69]There is no information available in the global level. However, The Global Ocean Science Report, launched in 2014, will provide a tool for mapping and evaluating the human and institutional capacity of States in terms of marine research, observations and data/information management, and provide a global overview of the main fields of interest, technological developments, capacity- building needs and overall trends, as well as information on research investments and the status of ocean research, see Report of the Secretary General, *Oceans and the law of the Sea*, A/70/74/Add. 1, 2015, par. 61.

[70]As far as the Mediterranean is concerned, the Mediterranean Science Commission database provides a list with resources and means of marine research institutions by country around the Mediterranean: http://www.ciesm.org/online/institutes/IndexInstituts.htm. Accessed: 9 Mar 2016.

[71]This term covers States conducting research themselves or whose private institutions are engaged in such research.

[72]Article 242.

[73]Article 243.

[74]For further information see http://www.argo.net. Accessed: 14 Mar 2016.

[75]For further analysis see Bork et al. (2008), p. 303.

projects as practice reveals. For instance, in the MEDITS survey program,[76] the research activities end at the boundary of the ecological and fishery protection zone claimed by Croatia.[77] Therefore, building mutual confidence is the very first step in launching cooperation for MSR activities.

4 Conclusion

The main objective of this chapter was to show how MSR can be conducted and promoted and to assess the current legal framework provided by UNCLOS. This framework establishes both general obligations and the legal basis for jurisdiction of the coastal States over MSR. Certainly, it does not resolve all problems satisfactorily and does not provide for any technical details. Being a product of a difficult compromise between the interests of the coastal and the researching States, it reflects the tension between appropriation and internationalization, which dominated the negotiations of the universal convention on the law of the sea. However, in the case of MSR, the balance seems to weigh more on the side of the coastal States. As it was eloquently noted: "freedom of MSR has ceased to exist in the law of the sea."[78] Admittedly, MSR is not yet free but largely controlled by the coastal States even in some parts of the high seas. This might explain why our knowledge on many issues concerning the role of the oceans is still limited.

This general regime provided by UNCLOS is unlikely to be changed, at least in the nearby future. Nevertheless, it could be further developed and the legal ambiguities clarified by regional cooperation and consistent State practice. Such cooperation could be undertaken by the coastal States themselves or in the framework of competent international organizations or even in the framework of the existing Regional Seas Programme under the auspices of UNEP as it has already been suggested.[79] Current technological developments in marine scientific research (e.g., remote sensing from satellites or collecting data through other means than ships) and their legal implications could be further discussed in such frameworks, and a code of conduct for MSR activities could be developed to diminish potential controversies. The need for a more integrated approach is more than evident. Instead of a strict balance of interests between coastal and researching States,

[76]The MEDITS survey programme intends to produce basic information on benthic and demersal species in terms of population distribution as well as demographic structure, on the continental shelf and along the upper slopes at a global scale in the Mediterranean sea through systematic bottom trawl surveys. For further information, see http://www.sibm.it/SITO%20MEDITS/principaleprogramme.htm. Accessed: 14 Mar 2016.

[77]*Cost and benefits arising from the establishment of maritime zones in the Mediterranean Sea, op. cit.*, p. 174.

[78]See de Marffy (1985), p. 957.

[79]See Oral (2014).

wider concerns need to be taken into account, such as issues of sustainability, as well as the necessity to know and better comprehend the marine environment.

References

Agnihotri KK, Agarwal SK (2009) Legal aspects of marine scientific research in exclusive economic zones: implications of the impeccable incident. Maritime Affairs 5(2):135–150

Bateman S (2009) Hydrographic surveying in exclusive economic zones: is it marine scientific research? In: Nordquist M et al (eds) Freedom of seas, passage rights and the 1982 law of the sea convention. Martinus Nijhoff Publishers, Leiden

Bork K et al (2008) The legal regulation of floats and gliders. In quest of a new regime? Ocean Dev Int Law 39:298–328

Caflisch L, Piccard J (1978) The legal regime of marine scientific research and the third united nations conference on the law of the sea. Zeitschrift fur auslandisches offentliches Recht und Volkerrecht 38:848–901

Cataldi G (2013) Prospects for the judicial settlement of the dispute between Croatia and Slovenia over Piran Bay. In: Boschiero N et al (eds) International courts and the development of international law: essays in honour of Tullio Treves. T.M.C. Asser Press, The Hague, pp 257–268

de Marffy A (1985) La recherche scientifique marine. In: Dupuy R-J, Vignes D (eds) Traité du nouveau droit de la mer. Economica/Bruylant, Paris/Bruxelles, pp 957–973

Dromgoole S (2010) Revisiting the relationship between marine scientific research and the underwater cultural heritage. Int J Mar Coast Law 25:33–61

Gavouneli M (2015) Offshore energy: troubled waters in the eastern Mediterranean sea. In: Scheiber HN, Kraska J, Kwon M-S (eds) Science, technology and new challenges to ocean law. Koninklijke Brill, Leiden, pp 253–279

Jarmache E (2003) Sur quelques difficultés de la recherche scientifique marine. In Mélanges offerts à Laurent Lucchini et Jean-Pierre Quéneudec. La mer et son droit, Pedone, Paris, pp 303–314

Jorem A, Tvedt MW (2014) Bioprospecting in the high seas: existing rights and obligations in view of a new legal regime for marine areas beyond national jurisdiction. Int J Mar Coast Law 29:321–343

Kirk E (2015) Science and the international regulation of marine pollution. In: Rothwell D, Elferink A, Scott K, Stephens T (eds) The Oxford handbook on the law of the sea. Oxford University Press, Oxford, pp 516–535

Leary DK (2007) International law and the genetic resources of the deep sea. Martinus Nijhoff Publishers, Leiden

Noussia K (2010) The decision of the Albanian Supreme Court Annulling the 2009 maritime delimitation agreement between Albania and Greece. Int J Mar Coast Law 25:601–608

Oral N (2014) The need for a regional framework for marine scientific research in the Black Sea and Mediterranean Sea. Mar Genomics 17:67–71

Pancracio J-P (2010) Droit de la mer. Dalloz, Paris

Roach JA (1996) Marine scientific research and the new law of the sea. Ocean Dev Int Law 27 (1–2):59–72

Roach JA (2001) La recherche scientifique marine dans la zone économique exclusive et sur le plateau continental: le problème du consentement de l'Etat côtier. Commission Océanographique d'experts, IOC/ABE-LOS I/11, http://unesdoc.unesco.org/images/0012/001226/122621F.pdf

Scovazzi T (2005) La zone de protection écologique italienne dans le contexte confus des zones côtières méditerranéennes. Annuaire du droit de la mer 10:209–222

Stephens T, Rothwell DR (2015) Marine scientific research. In: Rothwell DR, Oude Elferink AG, Scott KN, Stephens T (eds) The Oxford handbook of the law of the sea. Oxford University Press, Oxford, pp 559–581

Strati A (2012) Greek maritime zones and delimitation with neighboring States (in Greek). Nomiki Bibliothiki, Athens

Treves T (2012) Marine scientific research. In: Wolfrum R (ed) Max Planck encyclopedia of public international law. Oxford University Press, Oxford, accessible via: http://www.mpelpil.com

UN, DOALOS (2010) Marine scientific research, a revised guide to the implementation of the relevant provisions of UNCLOS. New York

Vidas D (2008) The UN Convention on the law of the sea, the European Union and the rule of law. What is going on in the Adriatic Sea? Fridthof Nansen Institute

Xue G (2009) Marine scientific research and hydrographic survey in the EEZs: closing up the legal loopholes? In: Nordquist M et al (eds) Freedom of seas, passage rights and the 1982 law of the sea convention. Martinus Nijhoff Publishers, Leiden, pp 209–225

Protecting Arctic Ocean Marine Biodiversity in the Area Beyond National Jurisdiction

Plausible Legal Frameworks for Protecting High Arctic Waters

Kamrul Hossain and Kathleen Morris

1 Introduction

An estimated 30% of the world's undiscovered gas and 13% of the world's undiscovered oil lies in the Arctic, most of it offshore.[1] Once impossible shipping routes—such as the Northern Sea Route and even a possible transarctic route—appear increasingly feasible due to sea ice melt.[2] Arctic shipping traffic as a whole is anticipated to increase in coming years due to ice melt allowing for longer shipping seasons (see footnote 2). Past sea ice melt indicates that the Arctic may be ice-free within decades, opening waters—and resources—previously sheltered by ice from mankind's exploits.[3] Once protected by ice, the vast northern landscape compares in size to Africa and exists as one of Earth's final pristine ecosystems.[4] Further, its wealth of resources includes more than oil, with living resources that include 5000 animal species; 2000 types of algae; and tens of thousands of ecologically critical microbes.[5]

[1]Eurasia Group for The Wilson Center (2013), p. 3.

[2]Emmerson and Lahn (2012), pp. 29–30.

[3]Wang and Overland (2012), pp. 4–5.

[4]Hossain (2014), pp. 18–20.

[5]Arctic Biodiversity Assessment Report (ABA) (2013), Chapter 14, p. 488.

K. Hossain (✉)
Northern Institute for Environmental and Minority Law, Arctic Center, University of Lapland, Rovaniemi, Finland
e-mail: khossain@ulapland.fi

K. Morris
Department of International Relations, University of Southern California, Los Angeles, CA, USA
e-mail: krmorris@usc.edu

G. Andreone (ed.), *The Future of the Law of the Sea*,
DOI 10.1007/978-3-319-51274-7_6

Present economic opportunity in the Arctic is only possible through a fundamental challenge to Arctic biodiversity: rapid climate change brought on by global warming.[6] The year 2015 proved to be the warmest year on record.[7] Global-warming-caused sea ice melt will not only pressure the Arctic marine ecosystem but also allow for anthropogenic activity and disturbances in previously unreachable parts of the Arctic Ocean. Opening waters promise to enable heightened resource extraction, tourism, shipping and navigation, and fishing activity. Arctic and non-Arctic states alike now express competitive economic stakes in the region in what has been termed the next "Great Game."[8] Thus, dual pressure of climate change and anthropogenic activity may threaten the living resources unique to the Arctic. Considering these economic and geopolitical shifts, it is worthwhile to consider whether current international law, such as the Law of the Sea, is sufficient to govern a changing region and protect its marine environment.

In anticipation of broader economic activity and environmental threat in the region—brought on by sea ice melt and the emergence of other states as stakeholders—the five Arctic coastal states reaffirmed a commitment to the Law of the Sea, including the United Nations Convention on the Law of the Seas (UNCLOS), in the Ilulissat Declaration of March 2008.[9] Prior to the Declaration, discourse over Arctic governance began to consider the need for a new international treaty to govern the region. Media, academia, and Arctic strategies of non-Arctic states expressed growing concern that the Law of the Sea, given its rudimentary framework, was not sufficient to govern the Arctic ecosystem in changing environmental and commercial contexts. The following text will consider: is a new international legal instrument necessary for the protection of the Arctic marine environment? We suggest that existing legal instruments, such as the UNCLOS, suffice to govern the region and enable more comprehensive environmental preservation, particularly in the form of marine protected areas (MPAs).

Our analysis examines current state-led efforts to designating large portions of the Arctic as "marine protected areas" (MPAs), which would limit human activity in protected waters in an effort to preserve marine biodiversity conforming to the approach known as ecosystem-based management (EBM). The principle of EBM emphasizes environmental cross-linkages that require *collaborative* ecosystem management transcending national jurisdictions and boundaries.[10] In contrast, current discourse around resource exploitation of the region describes a likelihood of geopolitical "scramble for the Arctic," drawing parallels to European imperial

[6]Arctic Climate Impact Assessment (ACIA) (2005), pp. 247, 1017.

[7]Hottest year on record according to surface temperature data from GISS Surface Temperature Analysis (GISTEMP) *See* GISTEMP Team (2016).

[8]Borgerson (2009).

[9]The Ilulissat Declaration provides the clear and firm statement that there is no need to develop a new arrangement for Arctic governance. *See Arctic Ocean Conference,* May 27–29, 2008, Ilulissat Declaration (May 28, 2008).

[10]UNEP/GPA (2006), p. 4.

competition over African resources.[11] In this scramble, state interests, including prospects for control over shipping lanes, oil resources, and fishing rights, depend on the lineation of exclusive economic zone boundaries based on seabed claims. Thus, while transboundary environmental protection is critical, the current Law of the Sea system emphasizes the importance of said boundaries, for delineating states' rights to respective coastal waters and resources. As such, the growing need for Arctic MPA creation in areas beyond national jurisdiction poses a challenge to the contemporary limitations of Law of the Sea.

Moreover, a considerable portion of the Arctic Ocean—2.8 million square kilometers—lies in an area beyond national jurisdiction (ABNJ), thus falling beyond the exclusive economic zone (EEZ) of any state actor.[12] Currently, global ocean management is hindered by the lack of a clear mechanism for MPA designation in ABNJs. Recent leadership by the Arctic Council established the goal of creating a pan-Arctic network of MPAs through piece-by-piece coordination of nation-led efforts to create MPAs within sovereign borders.[13] As such, the High Arctic cannot be designated as part of the network due to its status in ABNJ. Thus, in the following chapter drawing from our article "Legal Instruments for Marine Sanctuary in the High Arctic," we examine the limitations of MPA creation in the high seas under the contemporary Law of the Seas.[14] Our analysis will consider the ongoing process at the UN level to draft an internationally binding legal instrument on the conservation and sustainable use of marine biological diversity in the areas beyond national jurisdiction under the UN Convention on the Law of the Sea.[15] We, however, also consider various legal mechanisms that provide a mandate for MPA creation *even in ABNJ* that could allow for a plausible management regime to protect marine biodiversity.

No set universal legal mechanism is currently recognized to enable MPA creation in the High Arctic with acknowledgement of the UNCLOS. So we evaluate potential legal justification for the creation of an MPA in the High Arctic ABNJ, by way of precedent of mandates such as the UNCLOS and UN Convention on Biodiversity (UNCBD). We conclude that while the UNCLOS and UNCBD do offer a legal mandate for High Arctic MPA creation, a regional multilateral agreement offers the best solution for High Arctic MPA creation in the next decades.

[11]Grindheim (2009), p. 1.

[12]The Pew Charitable Trusts.

[13]PAME International Secretariat (April 2015), pp. 1–76.

[14]Morris and Hossain (2016).

[15]UN GA Res. 69/292 of 19 June 2015.

2 The Arctic Ocean: A Critical Intersection of Competing and Common Interests

According to the preamble of the United Nations Convention on Biological Diversity (CBD), there is an "intrinsic value of biological diversity and of the ecological, genetic, social, economic, scientific, educational, cultural, recreational and aesthetic values of biological diversity and its components."[16] Beyond this inherent value of Arctic biodiversity, human enjoyment and cultural valuation add another layer of importance to the Arctic environment. The economic value of marine biodiversity lies in its ecosystem services, such as ability to provide food, genetic resources, climate regulation through carbon sequestration, and a basis for local jobs.[17] Specific to the Arctic, critical industries such as environmental tourism depend on ecosystem health and preservation of biodiversity. Biodiversity is also critical to emerging medical science: biotechnology innovation depends on the genetic variability of marine species, with diverse physiological and biochemical properties resulting from evolution in extreme cold Arctic waters.[18] The benefits of Arctic biodiversity are felt globally, as they affect species adaptation to climate change, the global economy, future medical science, and overall species diversity on Earth.

Presently, as a result of the increase in the levels of CO_2 in the atmosphere, which accelerates melting of Arctic sea ice, Arctic biodiversity finds itself under the assault of anthropogenic climate change and ocean acidification. The Arctic Marine Shipping Assessment Report of 2009 concludes that sea ice melt will allow for ship navigation and resource extraction in areas previously covered by ice.[19] Tourism, shipping, oil and gas exploitation, fishing, and other industry-related maritime activity might therefore increase in the Arctic.[20] Thus, dual pressures—climate change and increased human activities—will stress the Arctic marine environment with dire implications for its unique biodiversity.

As the Arctic Ocean opens up to oil and gas extraction, countries have an economic interest in claiming sovereignty over the continental shelves underneath the Arctic Ocean. By virtue of Article 76 of the UNCLOS, such claim could extend to unlimited area of the ocean floor beyond 200 nautical miles, where the surface and subsurface waters above are still high seas subject to global commons (*see* Article 76).[21] As of August 2015, four out of five coastal states of the Arctic Ocean (the United States being the exception) had filed their submissions to extend the limits of their continental shelves into the Arctic Ocean.[22] Russia was the first

[16]United Nations, "Convention on Biological Diversity" (1992), p. 1.

[17]Fauria and Kettunen (2015). TEEB Report for the Arctic, pp. 32, 34, 36.

[18]Kattunen (2015), p. 11.

[19]Arctic Council. Arctic Marine Shipping Assessment 2009 Report, p. 4.

[20]Young (2010), pp. 165–166.

[21]United Nations Convention on the Law of the Sea. Article 76.

[22]Submissions To The CLCS (2015).

country to lodge its submission in 2001, claiming almost half of the Arctic Ocean seabed as its extended continental shelf.[23] In response to the request from the Commission on the Limits of Continental Shelf (CLCS) for resubmission with further data, Russia renewed its claims recently in 2015, claiming an even greater portion of seabed (see footnote 23). Norway submitted its claim in 2006 and received a final recommendation from the Commission on the Limits of Continental Shelf in 2009, whereas both Canada (partially) and Denmark submitted their claims in 2013 and 2014, respectively.[24] The United States is not yet a party to UNCLOS and is thus not yet able to lodge any formal submission to the CLCS. However, it is worth noting that most extractable resources fall within undisputed areas, within the Arctic littoral states' EEZs.

2.1 The Ilulissat Declaration: Reaffirming the Law of the Sea and Arctic Council Amid New Challenges

While competing claims may appear to signal competing interests in the region, regional actions have largely been evidenced as cooperative. At the March 2008 Arctic Ocean Conference in Ilulissat, Greenland, the five Arctic coastal states (Canada, Denmark, Russia, the United States of America, and Norway) convened and signed the resulting Ilulissat Declaration, which reaffirmed the commitment to the Law of the Seas and to an "orderly settlement of any possible overlapping claims."[25] The Ilulissat Declaration concluded that the Law of the Sea contained necessary provisions for responsible ecosystem management on the part of the five Arctic coastal states. Thus, the five Arctic coastal states reaffirmed their commitment to the existing legal framework for governance of the Arctic Ocean, under Law of the Sea. The Declaration also calls for heightened cooperation with the International Maritime Organization (IMO) to strengthen regulations to prevent pollution and accidents resulting from heightened ship traffic from shipping, tourism, resource development, and research vessels (see footnote 25). Despite appearing as an assertion of coastal state predominance in the region, the document reaffirms the five Arctic states' commitment to the Arctic Council, established in 1996, and other related forums (see footnote 25).

---

[23] Proelss and Müller (2008), pp. 651, 665–677.

[24] UN Commission on the Limits of the Continental Shelf (CLCS).

[25] *Arctic Ocean Conference*, Ilulissat Declaration (May 28, 2008).

2.2 Changing Tides: The Inclusion of Non-Arctic States as Arctic Council Observers

The largest challenge to the existing legal framework will instead likely come from states with no existing littoral rights in the region. The volume of applications for observer status in the Arctic Council increased notably in recent years.[26] Non-Arctic states increasingly perceive national interest in the region due to economic prospects made available by an opening Arctic. In particular, East Asian states such as Japan and China see potential commercial gains expanding fishing areas and faster shipping routes enabling trade.[27] Though previously lacking political status in the region, Asian influence has grown in the Arctic region by the way of gaining observer status on the Arctic Council. In 2013, China, India, Japan, Singapore, and South Korea all were granted observer status on the Arctic Council, along with Italy.[28] Though observers are void of decision-making power, the observer status allows for non-Arctic states to influence indirectly by way of being involved on participation in working groups, financial contributions, project proposals, and verbal and written statements.[29] China's Arctic strategy, as articulated by Chinese officials, cautiously reframes Arctic governance as being an international issue warranting recognition of non-Arctic states' interests due to common resources and shipping routes.[30] As noted by Jakobson (2010), the former Chinese Assistant Minister of Foreign Affairs Hu Zhengyue said:

> When determining the delimitation of outer continental shelves, the Arctic states need to not only properly handle relationships among themselves, but must also consider the relationship between the outer continental shelf and the international submarine area that is the common human heritage, to ensure a balance of coastal countries' interests and the common interests of the international community.[31]

Increased non-Arctic state involvement in the region may therefore challenge existing legal frameworks governing the high sea water column and complicate regional agreements in the Arctic. As previously mentioned, this contrasts with the Arctic 5 coastal states' bypass of the Arctic Council in favor of releasing the 2008 Ilulissat Declaration to reaffirm their commitment to using UNCLOS to settle any legal challenges in the Arctic, and suggesting the primacy of coastal states in the area. Amid growing interest of both observer states and coastal states, Oran Young

---

[26]The 2013 Kiruna Declaration welcomed China, India, Italy, Japan, Republic of Korea and Singapore as new Observer States. *See* Arctic Council Secretariat. Kiruna Declaration (2013).

[27]Jakobson (2010), p. 13.

[28]Arctic Council Secretariat. Kiruna Declaration (2013).

[29]Note that observer states' financial contributions cannot exceed those of Arctic States, except as directed by Senior Arctic Officials. *See* Arctic Council. Arctic Council Observer Manual for Subsidiary Bodies (2014).

[30]Jakobson (2010), pp. 9–10, 13.

[31]Jakobson (2010), p. 10.

(2010) suggests that the Arctic Council finds itself at a point of "state change" in which its existing governance structure is challenged.[32]

While international actors might acknowledge the High Arctic marine environment (in the international water column) as a common resource, notions of how the common resource should be used can be expected to differ greatly. States that stand to gain economically from fishing rights and navigational routes have the potential to conflict with state and non-state actors on their visions for the conservation of the Arctic Ocean. Further, a growing body of influential observer states have eventually outnumbered the Arctic states themselves. Admittedly, the observers' mandate does not hold a significant role in decision making in the Arctic Council. The Arctic Eight may, however, find their individual interests in the Arctic complicated by a growing number of actors exerting pressure by way of expertise, written opinions, and funding power. At the 2011 Nuuk Ministerial Meeting, one Russian diplomat voiced the concern that a growing majority of observers might demand more rights, perhaps to the extent of designating the Arctic as "universal humankind heritage" on the model of the Antarctic.[33] Thus, while present discourse often depicts non-Arctic observers as potential resource exploiters and *obstacles* to preservation, the growing influence of non-Arctic states in the region (particularly as Arctic Council observers) has the normative power to reframe the Arctic as a global commons of sorts mandating sweeping, collaborative protections. Thus, any observer state pressure to designate the Arctic as "universal humankind heritage" in need of a new legal regime would fundamentally challenge the status quo adherence to Law of the Seas for Arctic Ocean governance.

2.3 Heightened Industry-Caused Pressure on Marine Species

Market prices and technological barriers will likely temper any "rush for the Arctic," particularly in the case of oil.[34] Rather, short-term maritime activity is anticipated to consist primarily of destination shipping.[35] However, though potential grand-scale economic activity may be overstated, a lengthening navigational season (enabled by sea ice melt) in summer and spring creates greater potential for conflict between vessels and marine life.

Current late spring and summer month shipping generally takes place *after* marine mammals migrate through narrow choke points, such as the Bering Strait.[36] However, a lengthening shipping season risks more collisions between mammals and vessels during times of migration and reproduction in early spring months.

[32]Young (2010), p. 168.

[33]Graczyk and Koivurova (2012), p. 5.

[34]Anderson (2009), pp. 198, 207–215.

[35]Lawson (2010).

[36]Arctic Council AMSA Report (2009), pp. 135–136.

White Sea harp seals already undergo considerable pup mortality caused by collisions with marine traffic, as vessels often breach the ice in seal whelping groupings.[37] As shipping becomes possible earlier in the spring, there is increased risk of ship-caused disruptions along sections of water that are key to migration patterns and life cycles, such as feeding and nursery areas. Thus, heightened and highly adaptive conservation efforts will be particularly important in protecting marine mammal life cycles and migration patterns in coming years.

Migration patterns will also change, complicating the issue of protecting areas critical to marine life stages. Fish stocks are particularly sensitive to temperature and are predicted to continue to move toward the northern pole in search of cooler waters.[38] Sea-ice-dependent species in particular, such as polar bears and aquatic mammals, also continue to move pole-ward in the search of remaining sea ice. As sea ice continues to melt and species migration patterns and distributions shift northward, biologically significant populations will increasingly be found in the High Arctic. As the High Arctic also becomes vulnerable due to sea ice melt, species in an area beyond national jurisdiction will lie vulnerable, without the possibility of state-created marine protected areas to guard populations from ship traffic.

Other risks associated with heightened ship traffic and resource development include the introduction of non-native species, pollution, vessel collisions with marine life, noise pollution, and other disruptions to the Arctic marine ecosystem.[39] Not only are disruptions more likely as sea ice melts and northern routes are opened to ships, but these disruptions are increasingly likely to occur in waters that are beyond national jurisdiction and currently cannot be protected by a single state.

3 Need for Arctic Marine Protected Areas

In response to worldwide human-induced environmental damage, the UN CBD identifies designation of protected areas as an important strategy for biodiversity conservation. The International Union for the Conservation of Nature/World Commission on Protected Areas (IUCN/WCPA) defines a protected area as:

> clearly defined geographical space recognized, dedicated, and managed, through legal or other effective means, to achieve the long-term conservation of nature with associated ecosystem services and cultural values.[40]

The CBD's Aichi Target 11 established the goal of designating at least 10% of marine and coastal areas as protected areas by 2020.[41] Only a few years away from

[37]Vorontsova et al. (2008), pp. 586–592.

[38]Michel et al. (2009), pp. 487–518.

[39]Arctic Council AMSA Report (2009), p. 5.

[40]Dudley (2008), p. 8.

[41]Secretariat of the Convention on Biological Diversity (2010).

this deadline, merely 1.55% of Arctic marine and coastal waters were protected, OSPAR marine protected areas.[42] Arctic terrestrial habitats are well protected by comparison, with 17% considered protected by the year 2000.[43]

Though an enormous gap exists between terrestrial and marine protections, the Arctic Council did identify the creation of a pan-Arctic marine protected area (MPA) network as being a primary goal for the region in 2015. However, industrial activities in Arctic waters and coastal areas are increasing, while MPA creation still crawls behind. According to the Beaufort Sea Partnership (2009), MPA creation can take approximately 10 years.[44] Meanwhile, according to Wang and Overland's projections using 2007/2008 sea ice extant data, the Arctic is predicted to be nearly sea ice-free by 2037.[45] Under current legal infrastructure for MPA creation, glacial melt may outpace the creation of any comprehensive High Arctic marine reserve.

Terrestrial conservation is generally aided by clear jurisdiction, whereas the jurisdictional nature of the seas is much more complicated. Marine living resources move from one jurisdiction to another, including in the high seas. This contrasts with land-based protection, in which accepted borders generally allow for clear-cut, nation-led conservation efforts. Of course, the UNCLOS does allow for national claims to *some* waters as it designates areas within 200 nautical miles as being within respective countries' jurisdictions (in their exclusive economic zones or EEZs). Even so, a considerable portion of the Arctic Ocean—1.1 million square miles of ocean—falls beyond any country's exclusive economic zone.[46] Whereas the marine areas within national jurisdiction (falling within EEZ) can be protected by national regulations, as well as by regional agreements among coastal states, the high sea—an area beyond national jurisdiction—instead remains open to all states for free maritime use, following the general limits set by the UN Law of the Sea Convention.

No single state or governing body has sovereignty over the Arctic high seas, an ABNJ. So the discrepancy between terrestrial and marine reserve creation is further perpetuated by a perceived lack of any accepted legal instrument for multilateral protection efforts in the high seas. The UNCLOS protects states' sovereign rights in respective territorial waters, as well as rights to authority over the EEZ. The Convention, however, offers only rudimentary provisions in regard to high sea usage without having any concrete and adequate protection regime for the marine species occurring in the Arctic.

In lieu of a presently accepted legal framework for marine protected area creation in the ABNJ, the Arctic region presently adheres to a nation-led approach under the leadership of the Arctic Council and its working groups. This strategy

42 OSPAR Commission (2013), p. 29; OSPAR Commission (2009), pp. 7–8.

43 CAFF (2002), p. 4.

44 Beaufort Sea Partnership (2009).

45 Wang and Overland (2012), pp. 4–5.

46 The Pew Charitable Trusts: Arctic Ocean International.

notably restricts marine protections to areas already under national jurisdiction, in the exclusive economic zones of Arctic coastal states. As such, the Arctic high sea is left vulnerable. Warming waters will lead more species—and more ships—north while putting pressure on the existing ecosystem. Thus, we consider: what are the constraints and limits of the current initiative to build a pan-Arctic network of MPAs within EEZs? And if such an initiative cannot legally protect the Arctic high seas, what other legal pathways exist?

Specifically considering the challenge of protecting marine life in the High Arctic ABNJ, we will turn to examine potential legal mechanisms for MPA creation in Arctic areas beyond national jurisdiction: creation of an Antarctic-modeled Arctic sanctuary (*see* Sect. 5), an UNCLOS implementing agreement, an additional protocol to the UNCBD, and an Arctic specific regional agreement (*see* Sects. 6.1, 6.2, and 6.3, respectively). Our evaluation of both the limits and merits of the potential legal pathways leads us to suggest that a regional legal regime for MPA creation in the High Arctic offers the most politically feasible and expedient pathway to protection. This sort of legal regime, though limited in its capacity so long as non-Arctic states are not parties, would still hold normative power and set a critical precedent for the international community to recognize the importance of protecting the High Arctic's wealth of biodiversity.

However, we first turn to examine the existing pathway for protection: the state-led network coordinated by the Arctic Council.

4 The Arctic Council's State-Led Pathway for MPA Creation

Presently, under the UNCLOS, actors are constrained in their ability to create marine protected areas, particularly in the ABNJs, because MPAs cannot be established unilaterally in these areas.[47] Rather, human activity in areas that fall outside of national jurisdiction instead can only be governed by international arrangements, to some extent within the framework established by the IMO (see footnote 46).

Current action for MPA creation in the Arctic follows a nation-led approach, in which state actors designate MPAs within their particular EEZs. In 2015, the Arctic Council articulated a commitment to the coordinated creation of marine protected areas with its publication of the PAME Framework for a Pan-Arctic Network of Marine Protected Areas.

So what is the Arctic Council's vision for this pan-Arctic network? The 2015 PAME Framework envisions the following:

> An ecologically representative and well-connected collection of individual marine protected areas and other effective area-based conservation measures in the Arctic that operate cooperatively, at various spatial scales, and with a range of protection levels, in

[47]United Nations Convention on the Law of the Sea.

order to achieve the long-term conservation of the marine environment with associated ecosystem services and cultural values more effectively and comprehensively than individual sites could alone.[48]

The report plainly states that the network is not intended to be legally binding (see footnote 49). So while regional cooperation on the pan-Arctic MPA network will establish reciprocity for compliance among the Arctic states, it will be limited in enforcement capability for Arctic and non-Arctic states alike. PAME describes the Framework as offering guidance, which the Arctic states can each use to designate their own MPAs according to their respective timelines, goals, and authorities (see footnote 49). Individual state's goals and domestic processes will therefore affect the effectiveness of creating a pan-Arctic MPA network. The PAME Framework states that differing prioritization of MPA creation among the Arctic coastal states—not to mention their ruling parties and publics' opinions—will be a challenge.

Further, while the Framework acknowledges that "linkeages" exist between the pan-Arctic MPA network and the high seas, the Framework currently neglects to call for direct action in the ABNJ (see footnote 49). Particularly as warming temperatures and sea melt drive both marine species and human activity toward higher latitudes *in the ABNJ*, a clear legal instrument for MPA creation in the High Arctic ABNJ proves increasingly critical.

5 Arctic Sanctuary: Demand for an Antarctic Model Treaty

Following a media wave in 2008 depicting a "Wild West" type rush for the Arctic, nongovernmental organizations and academics alike began to articulate demand for a new treaty to govern the Arctic.[49] Greenpeace, for instance, calls for designation of an "Arctic Sanctuary" composed of 2.8 million square kilometers (out of a total Arctic Ocean size of over 14 million square kilometers), with strict regulation of shipping and prohibition of commercial fishing and hydrocarbon extraction.[50] Like others, Greenpeace suggested that this be done by way of a new Arctic Treaty, rather than through existing international law such as UNCLOS. Greenpeace further suggested that the Arctic Council's mandate to protect the region's environment might allow it to pursue such a treaty even beyond its member states' jurisdictions (see footnote 51). According to Greenpeace, the Arctic Council's past use of binding treaties in the case of the Agreement on Cooperation and Marine Oil Pollution Preparedness and Response and the Arctic Search and Rescue Agreement might then serve as precedent for another treaty beyond member state EEZs (so long as consistent with existing international law such as UNCLOS) (see

[48]PAME (2015), pp. 1–76.

[49]Holmes (2011).

[50]Hamilton (2014), pp. 4–15.

footnote 51). This notion of an "Arctic Sanctuary" of course invokes the existing Antarctic Treaty of 1959. The Antarctic Treaty was concluded in order to designate the area as a commons for scientific use, prohibiting mineral resource extraction, and for the peaceful use of the continent. The treaty further served to limit and regulate tourism in the area. Moreover, influential international bodies have also explicitly suggested that the Antarctic Treaty System (ATS) might serve as a model.[51] In October 2008, the European Parliament meeting on Arctic Governance held in Brussels articulated interest in such an international treaty:

> the Commission should be prepared to pursue the opening of international negotiations designed to lead to the adoption of an international treaty for the protection of the Arctic, having as its inspiration the Antarctic Treaty, as supplemented by the Madrid Protocol signed in 1991, but respecting the fundamental difference represented by the populated nature of the Arctic and the consequent rights and needs of the peoples and nations of the Arctic region; believes, however, that as a minimum starting-point such a treaty could at least cover the unpopulated and unclaimed area at the center of the Arctic Ocean.[52]

The Southern Pole conjures a similar image to the Arctic due to extreme cold conditions and a resulting fragile but critical ecosystem. Both regions have experienced the threat of expanding tourism and commercial activity, causing academics and the public alike to cite the Antarctic Treaty as a clear precedent for similar action in the Arctic. However, as the European Parliament conceded, the two regions differ in geographic nature as the Antarctic is a continental landmass surrounded by ocean while the Arctic is instead a body of water surrounded by landmass. Moreover, human settlements, including indigenous peoples having special interests, already populate the Arctic. The geopolitical significance of the Arctic is also not to be overlooked. Existing competing jurisdictional claims by coastal states may very well undermine such a treaty on the unclaimed center of the Arctic Ocean by the fact that more areas are under the process of being claimed, awaiting approval by the UN Commission on the Limits of the Continental Shelf. According to Oran Young and Paul Arthur Berkman's piece in Science Magazine (2009), these differences "rule out a similar treaty in the Arctic."[53]

While an international treaty might still be achieved with concessions for the notable differences between the two poles, it is worthy to consider whether such a treaty can be a feasible solution or if such a treaty best serves the interests of preserving the Arctic marine environment. In fact, an ATS-modeled international treaty may not be feasible or desirable. While a binding treaty like the ATS might possess greater normative power and pressure for compliance, the very strength of being legally binding also serves to hamper action.[54] Legally binding treaties notably take considerable lengths of time for negotiation and enactment; a minimum of four years is typical for negotiation alone (see footnote 55). The process of

[51]Young (2010), p. 168.

[52]European Parliament (2008), "Resolution of 9 October 2008 on Arctic Governance."

[53]Berkman and Young (2009), pp. 339–340.

[54]Young (2010), pp. 181–184.

ratification and negotiations might undermine the very intent of such a treaty, watering down its environmental protections until acceptable to all signatories. Moreover, the legal process—presumably requiring ratification of adjustments to the treaty—might hinder any such treaty's responsiveness to changes in scientific knowledge or Arctic environmental conditions (see footnote 55).

Of course, whether or not such an international treaty is desirable or expedient might not be the question of importance. Instead, is such a treaty necessary to protect the Arctic marine environment? Under existing international law, such as the UN Convention on Biodiversity and the UNCLOS, there already exists a mandate for the protection of the Arctic seas. Thus, MPA creation may not require a sweeping ATS-style treaty. Instead, clarification and extension of existing international legal frameworks may enable MPA creation.

6 Looking to the High Arctic: Legal Obligation for MPAs in ABNJs

Establishment of an MPA within the Arctic ABNJ would find its legal basis in UNCLOS Part XII, which establishes an obligation to protect the marine environment under Article 192. States also have a duty to protect rare and fragile ecosystems and the habitat of threatened species under UNCLOS Article 194 for the prevention of pollution of the marine environment and the duty to cooperate under Article 107.[55] According to Molenaar and Elferink (2009), MPAs can be established in ABNJ within the framework of UNCLOS on the basis of Article 194(5).[56] Article 194 (5) articulates the obligation to "protect and preserve rare or fragile ecosystems as well as the habitat of depleted, threatened or endangered species and other forms of marine life." The other articles of UNCLOS also lend support to the creation of MPAs in ABNJ for the purpose of, for example, conservation and management of living resources in Article 61.[57] Finally, Article 145 establishes obligation for "the protection and conservation of the natural resources of the Area and the prevention of damage to the flora and fauna of the marine environment" (see footnote 19). The Area is defined by the Convention as the seabed and ocean floor and subsoil thereof, beyond the limits of national jurisdiction. The marine environment in the water column above the "Area" therefore also would fall beyond boundaries of national jurisdiction.[58] Though there apparently exists a legal basis for the creation of an MPA in an ABNJ, there presently exists no accepted comprehensive legal framework for execution of such an obligation.

---

[55]United Nations Convention on the Law of the Sea. Article 76.

[56]Molenaar and Oude Elferink (2009), pp. 5, 9.

[57]United Nations Convention on the Law of the Sea. Article 61.

[58]United Nations Convention on the Law of the Sea. Article 1.

In lieu of an existing legal *framework* to fulfill the obligation to protect the High Arctic marine environment, we suggest three other potential approaches for MPA creation in an ABNJ: an UNCLOS implementing agreement or a protocol under UNCBD or a regional agreement to be concluded among the Arctic states. The aforementioned strategies are drawn from the recommendations from experts at the international seminar "Towards a legal framework for the creation and management of cross-sectoral marine protected areas in areas beyond national jurisdiction" in Boulogne-sur-Mer, France, in 2011.[59] The Boulogne-sur-Mer international seminar was organized by IUCN and Institute for Sustainable Development and International Relations (IDDRI) in partnership with the Agency for Marine Protected Areas, University of the Littoral – Opal Coast, the European Office for Conservation and Development, and the Nausicaá (National Sea Centre – Boulogne-sur-Mer) in an effort to identify avenues for the creation of MPAs in ABNJs. Twenty international experts collaborated in applying precedents of international law to create viable scenarios for the creation and management of MPAs in high seas by 2030—the decade in which an ice-free Arctic summer is predicted. Findings from the seminar were to inform the United Nations at the Rio+20 Conference and the IUCN World Congress in 2012.[60]

6.1 United Nations Convention on the Law of the Sea (UNCLOS) Implementing Agreement

The UNCLOS, while providing legal mandate for conservation measures in ABNJs such as MPAs, lacks an implementing agreement to provide a legal framework for MPA creation in an area beyond national jurisdiction. Part XII of the Convention provides that states' sovereign rights in the marine area coexist with the duty to protect and preserve the marine environment (Article 193).[61] UNCLOS establishes the general obligations of all states to safeguard the marine environment in its entirety while offering a structure for dealing with all sources of pollution at sea. The Rio+20 UN Conference on Sustainable Development in 2012 expressed particular interest in reforming the institutional framework for ocean governance, with particular attention to an amendment to UNCLOS that would address gaps in the current framework for conservation in ABNJs.[62] The Rio+20 proposal referenced "possible development of a multilateral agreement" under UNCLOS specifically (see footnote 63). The proposal also necessitated the designation of a lead UN agency to manage MPA designation. With growing demand for an UNCLOS implementing agreement from within the United Nations, it appears that internal

[59] Druel et al. (2011), pp. 1–28.

[60] IOC/UNESCO, IMO, FAO, UNDP (2011), pp. 39–40.

[61] United Nations Convention on the Law of the Sea.

[62] IOC/UNESCO, IMO, FAO, UNDP (2011), pp. 39–40.

pressure and public opinion exist. As a result, at the UN level the General Assembly established a Preparatory Committee to produce a draft text for an internationally binding legal instrument on the conservation and sustainable use of marine biological diversity in the ABNJ under UNCLOS.[63] However, negotiating a treaty as such is a far-reaching process since the Committee has just had its first meeting from March 28 to April 8, 2016, with a timeline to report to the General Assembly on the draft by the end of 2017, at which point the Assembly will take effort to convene an international conference to negotiate the agreement. Even if the agreement were eventually materialized, the general scope of the potential agreement would not be capable of dealing with Arctic-specific critical conditions unless an Arctic chapter in it is agreed upon.[64] Consequently, despite the Preparatory Committee's endeavor to address area-based management tools (ABMTs), including MPAs,[65] a universal nature of the scope of the agreement would probably fall short of addressing critical specificities prevailing in the Arctic. Nevertheless, the creation of an implementing agreement through UNCLOS offers clearly a legitimate avenue for the creation and management of MPAs in ABNJs, enforceable for all signatories. Compared to regionally authorized MPAs, an UNCLOS implementing instrument would make all UNCLOS state parties to the agreement, following ratification on behalf of each signatory. By explicitly forging an implementation agreement for governance of the high seas under UNCLOS, all the state parties to UNCLOS would also be mandated to recognize an MPA in the Arctic high seas. While perhaps less expedient than a regional agreement, such expansion and clarification of the UNCLOS would mandate greater participation. Notably, this arrangement would also provide a legitimate avenue for enforcement under the International Tribunal for the Law of the Sea.[66] However, an UNCLOS implementing agreement would require a complicated process of renewed ratification—possibly with less clarity as concerns Arctic-specific conditions. Thus, like Antarctic-modeled treaty making, this legal pathway presents similar complexity in the *effective* realization of such an implementing agreement, even under the UNCLOS.

[63]UN GA Res. 69/292 of 19 June 2015.

[64]Hossain (2016).

[65]See Chair's overview of the first session of the Preparatory Committee, Preparatory Committee established by General Assembly resolution 69/292: Development of an international legally binding instrument under the United Nations Convention on the Law of the Sea on the conservation and sustainable use of marine biological diversity of areas beyond national jurisdiction.

[66]Druel et al. (2011), p. 16.

6.2 United Nations Convention on Biological Diversity (UNCBD) Additional Protocol

Though present international legal discourse gives prominence to the UNCBD, a mandate for the protection of the global marine environment might allow for a legal framework by way of an additional protocol. Article 5 establishes the obligation of contracting parties to cooperate with other contracting parties either directly or through international organizations in the interest of conserving biodiversity in areas beyond national jurisdiction.[67] In the UNCBD's preamble, the Convention instructs its parties to use the precautionary approach in environmental conservation. The Conference of the Parties (COP) Decision II/10 reiterates the direction to use the precautionary approach in regard to marine environments.[68] Given that all the Arctic states, excluding the United States, are party to the UNCBD, they are all subject to the mandate to use precautionary approach in marine governance.

The UNCBD possesses particular validity as a means for a legal framework for MPA creation due to the CBD's precedent of offering scientific insight on the designation of Ecologically or Biologically Significant Areas (EBSAs), criteria also used to create MPAs.[69] The UNCBD already possesses significant expertise in the area, evidenced by the existing Working Group on Protected Areas. The Working Group's mandate explicitly calls for an effort to identify methods for MPA creation in areas beyond national jurisdiction that are consistent with existing international law.[70]

The UNCBD has already created two additional protocols that serve as supplementary agreements to the original convention: the Cartagena Protocol and the Nagoya Protocol.[71] The Nagoya Protocol in particular was agreed upon in order to create a clear legal framework for the satisfaction of the CBD objective of fair and equitable sharing of benefits from the use of genetic resources (see footnote 71). With these two protocols serving as precedent, a similar protocol might be enacted to create a transparent and formalized legal framework for the creation of MPAs even in areas beyond national jurisdiction, as the mandate for such action is provided in the UNCBD. And yet again, this type of framework requires a similarly lengthy process as is required for an implementing agreement under UNCLOS, making its realization both time consuming and complicated.

[67]UNCBD (1992).

[68]Conference of the Parties to the Convention on Biological Diversity, Jakarta, Indonesia, Nov. 6–17, 1995. *Conservation and Sustainable Use of Marine and Coastal Biological Diversity*, Decision II/10, Part XI, UNEP/CBD/COP/DEC/II/10.

[69]The EBSA criteria has been used in the past to aid in conservation targets by the Sargasso Sea Alliance (SSA) and OSPAR. *See* Freestone et al. (2014).

[70]Ad Hoc Open-ended Working Group on Protected Areas (2004).

[71]The Cartagena Protocol on Biosafety (2003).

6.3 A Regional Arrangement

Marine protected areas in ABNJs already have been established in other areas of the globe under the auspices of regional arrangements. Such regional arrangements have been used to establish MPAs in high seas of the Northern Ocean, the Northeast Atlantic, the Mediterranean, and the South Pacific, though no such arrangement has been used comprehensively in the Arctic.[72] The Convention for the Protection of the Marine Environment of the North-East Atlantic ("OSPAR Convention") of 1992 established the OSPAR Commission in order to foresee the conservation and protection of the North-East Atlantic marine area. The OSPAR Convention's preamble cites UNCLOS Article 197 as providing a mandate for global and regional cooperation of the marine environment.[73] It is suggested that cooperation between OSPAR Convention, the regulatory body and legal instrument tasked with preserving environment and resources of the North-East Atlantic, and sectoral regulatory organizations, such as Regional Fisheries Management Organizations (RFMOs), offers a comprehensive model for cooperation to create a similar regional agreement.[74] Though limited in its Arctic claims, OSPAR's work in Arctic ABNJs shows that successful precedent exists for a coalition of regional actors to approach MPA creation in the high seas.[75] Similar models for MPA creation in ABNJ under the auspices of regional arrangements include the nonbinding Hamilton Declaration on Collaboration for the Conservation of the Sargasso Sea and under the Convention for the Protection of the Marine Environment and the Coastal Region of the Mediterranean (the ammended Barcelona Convention).[76]

Though no regional sea management organization exists within the Arctic, the Arctic Council might serve as a venue for the creation of similar regional organization to facilitate MPA creation within an ABNJ. The 1996 Ottawa Declaration established the Arctic Council of the eight Arctic states—Sweden, the Russian Federation, the United States of America, Finland, Norway, Canada, Iceland, and Denmark—to ensure circumpolar cooperation with particular focus on environmental protection and sustainable development.[77] Greenpeace suggests that the Arctic Council's mandate to protect the Arctic environment might give it grounds for greater action, extending even beyond its respective countries' borders.[78] The

[72]Rochette et al. (2014), pp. 109–117.

[73]OSPAR Convention For The Protection Of The Marine Environment Of The North-East Atlantic (1992), 1–33.

[74]Druel et al. (2011), pp. 10–11.

[75]OSPAR in particular has pioneered efforts for MPA creation in ABNJs, in the North-East Atlantic Ocean. *See* Molenaar and Oude Elferink (2009), pp. 5, 9.

[76]The EBSA criteria has been used in the past to aid in conservation targets by the Sargasso Sea Alliance (SSA) and OSPAR. *See* Freestone et al. (2014).

[77]Arctic Council, "Declaration on the Establishment of the Arctic Council" (Ottawa, Canada, 1996).

[78]Hamilton (2014), pp. 4–15.

Arctic Council has taken initiative to negotiate even *binding* regional agreements in areas beyond national jurisdiction through past efforts such as the Agreement on Cooperation and Marine Oil Pollution Preparedness and Response and the Arctic Search and Rescue Agreement.[79] The precedent of binding treaties in waters beyond members' EEZs could serve as grounds for a regional treaty creating MPAs even beyond EEZs, though only binding for the eight Arctic states and other parties that choose to accede to the treaty through future signature and ratification.

Such a regional legal arrangement may prove to be more politically expedient compared to a UNCBD additional protocol or an UNCLOS implementing agreement. Common needs in the Arctic have resulted in notable past regional cooperation and governance arrangements. The Arctic Council possesses a mandate to protect the Arctic environment by way of the Ottawa Declaration, reaffirmed by the Arctic coastal states in the 2008 Ilulissat Declaration. Greenpeace suggests that the Arctic Council's mandate to protect the Arctic environment might give it grounds for greater action, extending even beyond its respective countries' borders. Past Arctic Council initiatives resulted in a number of binding and nonbinding agreements relating to the marine environment, creating precedent for regional approaches to Arctic Ocean governance. A binding agreement, to be concluded within the auspices of the Arctic Council, among the Arctic Eight might be used to coordinate MPA creation in the Arctic ABNJ. Further, PAME's existing work to create a pan-Arctic MPA network among the Arctic EEZs provides a reasonable foundation for an extension of the network into the high seas.

A regional-agreement-based MPA does not legally bind a nonstate party, thus remaining compliant with international law and UNCLOS. Further, coordination with international bodies such as the IMO is considered necessary to ensure that creation of an MPA in the high seas does not violate existing international maritime law.[80] Compliance with UNCLOS does limit the ability of such a regional agreement to enforce high sea MPA observance among nonmembers. However, such regional agreement can establish powerful normative guidance and also achieve nonparty compliance through mutual observer status, which would establish reciprocity for the mutual acknowledgement of participating parties' protected areas. For instance, the OSPAR Convention cannot enforce MPAs for international actors not party to the Convention (abiding by UNCLOS), but the Convention has mechanisms to engage nonparty states operating in OSPAR waters. The 17 contracting parties can invite nonparty states to accede to the Convention, often through arrangements of mutual observer status or memorandums of understanding (MoUs), thus making the MPA in ABNJ more enforceable while acknowledging freedom of the high seas. Thus, an Arctic-style regional agreement modeled after OSPAR could provide a robust marine environmental governance regime.

[79] Arctic Council: Agreements. (2015, September 16).

[80] OSPAR in particular has noted the need for cooperation with international bodies such as the IMO. *See* OSPAR Commission (2012), p. 19.

Thus, lacking a new legal framework under the UNCLOS and UNCBD, a regional agreement is arguably the most suitable tool in its enforcement capacity and ability to set a norm for nonparties to comply.

7 Conclusion: A Regional Arrangement in Wait of a Formal Legal Framework

A regional agreement will only enforce recognition of a High Arctic MPA among ratifying signatory member states, with limited capacity for enforcement. By comparison, a legal framework through an implementing agreement under UNCLOS would allow for wider reach, extending to all ratifying state parties. Further, such an arrangement would allow for an enforcement mechanism by way of the International Tribunal for Law of the Sea. The potential for an additional protocol to the UNCBD also should not be ignored, given past precedent, international scope, and existing expertise in the creation of protected areas. Both instruments would lend international scope and recognition for a High Arctic MPA while addressing a critical gap in existing global marine governance. Though increasing public pressure exists for such a framework, the near horizon for an ice-free Arctic obliges immediate and effective action. The complicated lengthy process in reaching a large-scale consensus, either for UNCLOS implementing agreement or for UNCBD Protocol, is challenging. In lieu of such an international agreement, a regional agreement offers to begin MPA designation in the Arctic high seas under existing legal mandates, and among only the eight Arctic states. Drawing upon the expertise and collaboration of the pan-Arctic MPA network within members' EEZs, the Arctic states under the auspices of the Arctic Council possess proficiency to begin MPA designation in the Arctic ABNJ. Increased use of such regional agreements to protect the global commons might provide the impetus for broader protection under the UNCLOS or UNCBD.

References

ACIA (2005) Arctic climate impact assessment. ACIA Overview report. Cambridge University Press

Ad Hoc Open-ended Working Group on Protected Areas (2004) Convention on biological diversity. https://www.cbd.int/convention/wgpa.shtml. Accessed 5 Apr 2016

Anderson A (2009) After the Ice: Life, Death, and Geopolitics in the New Arctic: "Too Many Ships, Too Soon?". Smithsonian, New York

Arctic Council (2009) Arctic Marine Shipping Assessment 2009 Report. http://www.arctic.noaa.gov/detect/documents/AMSA_2009_Report_2nd_print.pdf

Arctic Council (2014) Arctic Council observer manual for subsidiary bodies. https://oaarchive.arctic-council.org/handle/11374/939

Arctic Council (2015, September 16) Agreements. http://www.arctic-council.org/index.php/en/our-work/agreements. Accessed 4 Apr 2016

Arctic Council (1996) Declaration on the establishment of the Arctic Council. Ottawa, Canada. https://oaarchive.arctic-council.org/bitstream/handle/11374/85/00_ottawa_decl_1996_signed%284%29.pdf?sequence=1&isAllowed=y

Arctic Council Secretariat (2013) Kiruna Declaration: on the occasion of the Eighth Ministerial Meeting of the Arctic Council. Kiruna, SW. https://oaarchive.arctic-council.org/handle/11374/93?show=full

Arctic Ocean Conference, May 27–29, 2008, Ilulissat Declaration (May 28, 2008). http://www.oceanlaw.org/downloads/arctic/Ilulissat_Declaration.pdf

Beaufort Sea Partnership (2009) Beaufort Sea Partnership Integrated Ocean Management Plan (IOMP) for the Beaufort Sea: 2009 and Beyond. Beaufort Sea Planning Office, Inuvik, NT

Berkman PA, Young OR (2009) Governance and environmental change in the arctic ocean. Science 324:339–340. doi:10.1126/science.1173200

Borgerson SG (2009, March 25) The Great Game Moves North. Retrieved via Foreign Affairs website. https://www.foreignaffairs.com/articles/global-commons/2009-03-25/great-game-moves-north. Accessed 1 Apr 2016

CAFF (Conservation of Arctic Flora and Fauna) (2002) Protected areas of the arctic: conserving a full range of values. Ottawa, Canada

Chair's overview of the first session of the Preparatory Committee, Preparatory Committee established by General Assembly resolution 69/292: Development of an international legally binding instrument under the United Nations Convention on the Law of the Sea on the conservation and sustainable use of marine biological diversity of areas beyond national jurisdiction, http://www.un.org/depts/los/biodiversity/prepcom_files/PrepCom_1_Chair's_Overview.pdf

Conference of the Parties to the Convention on Biological Diversity, Jakarta, Indonesia, Nov. 6–17, 1995. Conservation and Sustainable Use of Marine and Coastal Biological Diversity, Decision II/10, Part XI, UNEP/CBD/COP/DEC/II/10. http://www.cbd.int/decision/cop/?id=7083

Druel E, Billée R, Treyer S (2011) A legal scenario analysis for marine protected areas in areas beyond national jurisdiction. Report from the Boulogne-Sur-Mer Seminar, 19–21 September 2011

Dudley N (2008) Guidelines for applying protected area management categories. IUCN, Gland, Switzerland. https://portals.iucn.org/library/efiles/documents/PAPS-016.pdf

Emmerson C, Lahn G (2012) Arctic opening: opportunity and risk in the high North (Rep.). Available via Lloyd's of London website. https://www.lloyds.com/~/media/files/news and insight/360 risk insight/arctic_risk_report_webview.pdf. Accessed 12 Mar 2016

Eurasia Group for The Wilson Center (2013) Challenges for arctic oil and gas development. https://www.wilsoncenter.org/sites/default/files/Artic Report_F2.pdf. Accessed 15 Feb 2016

European Parliament (2008) Resolution of 9 October 2008 on Arctic Governance. http://www.europarl.europa.eu/sides/getDoc.do?type=TA&language=EN&reference=P6-TA-2008-474. Accessed 12 Feb 2016

Fauria M, Kettunen M (2015) The Economics of Ecosystems and Biodiversity (TEEB) for the Arctic: a scoping study. CAFF International Secretariat, Akureyri, Iceland

Freestone D et al (2014) Can existing institutions protect biodiversity in areas beyond national jurisdiction? Experiences from two on-going processes. Mar Policy 49. doi:10.1016/j.marpol.2013.12.007

GISTEMP Team (2016) GISS Surface Temperature Analysis (GISTEMP). NASA Goddard Institute for Space Studies. http://data.giss.nasa.gov/gistemp/. Accessed 9 Mar 2016

Graczyk P, Koivurova T (2012) A new era in the arctic council's external relations? Broader consequences of the Nuuk observer rules for arctic governance. Polar Record. doi:10.1017/S0032247412000824

Grindheim A (2009) The scramble for the Arctic? A discourse analysis of Norway and the EU's strategies towards the European Arctic. Fridtjof Nansen Institute, Lysaker, Norway. ISBN: 978-82-7613-566-4

Hamilton N (2014) Arctic Sanctuary. Greenpeace, June (2014). http://www.greenpeace.org/international/Global/international/publications/oceans/2014/Arctic%20Sanctuary.pdf

Holmes JR (2011) The Arctic Sea-a New Wild West? The Diplomat. http://thediplomat.com/2011/04/the-arctic-sea-a-new-wild-west/. Accessed 04 Feb 2016

Hossain K (2014) An arctic challenge: areas beyond national jurisdiction. The Circle

Hossain K (2016) Arctic marine biodiversity in areas beyond national jurisdiction: framing a legally binding MPA regime? ASIL Insights 20(14). https://www.asil.org/insights/volume/20/issue/14/arctic-marine-biodiversity-areas-beyond-national-jurisdiction-framing

IOC/UNESCO, IMO, FAO, UNDP (2011) A blueprint for ocean and coastal sustainability. Paris, France. http://www.uncsd2012.org/content/documents/interagency_blue_paper_ocean_rioPlus20.pdf

Jakobson L (2010) China prepares for an ice-free arctic. SIPRI Insights on Peace and Security, no. 2010/2. http://books.sipri.org/files/insight/SIPRIInsight1002.pdf

Kattunen M (2015) Business and ecosystem services: taking responsibility, identifying dependencies, seizing opportunities. The Circle 02.15

Lawson BW (2010) Think Again: The Arctic. Foreign Policy. http://foreignpolicy.com/2010/08/06/think-again-the-arctic/. Accessed 04 Jan 2016

Meeting of the OSPAR Commission (2009) OSPAR's Regulatory Regime for establishing Marine Protected Areas (MPAs) in Areas Beyond National Jurisdiction (ABNJ) of the OSPAR Maritime Area

Meltofte H (ed) (2013) Arctic Biodiversity Assessment (ABA). Status and trends in Arctic biodiversity. Ch. 14. Conservation of Arctic Flora and Fauna, Akureyri, Iceland

Michel C et al (2009) Arctic Biodiversity Assessment: Marine Ecosystems. Chapter 14, pp 487–518. http://www.caff.is/assessment-series/10-arctic-biodiversity assessment/219-arctic-biodiversity-assessment-2013-chapter-14-marine-ecosystems

Molenaar EJ, Oude Elferink A (2009) Marine protected areas in areas beyond national jurisdiction-the pioneering efforts under the OSPAR convention. Utrecht Law Rev

Morris K, Hossain K (2016) Legal instruments for marine sanctuary in the high arctic. Laws 5 (2):20. doi:10.3390/laws5020020

OSPAR Commission (2013) 2012 Status Report on the OSPAR Network of Marine Protected Areas - Biodiversity Series. http://www.ospar.org/ospar-data/p00618_20

OSPAR, Convention for the Protection of the Marine Environment of the North-East Atlantic, 1992. http://www.ospar.org/convention/text

PAME International Secretariat (April 2015) Framework for a Pan-Arctic Network of Marine Protected Areas: 1-76. Akureyri, Iceland. http://www.pame.is/images/03_Projects/MPA/MPA_Report.pdf. Accessed 10 Dec 2015

Proelss A, Müller T (2008) The legal regime of the Arctic Ocean. Heidelberg J Int Law 68, 651, 665–677

Rochette J et al (2014) The regional approach to the conservation and sustainable use of marine biodiversity in areas beyond national jurisdiction. Mar Policy 49:109–117. doi:10.1016/j.marpol.2014.02.005

Secretariat of the Convention on Biological Diversity (2010) Strategic Plan for Biodiversity 2011–2020 and the Aichi Targets. Montreal, Canada. http://www.cbd.int/doc/strategic-plan/2011-2020/Aichi-Targets-EN.pdf. Accessed 16 Feb 2016

The Cartagena Protocol on Biosafety. Convention on Biological Diversity: http://bch.cbd.int/protocol/ (2003). Accessed 05 Apr 2016

The Pew Charitable Trusts. Why an International Fisheries Agreement Is Needed in the Central Arctic Ocean FAQ. http://www.pewtrusts.org/en/projects/arctic-ocean-international/solutions/faqs. Accessed 15 Feb 2016

UNCBD (1992) Convention on Biological Diversity. https://www.cbd.int/doc/legal/cbd-en.pdf. Accessed 05 Apr 2016

UN Commission on the Limits of the Continental Shelf (CLCS). http://www.un.org/Depts/los/clcs_new/commission_submissions.htm

UNEP/GPA (2006) Ecosystem-based management: markers for assessing progress. UNEP/GPA, The Hague. ISBN: 92-807-2707-9

UN GA Res. 69/292 of 19 June 2015. http://www.un.org/en/ga/search/view_doc.asp?symbol=A/RES/69/292

United Nations Convention on Biological Diversity (1992). https://www.cbd.int/doc/legal/cbd-en.pdf

United Nations Convention on the Law of the Sea. http://www.un.org/depts/los/convention_agreements/texts/unclos/unclos_e.pdf

UN News Center (3 August 2015) Submissions To The CLCS. http://www.un.org/Depts/los/clcs_new/commission_submissions.htm. Accessed 25 Oct 2015

Vorontsova MN, Chernook VI, Glazov DM, Filipova AV (2008) Current threats to the survival of the harp seal (Phoca groenlandica) White Sea population. In: Proceedings of the Marine Mammals of the Holarctic. Odessa, October 2008. http://2mn.org/downloads/bookshelf/mmh5_book/1-18.pdf

Wang M, Overland JE (2012) A sea ice free summer arctic within 30 years: an update from CMIP5 models. Geophys Res Lett 39(18). doi:10.1029/2012gl052868

Young OR (2010) Arctic governance - pathways to the future. Arct Rev Law Polit 1(2). https://arcticreview.no/index.php/arctic/article/view/15. doi:10.17585/arctic.v1.15

The Environmental Legal Framework for the Development of Blue Energy in Europe

Enrique J. Martínez Pérez

1 Introduction

As agreed by the Heads of State and Government of the Member States of the European Council in March 2007, the European Union set itself the target of using energy from renewable sources to cover 20% of the European Union's total energy consumption and 10% of energy consumption in the transport industry by 2020.[1] According to the latest European Commission reports, these targets are well on the way to being reached, for in 2014 renewable energy covered an estimated share of 15.3% of gross final consumption, close to 8.3% more than in 2004.[2] Hydropower is still the production leader, but it is losing ground to wind power (27.5%), biomass and biogas (16.2%), and solar power (10%).[3] The latter accounts for only 0.5% of

This article was undertaken within the framework of the research project "La Unión Europea y el Derecho del Mar" (DER2013-45995-R) funded by the Spanish Ministry of Economy and Competitiveness and the COST Action IS 1105, supported by COST (European Cooperation in Science and Technology).

[1]Directive 2009/28/EC of the European Parliament and of the Council of 23 April 2009 on the promotion of the use of energy from renewable sources and amending and subsequently repealing Directives 2001/77/EC and 2003/30/EC, OJ L 140, 5 June 2009.

[2]Report from the Commission to the European Parliament, the Council, the European Economic and Social Committee and the Committee of the Regions. Renewable energy progress report. COM (2015) 293 final, 15 June 2015.

[3]There are at present 128.8 GW of installed wind power capacity, of which 120.6 GW are at onshore wind farms, and 8 GW, at offshore wind farms (European Wind Energy Association (2015). Wind in power: 2014 European statistics. http://www.ewea.org/fileadmin/files/library/publications/statistics/EWEA-Annual-Statistics-2014.pdf. Accessed 19 Nov 2015).

E.J. Martínez Pérez (✉)
Faculty of Law, University of Valladolid, Valladolid, Spain
e-mail: enriquejesus.martinez@uva.es

G. Andreone (ed.), *The Future of the Law of the Sea*,
DOI 10.1007/978-3-319-51274-7_7

the EU's total electricity consumption,[4] although by 2020 installed capacity is anticipated to reach 43 GW, which would be 3% of total consumption.[5] Nevertheless, the seas offer us other sources of clean energy, sources that are still in an embryonic stage yet can, with public support and technological improvements, achieve the same kind of development as wind power.

Waves, tides, and temperature and salinity differences can be tapped for energy. These new ocean energy sources enjoy the same advantages as wind energy: they help reduce greenhouse gas emissions, they boost energy security, they favor industrial and technological developments, and they are a major source of jobs in high-unemployment areas. But they also face important challenges, such as the high costs of technology, the development of grid connections for renewable marine energy, and the issue we will address here, uncertainty over the environmental impact of the new installations and their compatibility with other maritime activities. So when projects of this kind are introduced, rigorous assessments of their environmental effects must be run to identify the impacts of projects on protected areas, on plants and animals, and on other uses, such as navigation. These assessments must take account of EU law in the framework of biodiversity policy and integrated maritime policy, pay special attention to the rules of maritime spatial planning and marine strategy, and not overlook the international legal obligations established by international environmental law and marine law.

2 The Impact of the Law of the Sea: Maritime Safety Issues

"Ocean energy" refers to energy that comes from exploiting waves, tides, and temperature and salinity differences.[6] Most installations and projects being tested are located in maritime zones under the sovereignty of coastal States (encompassing internal waters, archipelagic waters, and territorial seas), although technological strides such as those made with wind energy have enabled ocean

[4]There are just 84 offshore wind farms scattered over 11 European countries (European Wind Energy Association (2016). The European offshore wind industry -key trends and statistics 2015. http://www.ewea.org/fileadmin/files/library/publications/statistics/EWEA-European-Offshore-Statistics-2015.pdf. Accessed 15 Feb 2016).

[5]Communication from the Commission to the European Parliament, the Council, the European Economic and Social Committee and the Committee of the Regions. *Blue Energy. Action needed to deliver on the potential of ocean energy in European seas and oceans by 2020 and beyond*, COM (2014) 8 final, 20 January 2014, p. 4.

[6]The Commission has identified four forms of ocean energy: "Wave energy depends on wave height, speed, length, and the density of the water. Tidal stream energy is generated from the flow of water in narrow channels whereas tidal range technologies (or 'tidal barrages') exploit the difference in surface height in a dammed estuary or bay. Ocean energy can also be generated from temperature differences between surface and sub-surface water while salinity gradient power relies on the difference in salinity between salt and fresh water" (European Commission, *supra* note 5, at 2).

energy installations to be developed in areas much further away from the coast, where States does not enjoy sovereignty as such but a more limited set of "sovereign rights."[7]

The United Nations Convention on the Law of the Sea (UNCLOS) does not set many conditions on the development of blue energy in areas under the sovereignty or jurisdiction of States. On the one hand, a State extends its sovereignty across a belt of sea adjacent to its territory out to a maximum distance of 12 miles (territorial sea), and therefore, although no such express mention is made, a State may establish marine installations there by virtue of its sovereignty.[8] On the other, States have the same right in the exclusive economic zone, an area adjacent to the territorial sea and measuring a maximum of 200 nautical miles from the baselines from which the breadth of the territorial sea is measured. Under article 56, a coastal State is expressly acknowledged as having sovereign rights in its exclusive economic zone to perform "other activities for the economic exploitation and exploration of the zone, such as the production of energy from the water, currents and winds," plus, according to article 60, "the exclusive right to construct and to authorize and regulate the construction, operation and use of . . . installations and structures for the purposes provided for in article 56."

In the exclusive economic zone, unlike in other maritime zones, exercising such rights requires an express proclamation. Moreover, territorial or material limitations can be placed on the rights (called *minoris generis* or *sui generis* zones).[9] Spain, for example, initially limited its exclusive economic zone to the waters of the Atlantic Ocean and the Bay of Biscay.[10] In the Mediterranean Sea, however, due to the special characteristics of that area, Spain, like other countries such as Algeria, Libya, and Malta, established a Fishing Protection Zone in which the country only claimed sovereign rights for the preservation of living marine resources and the management and control of fishing activities.[11] Then there is the case of France and Italy, which established Ecological Protection Zones with powers for the preservation of the marine environment.[12] Less usual in practice is to find state declarations limiting a state's power to energy activities. One of the few examples was the Renewable Energy Zone declared by the United Kingdom in section 84 of the Energy Act 2004, in which the State vests itself with exclusive rights for the production of water and wind energy under Part V of the Montego Bay

[7]See Cottier (2015), p. 133.

[8]Article 2.

[9]See Andreone (2015), p. 163.

[10]Act 15/1978, of 20 february 1978, on Economic Exclusive Zone (BOE núm. 46, 23 February 1978), first final provision.

[11]Royal Decree 1315/1997, of 1 August 1997, establishing a Fisheries Protection Zone in the Mediterranean Sea (BOE núm. 204, 26 August 1997), article 2.

[12]See Papanicolopulu (2007), pp. 381–398.

Convention.[13] Recently, however, many of these States have transformed their *minoris generis* zones to economic exclusive zones.[14]

States have freedom to construct installations on the high seas as well, albeit subject to the provisions of Part VI of the Convention, which establishes the legal regime governing the continental shelf. On the continental shelf, unlike in the exclusive economic zone, the rights of the coastal State exist *ipso facto* and *ab initio*, as may be gathered from article 77 of the Convention. As a consequence, States likewise exercises sovereign rights for the exploitation of the natural resources on the seabed and marine subsoil, without the need of occupation or express proclamation. But here, unlike in the exclusive economic zone, the natural resources only include mineral resources and other nonliving resources. Wind and water are not mentioned. It is true, however, that article 80 contains a clause referring back to article 60, allowing the construction of installations and structures on the continental shelf, but the referral in article 80 includes the expression *mutatis mutandis*, namely "with the necessary changes," which in our opinion means that the right is limited to the construction of installations for the stated purposes on the continental shelf, not in the exclusive economic zone. Although a broader interpretation could be argued, the point would be moot because, as we have just said, any State can declare an exclusive economic zone as such or a limited exclusive economic zone.

At any rate, the rights that States exercise in these marine areas are not absolute but are subject to certain limitations, especially where navigation is concerned. And the fact is that the energy installation can endanger maritime navigation safety if they are located or lie near regular routes or maritime traffic separation schemes. As regards the territorial sea, States can establish safety zones prohibiting or restricting navigation around power plants or structures.[15] Nonetheless, the right of innocent passage of all ships must be guaranteed as well.[16] Alternative sea lanes must therefore always be ensured[17] because otherwise the exercise of the right of innocent passage would be denied or hindered, and article 24 of the Montego Bay Convention would be violated. At all events, coastal States may adopt navigation laws and regulations and regulate maritime traffic under very few limitations; they need only give due publicity and take into account the recommendations of the

[13]See Scott (2006), pp. 89–118.

[14]See Andreone and Cataldi (2014), pp. 226–230.

[15]Article 21. Spain, for example, issued a prohibition in advance against marine energy farms in areas where there are maritime traffic separation schemes and zones adjacent thereto, via Royal Decree 1028/2007 of 20 July concerning the procedure for processing applications for authorization for electricity generation facilities in the territorial sea (BOE núm. 183, 1 August 2007, second additional provision).

[16]Article 17.

[17]Article 22.

competent international organization.[18] In this respect, Regulation V/8 (Routeing) of the International Convention for the Safety of Life at Sea (London, 1 November 1974)[19] acknowledges that the International Maritime Organization (IMO) is the only one recognized as creating guidelines, criteria, and rules applicable to maritime traffic routeing, although the governments concerned hold the responsibility of taking the initiative,[20] and rule 1 paragraph d) of the International Regulations for Preventing Collisions at Sea (London, 20 October 1972) likewise establishes that the IMO is the organization in charge of adopting traffic separation schemes (TSS). The main rule on this subject is IMO Resolution A.572 (14) on "General Provisions on Ships' Routeing," which recommends following IMO guidelines on TSS establishment and even submitting schemes to the IMO for approval.[21] Otherwise, traffic separation schemes must at least be made known in nautical publications and charts.[22]

In the exclusive economic zone, States can also establish safety zones around installations to safeguard navigation when they see fit. All ships must respect these safety zones.[23] Their breadth depends on the nature and functions of the installations but shall not exceed a distance of 500 meters around them, except as authorized by generally accepted international standards or the recommendation of the competent international organization.[24] Safety zones are set in Resolution A.671(16) on "Safety Zones and Safety of Navigation Around Offshore Installations and Structures," which includes an annex giving a series of guidelines for the correct reporting of all information about safety zones.[25] However, in no case can installations or structures be established, nor can safety zones be established around installations or structures, when they can interfere with the use of recognized sea lanes that are essential to international navigation.[26] IMO's Resolution A.572 (14) also recommends not emplacing structures inside or near traffic separation schemes. Should no other emplacement be possible, permanent modifications of the schemes must nonetheless be submitted to the IMO for approval.[27]

[18]Article 22.3. In that regard, Spain recently updated its legislation on navigation, establishing that the use of the maritime traffic systems "shall be mandatory for all ships once they have obtained the international approval and publication that may be necessary as appropriate. In any event, use of the maritime traffic systems may only be mandatory when located in internal waters or in the territorial sea and, in the event of approval by the International Maritime Organisation, within the exclusive economic zone" (Act 14/2014, dated 24 July, on maritime navigation, BOE núm. 180, 25 July 2014, article 30).

[19]Resolution MSC. 46 (65), adopted on 16 May 1995, annex 2.

[20]See Birnie (1997), p. 34.

[21]Para. 3.12.

[22]Para. 3.13.

[23]Article 60.6.

[24]Article 60. 4–5.

[25]Adopted on 19 October 1989.

[26]Article 60.7.

[27]Para. 3.11.

3 The Integration of Ocean Energy in Maritime Spatial Plans

Ocean energy has to compete with other maritime interests and activities, including classic pursuits (such as fishing, navigation, maritime shipping, and oil and gas extraction) and more innovative activities (such as aquaculture). Many States have drawn up maritime space management plans where the different uses of the sea are regulated. However, most of them fail to include renewable energy activities.[28]

Recently, however, has been adopted Directive 2014/89/EU of 23 July 2014[29] establishing a framework for maritime spatial planning, which will assist Member States to identify compatible uses within a given maritime space, thus precluding future conflicts, although they enjoy a broad margin of discretion to implement the obligations deriving from this directive. Most obligations are only procedural, not substantive.[30] In addition, the minimum requirements that all Member States must meet are in fact very few and highly abstract. In comparison with the initial proposal, which made it obligatory to carry out a clear *demarcation* of the marine space reflecting the actual and potential spatial and temporal distribution of activities,[31] Member States are now only required to determine uses and activities on their maritime spatial plans (art. 8). That said, when Member States draw up their maritime spatial plans, they must always take account of land–sea interactions and environmental, economic, and social aspects and guarantee coherence between maritime planning and integrated coastal management strategies (art. 6).

In any case, as stated in article 1, the directive will contribute to "promoting the sustainable growth of maritime economies, the sustainable development of marine areas and the sustainable use of marine resources." Moreover, it has a very wide scope of application since it applies to the marine waters of Member States, except for coastal waters. Therefore, as recognized in Directive 2008/56/EC, it includes waters, the seabed, and the subsoil where Member States exercise jurisdictional rights; as we have just seen, this is the territorial sea, the contiguous zone, the continental shelf, and the exclusive economic zone.[32] But above all, and most importantly for our study, the uses and activities that must be taken into account include installations and infrastructure for the production of energy from renewable sources and undersea cable and pipeline routes.

[28]See Long (2013), p. 37.

[29]Council Directive (EU) 2014/89 establishing a framework for maritime spatial planning [2014] OJ L 257/135, 28 August 2014.

[30]See Zervaki (2015), p. 106.

[31]COM (2013) 133, article 7.1.

[32]Council Directive (EC) 2008/56 establishing a framework of Community action in the field of marine environmental policy (Marine Strategy Framework Directive) [2008] OJ L164/19, 25 June 2008, article 3. 1. a.

4 Environmental Impact Assessment

One of the most complicated stages in the development of renewable energies occurs at the assessment of the potential impact on the marine environment. The obligation to conduct an environmental impact assessment derives from different international law sources (treaty and custom), which determines the specific content and the spatial scope. Simultaneously, EU law has established further requirements to carry out an environmental assessment, as discussed below.

4.1 At the EU Level

Strategic planning is the first preventive instrument for reducing negative environmental impacts since it enables the States to decide on the capacity and location of renewable ocean energy projects. In Europe, strategic environmental assessment, which is regulated in Directive 2001/42/EC, is compulsory for all plans and programs concerning agriculture, forestry, fisheries, industry, transport, waste management, water management, telecommunications, tourism, town and country planning or land use, and, more to the point for us, energy.[33] The deliverable is an environmental report that assesses aspects such as biodiversity, population, human health, fauna, flora, land, water, air, cultural heritage, and landscape.[34]

The effects of installations on the population and human health are minimum since any site would lie far from populated areas and the energy would be clean. Nevertheless, people do not like how some energy projects change the landscape. Accordingly, although visual impact cannot be considered a strong enough argument to rule out offshore energy projects entirely along the coast, a wide strip along the coastline could be established as being area suitable with environmental restrictions.[35] Any project inside this area is required to undergo a further assessment of its environmental feasibility if there is any possibility that it might have certain negative effects. This assessment mandatorily entails a specific visual impact analysis for each project within the marine strip.[36] Unlike other energy projects,

[33]Directive No 2001/42/EC of the EP and the Council on the assessment of the effects of certain plans and programmes on the Environment, [2001] OJ L 197/30, 21 July 2001, article 3 (2) (a). See Vazquez Gomez (2012), pp. 146–159.

[34]Annex I.

[35]In Spain, for example, identified some of the negative impacts of ocean energy installations in the *Strategic Environmental Study of the Spanish Coast for the Installation of Marine Wind Farms*, an assessment focusing on finding areas in the maritime public domain that qualify as marine installation sites. Zoning was done bearing in mind the potential perception that marine wind farms visible from the coast alter the landscape. Available at http://www.mityc.es/energia/electricidad/RegimenEspecial/eolicas_marinas/Documents/EEAL_parques_eolicos_marinos_Final.pdf. Accessed 8 Mar 2016.

[36]In that regard, see OSPAR *Guidance on Environmental Considerations for Offshore Wind Farm Development*, ref. 2008-3, available at http://www.ospar.org. Accessed 7 Jul 2016.

such as offshore wind farm, ocean energy will in all probability be found more acceptable because many ocean energy devices (like underwater tidal power turbines) are entirely or partly submerged. Only some floating structures and installations requiring the construction of landscape-changing barriers may prove less welcome.

Another item to bear in mind at the start of any planning effort is where the grid access points are because if new infrastructure proves necessary, its environmental impact will have to be analyzed as well. Lack of cross-border grid interconnections is one of the reasons why there is so little harnessing of ocean energy. For that reason, the European Union has encouraged the development of cross-border grid connections to ensure a stable total supply of renewable energy to the grid and to enable this supply of energy to be marketed, thus improving its efficiency.[37] Infrastructure may cut across different maritime zones belonging to different States, so planning, authorization, and regulation issues remain in the hands of each Member State. Under UNCLOS, all States are entitled to lay submarine cables and pipelines in the exclusive economic zone (article 58) and on the continental shelf (articles 87 and 79.1). Nevertheless, the exercise of this right may be subject to some restrictions.

The coastal State can take measures for the prevention, reduction, and control of pollution from pipelines but not from submarine cables, nor can the coastal State's consent be required for the laying of submarine cables.[38] This difference in standards is due to the low environmental impact of damage at submarine cable installations.[39] Nevertheless, as the Convention does allow coastal States to take *reasonable measures* for the exploration of the continental shelf and the exploitation of their natural resources, some States argued that they can impose certain conditions on cable laying.[40] Furthermore, all States must comply with the laws and regulations adopted by the coastal State in the exclusive economic zone regarding the exploitation of natural resources and the protection and preservation of the marine environment, and these laws and regulations must in their turn respect the rights and duties of other States and be in accordance with the provisions of the Convention and other rules of international law.[41] In the light of these provisions, many States (some of them EU Member States) have of late adopted legislation under which the legal procedure for cables and pipelines is the same, so that prior permits have to be obtained for cables as well, and there may even be fees or taxes to be paid.[42]

---

[37] Communication from the Commission to the European Parliament, the Council, the European Economic and Social Committee and the Committee of the Regions—*Offshore Wind Energy: Action needed to deliver on the Energy Policy Objectives for 2020 and beyond*, COM (2008) 768 final, 13 November 2008, para. 2.1.

[38] Article 79 (2).

[39] See Roeben (2013), p. 847.

[40] See Ford-Ramsden, Davenport (2014), p. 148.

[41] Article 58 (3).

[42] See Ford-Ramsden, Davenport (2014), pp. 148–151.

In accordance with article 79 paragraph 4 of the Montego Bay Convention, in contrast, before cables may enter the territory or territorial sea of a coastal State, authorization must be obtained from the coastal State, which can set conditions regarding the route of the cable. The coastal State also has jurisdiction over cables used with respect to structures built to tap ocean energy. Because the sovereignty of a State also extends to its territorial sea, the coastal State can demand compliance with national legislation before it grants permits and licenses. Lastly, in accordance with paragraph 5, when cables are laid, account must be taken of the cables already installed, and the possibility of their repair must not be hampered.

The second of the essential instruments of EU legislation is the environmental impact assessment regulated in Directive 2011/92/EU, applicable to public and private projects.[43] Renewable energy installations were not listed as such in the category of projects within the scope of Directive 85/337/EEC,[44] but much of the necessary construction work (that, by its nature, dimension or location has a major impact on the environment) was required to be assessed anyway because it qualified as a project in Annex II of the Directive.[45] This situation changed with the entry in force of the new Directive 97/11/EC, which expressly includes installations harnessing wind power for energy production in Annex II.[46] In these cases, Member States still enjoy a broad margin of discretion to decide if they have to carry out an environmental impact assessment, but they are obligated to determine whether the project is likely to have significant effects on the environment. This requirement extends to any change or expansion of an installation that is already authorized, executed, or in the process of being executed.[47]

In any case, when Member States conduct *screening* (the assessment process to determine whether or not there are any significant effects on the environment, so as to decide whether or not a particular project requires an environmental impact assessment), they must always take account of the criteria established in Annex III, *inter alia*, the environmental sensitivity of the geographic areas that the installations may affect. Harm to the landscape would be a point in favor of the existence of environmental effects, but only, as Annex III expressly states, if it affects landscapes of historical, cultural, and/or archaeological significance. So this would seem to exclude subjective, aesthetic ideas about the beauty of the landscape of a given area.[48]

---

[43]Directive 2011/92/EU of the European Parliament and of the Council of 13 December 2011 on the assessment of the effects of certain public and private projects on the environment, OJ L 26, 28 December 2012 (amended by directive 2014/52/EU, OJ L 124, 25 April 2014).

[44]OJ L 175, 5 July 1985.

[45]Case C-215/06, Commission v Ireland [2008] ECR I-4911, para 94.

[46]Council Directive 97/11/EC of 3 March 1997 amending on the assessment of the effects of certain public and private projects on the environment, OJ L 73, 14 March 1997.

[47]Case C-215/06, para. 108.

[48]Opinion dated 22 January 2009 of Advocate-General J. Kokott in *Mellor* (C-75/08), para. 48–55.

4.2 In a Cross-Border Context: The Obligation of Due Diligence

New legal obligations arise when the planned activities may have cross-border impact. The Maritime Spatial Planning Directive has already endeavored to improve cross-border cooperation to harness the oceans as an energy source. It requires Member States whose waters are adjacent to consult and coordinate their plans with one another and with third countries.[49] True, the obligations in that respect are not given in any great detail.[50] The directive only states that such cooperation may take the form of existing regional institutional cooperation structures, networks, or structures of competent authorities or any other method, such as taking advantage of the framework for sea-basin strategies.[51] And all that the directive says on cooperation with third countries is that regional institutional cooperation or existing international forums may be used.[52]

Furthermore, as the International Court of Justice declared in the case of *Pulp Mills on the River Uruguay (Argentina v Uruguay)*, a State would fail to comply with its obligation of due diligence, and the duty of vigilance and prevention that it implies, if the State did not undertake an environmental impact assessment on the potential effects of the projects. The Court considered it an obligation enshrined in general international law to carry out an environmental assessment whenever there is a risk that an industrial activity may have significant adverse impact in a cross-border context.[53] The Court moreover observed that such an assessment must be carried out before the activity goes into operation, although the activity's effects on the environment also have to be subjected to continuous monitoring throughout the project's life. However, the Court did recognize, as the International Law Commission did earlier in the 2001 *Draft Articles on Prevention of Transboundary Harm from Hazardous Activities*,[54] that general international law, as reflected in most prevailing international conventions, does not specify the scope or content of impact assessments. Thus, it falls to each State to determine the specific content of

[49]Preamble, para 20, articles 11–12.

[50]See Soininen (2015), pp. 193–195.

[51]Article 11.

[52]Article 12.

[53]*Pull Mills on the River Uruguay (Argentina v. Uruguay)*, Judgement, 20 April 2010, ICJ Reports 2010, para. 204.

[54]*Draft Articles on Prevention of Transboundary Harm from Hazardous Activities with commentaries*, Report of the International Law Commission on the work of its fifty-third session, Yearbook of the International Law Commission (2001-II), Part. 2, UN Doc. A/56/10, commentary on article 7. However, «such an assessment should contain an evaluation of the possible transboundary harmful impact of the activity. In order for the States likely to be affected to evaluate the risk to which they might be exposed, they need to know what possible harmful effects that activity might have on them. » (p. 159).

impact assessments through its own domestic law, taking account of the nature and magnitude of the proposed project and its possible adverse environmental impact.[55] The International Tribunal for the Law of the Sea took a step farther in its *Advisory Opinion on Responsibilities and Obligations of States Sponsoring Persons and Entities with Respect to Activities in the Area*, considering that "The Court's reasoning in a transboundary context may also apply to activities with an impact on the environment in an area beyond the limits of national jurisdiction; and the Court's references to 'shared resources' may also apply to resources that are the common heritage of mankind."[56] Consequently, this opinion opens up the possibility of extending said obligation beyond the mere cross-border sphere.

In the European context, there is a legal instrument containing more detailed procedural rules, the Convention on Transboundary Environmental Impact Assessment – the "Espoo (EIA) Convention" (Espoo, 25 February 1991).[57] An environmental assessment must be undertaken prior to any decision to authorize or undertake an activity, and each environmental assessment must contain at least the information in Appendix II (description of the planned activity, alternative solutions, corrective measures, etc.). In principle, this obligation concerns only the activities listed in Appendix I, which for now does not generically mention ocean energy. However, in their Decision III/7 (2004), the Parties to the Convention agreed to a second amendment to the Convention and revising the activities listed in Appendix I to include, *inter alia*, installations that harness wind power for energy production (wind farms).[58] The possibility of applying the terms of the Convention is also envisioned in any case where the parties involved agree to do so, if the proposed activities have a harmful transboundary impact due to their breadth, location (closeness to an international border), and long-distance effects.[59] So, for example, although cable laying is not one of the activities listed in Appendix I, the North Seas Countries' Offshore Grid (NSCOGI) initiative, a forum for regional cooperation in energy matters whose collaboration was formalized in a Memorandum of Understanding in 2010,[60] considers that coordination of national processes to authorize transboundary infrastructure should be guided by the principles of the Espoo Convention.[61]

---

[55] Para. 205.

[56] *Responsibilities and obligations of States with respect to activities in the Area*, Advisory Opinion, 1 February 2011, ITLOS Reports 2011, para.148.

[57] OJ C 104, 24 April 1992.

[58] Text available at http://www.unece.org/env/eia/about/amendment2.html. Accessed 10 Apr 2016.

[59] Article 2 (5) in conjunction Annex III.

[60] The North Seas Countries' Offshore Grid Initiative, Memorandum of Understanding, 3 December 2010. http://www.benelux.int/files/8113/9625/9202/MoU_NSCOGI.pdf. Accessed 10 Apr 2016.

[61] See Roeben (2013), p. 861.

5 Protection of Flora and Fauna

Another of the effects of the construction of installations of this type is the loss of marine habitat. Studies suggest that various species of marine animals and fish may be particularly vulnerable. The type and degree of impact are very much dependent upon a range of factors, such as location and design of the individual ocean energy developments. There are many international agreements that seek to protect and preserve marine ecosystems. In matters of ocean energy, the Convention on Wetlands of International Importance Especially as Waterfowl Habitat (Ramsar, 2 February 1971)[62] is extremely relevant because its Parties accept the responsibility of safeguarding the coastal wetland areas used by waterfowl in their seasonal migration.[63] Account was also taken, although at another level of protection, of other areas protected by the Convention for the Protection of the Marine Environment of the North-East Atlantic, OSPAR (Paris, 22 September 1992),[64] the Convention on the Protection of the Marine Environment in the Baltic Sea Area, HELCOM (Helsinki, 9 April 1992), the Convention for the Protection of the Marine Environment and the Coastal Region of the Mediterranean (Barcelona, 16 February 1976[65]), and the Protocol Concerning Specially Protected Areas and Biological Diversity in the Mediterranean (Barcelona, 10 June 1995[66]).

Other marine species besides waterfowl, such as cetaceans, can be affected by installations built in marine zones that they inhabit or cross on their regular migration route. The Convention on the Conservation of Migratory Species of Wild Animals (Bonn, 23 June 1979) has the objective of conserving migratory species throughout their area of natural distribution.[67] Under article V, States are obligated to make complementary agreements covering the whole of the area throughout which migratory species are distributed. One of them is the Agreement on the Conservation of Cetaceans in the Black Sea, Mediterranean Sea, and Contiguous Atlantic Area (Monaco, 24 November 1996), whose objective is to maintain a favorable state of conservation in a set of threatened species.[68]

---

[62]Text available at http://www.ramsar.org/sites/default/files/documents/library/scan_certified_e.pdf. Accessed 8 Mar 2016.

[63]Naturally, the Spanish authorities bore the Ramsar Convention strongly in mind when drawing up the procedure for the strategic assessment of the Spanish coast for the installation of marine wind farms. The Spanish authorities established a six-mile strip along the coastline around wetlands of international importance and catalogued it as a "no-go" zone, that is, a coastal zone not suitable for the installation of wind farms, because there the authorities have identified potential environmental effects incompatible with other marine environment uses that are considered to take priority (*Strategic Environmental Study of the Spanish Coast for the Installation of Marine Wind Farms, supra* note 35).

[64]OJ L 104, 3 April 1998.

[65]OJ L 240, 19 September 1977.

[66]OJ L 322, 14 December 1999.

[67]OJ L 210, 19 July 1982.

[68]Text available at http://www.accobams.org. Accessed 20 May 2016.

There are also strict obligations arising under EU law. Directive 92/43/EEC on Habitats[69] and Directive 79/409/EEC on Birds[70] established a network of protected marine areas of Community importance (Natura 2000) resembling protected terrestrial areas but with less intensity. Both directives[71] explicitly stated that they were to apply in the European territory of the Member States. Against the opinion of the European Commission, this stipulation was at first interpreted restrictively by some States that considered that their obligations should be limited to their internal waters and territorial seas only. However, the 2001 Council Meeting on Fisheries in Luxembourg[72] urged Member States to apply the directives in the exclusive economic zone, as some domestic courts had already instructed them to do. Years later, this position was also backed by the Court of Justice itself in case *Commission/United Kingdom* (C-6/04) of 20 October 2005.[73]

Actually, the marine component of the Natura 2000 network is not yet complete, due fundamentally to the fact that scientific knowledge of marine species and their habitats is less abundant. In December 2013, there were only 2292 sites of Community importance (SCI) and 983 special protected areas (SPA) in marine waters, which contrasts sharply with the 26,410 zones in the terrestrial Natura network.[74] In this sense, it is important for the Member States to designate protected marine areas as soon as possible and to approve their management plans, to put an end to the legal uncertainty about the suitability of ocean energy installations.

---

[69]Council Directive 92/43/EEC of 21 May 1992 on the conservation of natural habitats and of wild fauna and flora, OJ L 206, 22 July 1992. The Strait of Gibraltar, which is included in the geographical scope of the agreement, is one of those protected areas. Spanish authorities are aware of this and thus classified the strait as a "wind no-go area" in the Strategic Environmental Study mentioned above. Not so other zones, such as the Mediterranean; although extremely important for cetaceans and other marine species, they are difficult to exclude in the strategic phase, since migration routes and critical areas were established on the basis of very extensive delimitations. Thus, the most advisable course there is to postpone environmental viability and authorization to the project impact assessment phase (*Strategic Environmental Study of the Spanish Coast for the Installation of Marine Wind Farms, supra* note 35).

[70]Council Directive 79/409/EEC of 2 April 1979 on the conservation of wild birds, OJ L 103, 25 April 1979.

[71]Article 1 and 2.

[72]Annex, 2344th Council meeting- fisheries -Luxembourg, 25 April 2001, doc. 8077/01, para 15.

[73]Para. 115–117. As indicated in detail by the Opinion of Advocate General Kokott: «While the Habitats Directive admittedly contains no express rule concerning its territorial scope, it is consonant with its objectives to apply it beyond coastal waters. In accordance with Article 2(1), the directive is meant to contribute towards ensuring bio-diversity through the conservation of natural habitats and of wild fauna and flora in the European territory of the Member States to which the Treaty applies. This objective supports the conclusion that the area within which the directive applies coincides with that of the Treaty. In accordance with the aforementioned case-law, the area within which the Treaty applies is not limited to the territorial waters. Also, the directive protects habitats such as reefs and species such as sea mammals which are frequently, in part even predominantly, to be found outside territorial waters» (Opinion dated 9 June 2005 of Advocate-General J. Kokott in *Commission/United Kingdom* (C-6/04) para 132).

[74]European Commission (2013) *Natura 2000 Barometer*. http://ec.europa.eu/environment/nature/natura2000/barometer/index_en.htm. Accessed 22 Jun 2016.

If a natural habitat is eventually protected by the EU under the Natura 2000 network, either as a special area of conservation under Directive 92/43/EEC on Habitats or as a special protection area for birds under Directive 79/409/EEC on Birds, Member States are obligated to take the necessary measures to avoid natural habitat degradation and disturbances to species in the area. The directives do not in principle prohibit new projects or activities (such as energy-producing facilities) in Natura 2000 network. If the infrastructure could affect protected sites, however, the appropriate assessment would have to be carried out.[75] To this effect, the European Commission has published a series of instructions giving interpretative and methodological guidance on how to conduct the assessment called for in article 6(3) and (4) of Directive 92/43/EEC on Habitats. The process consists basically of four phases: description of the elements of the project, the conservation objectives, the effects on the main species and habitats, and the possible corrective measures.[76] During the process, it is quite normal for scientific doubts and other uncertainties to arise about the effects of the new installations (e.g., the effects of the noise they make). If so, the precautionary principle has to be applied, as advised in some international recommendations on the subject, such as those prepared by the Convention on the Conservation of Migratory Species of Wild Animals.[77] But even when it is concluded that environmental damage does exist, Member States may authorize projects anyway if there are no alternative solutions or if there are overriding reasons of public interest, although in that event Member States are obligated to create or improve another habitat elsewhere as a compensating measure.[78]

6 Marine Environmental Protection

Part XII of the United Nations Convention on the Law of the Sea establishes a general obligation on all States to protect and preserve the marine environment,[79] although, as we have just seen, other principles and general rules can be found

---

[75]See European Commission (2011), *Wind energy developments and Nature 2000* (Guidance document).

[76]See European Commission (2001). *Assessment of plans and projects significantly affecting Natura 2000 sites. Methodological guidance on the provisions of Article 6(3) and (4) of the Habitats Directive 92/43/EEC*. http://ec.europa.eu/environment/nature/natura2000/management/docs/art6/natura_2000_assess_en.pdf. Accessed 22 Jun 2016.

[77]Resolution 7.5, *Wind Turbines and Migratory Species*, adopted by the Conference of the Parties at its Seventh Meeting (Bonn, 18-24 September 2002), UNEP/CMS/Res. 7.5.

[78]Article 6. 4. See also European Commission (2007), *Guidance document on Article 6(4) of the 'Habitats Directive' 92/43/EEC clarification of the concepts of: alternative solutions, imperative reasons of overriding public interest, compensatory measures, overall coherence, opinion of the commission*, available at http://ec.europa.eu/environment/nature/natura2000/management/docs/art6/new_guidance_art6_4_en.pdf.

[79]Article 192.

throughout the entire Convention. However, States must avoid all unjustifiable interference with the activities carried out by other States in the exercise of their rights.[80] Moreover, States must take "all measures necessary to ensure that activities under their jurisdiction or control are so conducted as not to cause damage by pollution to other States and their environment"; hence, even though the Convention does not impose obligations on the States beyond their national jurisdiction, they are obligated to take measures with respect to renewable energy installations that are under their control, even on the high seas.[81]

One way of complying with this obligation is by designating "marine protected areas," which, as defined by the Convention on Biological Diversity, are understood to be "any defined area within or adjacent to the marine environment, together with its overlying waters and associated flora, fauna and historical and cultural features, which has been reserved by legislation or other effective means, including custom, with the effect that its marine and/or coastal biodiversity enjoys a higher level of protection than its surroundings. Areas within the marine environment include permanent shallow marine waters; sea bays; straits; lagoons; estuaries; subtidal aquatic beds (kelp beds, seagrass beds; tropical marine meadows); coral reefs; intertidal muds; sand or salt flats and marshes; deep-water coral reefs; deep-water vents; and open ocean habitats."[82] While we cannot find an express legal basis in any international legal instrument allowing the creation of marine protected areas, there are around 5000 protected sites, of which 10% are established in waters beyond national jurisdiction.[83] However, marine protected areas could be justified under article 194(5), which requires, among the measures for conserving the marine environment, "those necessary to protect and preserve rare or fragile ecosystems as well as the habitat of depleted, threatened or endangered species and other forms of marine life."[84]

The regulation of protected areas is addressed in different legal instruments both within areas under national jurisdiction and in the high seas.[85] Some regional organizations also foresee to take measures to protect and preserve the marine environment. Their great challenge is to reconcile the interests of States that wish to establish conservation measures with the interests of States that prefer other kinds of uses, which could include harnessing renewable energies. The point is not to

[80]Article 194(4).

[81]See Abad Castelos (2014), p. 227.

[82]Decision VII/5 (2004), Seventh meeting of the Conference of the Parties (COP-7), 9–20 February 2004—Kuala Lumpur, Malaysia, doc. UNEP/CBD/COP/DEC/VII/5, 13 April 2004.

[83]See Sands et al. (2012), p. 442.

[84]See Scovazzi (2004), p. 5.

[85]In this regards, as we know a new implementing agreement of UNCLOS is being negotiated on the conservation and sustainable use of marine biological diversity in areas beyond national jurisdiction which addresses marine protected areas as one of its elements (UN General Assembly adopted, on 19 June 2015, Resolution 69/292).

prohibit a set of activities unnecessarily but to set up a wide range of protective measures to ensure that conservation targets are met. The Protocol Concerning Specially Protected Areas and Biological Diversity in the Mediterranean (Barcelona, 10 June 1995) is a good example. It allows each of its Parties to regulate (taking account of the characteristics of each protected area) a set of activities (including ocean energy projects[86]), which can endanger the state of conservation of the ecosystems or species.[87]

Finally, it should be recalled that the Convention on the Law of the Sea also stipulates a set of general obligations with respect to power grids,[88] *inter alia*, the obligation to take measures to control marine pollution from the use of technologies under jurisdiction or control of States[89] and the obligation to minimize pollution from the construction and operation of installations and devices operating in the marine environment.[90]

7 Final Remarks

The European Commission has recognized that the environmental effects of ocean energy installations have not yet been identified, nor how environmental legislation in the different phases of projects should be applied. Nonetheless, experience gained from other activities, such as marine wind energy, can act as a guide for the implementation of new initiatives. The assessment carried out by the Spanish authorities for the development of marine wind energy demonstrated the need to accommodate some very different legal obligations arising not only from domestic law and EU law but also from international law. Even so, the existing regulatory framework has many gaps. States are ultimately forced to seek new forms of cooperation according to their needs. We will also have to stay attentive to the work of the International Renewable Energy Agency (IRENA). Although the agency's powers are limited, its objective is the widespread introduction of all forms of renewable energy, *inter alia*, marine energy, which includes tidal power, wave power, and ocean thermal energy.[91]

The European Union must also urge Member States to approve all legislative instruments that can hinder ocean energy development, such as maritime spatial

[86] See Scovazzi (2014), p. 427.

[87] Article 6.

[88] See Roeben (2013), p. 850.

[89] Article 196 (1).

[90] Article 194(3)(d).

[91] International Renewable Energy Agency (IRENA), Statute, 26 January 2009, available at http://www.irena.org. Accessed 23 Jun 2016.

planning instruments and designations of Natura 2000 marine protected areas. In addition, although the EU has already adopted the basic principles for cross-border grid cooperation, the establishment of regional structures needs to be fostered as well, to harmonize the requirements set for each individual project. This is a task that falls essentially within the competence of the Member States.

References

Abad Castelos M (2014) Marine renewable energies: opportunities, law, and management. Ocean Dev Int Law 45:221–237

Andreone G (2015) The exclusive economic zone. In: Rothwell D, Oude Elferink A, Scott K, Stephens T (eds) The Oxford handbook of the law of the sea. Oxford University Press, Oxford, pp 159–180

Andreone G, Cataldi, G (2014) Sui Generis Zones. In Attard D, Fitzmaurice M, Martínez Gutiérrez N (eds) The IMLI manual on international maritime law: Volume I: the law of the sea. Oxford University Press, Oxford, pp 217–238

Birnie P (1997) The status of environmental 'Soft Law': trends and examples with special focus on IMO norms. In: Ringbom H (ed) Competing norms in the law of marine environmental protection: focus on ship safety and pollution prevention. Kluwer, London, pp 31–57

Cottier T (2015) Equitable principles of maritime boundary delimitation. Cambridge University Press, Cambridge

Ford-Ramsden K, Davenport T (2014) The manufacture and laying of submarine cable. In: Burnett DR, Beckman RC, Davenport TM (eds) Submarine cables. The handbook of law and policy. Martinus Nijhoff, Leiden/Boston, pp 123–154

Long R. (2013) Offshore wind energy and ecosystem-based management: are the EU regulatory answers really blowing in the wind?. In: Nordquist MH, Moore JN, Chircop AE, Long R (eds) The regulation of continental shelf development: rethinking international standards. Martinus Nijhoff, Leiden/Boston, pp 15–52

Papanicolopulu I (2007) A note on maritime delimitation in a multizonal context: the case of the Mediterranean. Ocean Dev Int Law 38:381–398

Roeben V (2013) Governing shared offshore electricity infrastructure in the Northern Seas. Int Comput Law Q 62(04):839–864

Sands P, Peel J, Fabra A, MacKenzie R (2012) Principles of international environmental law, 3rd edn. Cambridge University Press, Cambridge/New York

Scott KN (2006) Tilting at offshore windmills: regulating wind farm development within the renewable energy zone. J Environ Law 18:89–118

Scovazzi T (2004) Marine protected areas on the high seas: some legal and policy considerations. Int J Mar Coast Law 19(1):1–17

Scovazzi T (2014) Marine protected areas in the Mediterranean. In: Juste J, Bou V (eds) Derecho del mar y sostenibilidad ambiental en el Mediterráneo. Tirant lo Blanch, Valencia, pp 425–441

Soininen N (2015) Marine spatial planning in the European Union. In: Hassan D, Kuokkanen T, Soininen N (eds) Transboundary marine spatial planning and international law. Routledge, New York, pp 189–120

Vazquez Gomez E (2012) The ecosystem approach to fisheries in environmental emergencies. In: Andreone G, Caligiuri A, Cataldi G (eds) Law of the sea and environmental emergences. Editoriale Scientifica, Naples, pp 146–159

Zervaki A (2015) Introducing maritime spatial planning legislations in the EU: fishing in troubled waters? MarSafeLaw J 1:95–114

The Black Sea and Blue Energy: Challenges, Opportunities and the Role of the European Union

Montserrat Abad Castelos

1 Introduction

The Black Sea enjoys enormous importance from a number of standpoints: if we consider all the economic, political, social and environmental factors that come together, its strategic nature for the world as a whole, not just Europe, is immediately apparent. In this regard, mention is often made of it serving as a bridge between Europe and Asia since it connects Europe with the Caspian Sea area, Central Asia, the Middle East and, going further, with South-East Asia and China.[1] Its strategic nature is also due to its connection with certain wide-ranging threats of

This chapter has been written within the framework of the following research projects: "NETwork of experts on the Legal aspects of MARitime SAFEty and security (www.marsafenet.org), awarded by the European Union, COST Action IS1105 (Working Group IV, "Protection of Fragile and Semi-Enclosed Seas"); "Alianza Público-Privada en la Cooperación para el Desarrollo en el Sector Pesquero: Las Empresas Pesqueras Españolas en los Países en Desarrollo" [Public-Private Partnership in Cooperation for Development in the Fisheries Sector: The Spanish fishing companies in Developing Countries], awarded by the Spanish Ministry of Economy and Competitiveness (DER 2013-45995-R); "Actores económicos internacionales y derechos humanos. Especial relevancia para España" [International Economic Actors and Human Rights. Particular relevance for Spain], awarded by the Spanish Ministry of Economy and Competitiveness (DER2014-55484-P); and "Nuevos escenarios jurídicos marítimo-pesqueros y la protección de las gentes del mar" ["New legal scenarios in the maritime and fisheries sector and the protection of seafarers"] (Network of Excellence), awarded by the Spanish Ministry of Economy and Competitiveness (DER2015-70965-REDT).

[1] *High-level Black Sea Stakeholder Conference* (2014), Sustainable Development of the Blue Economy of the Black Sea, *Background paper for the stakeholders conference*, 30 January 2014, Bucharest, Romania, p. 4.

M. Abad Castelos (✉)
Faculty of Social Science and Law, University Carlos III, Madrid, Spain
e-mail: montserrat.abad@uc3m.es

G. Andreone (ed.), *The Future of the Law of the Sea*,
DOI 10.1007/978-3-319-51274-7_8

a more global nature, such as human trafficking as a ramification of illegal migration, terrorism or drug trafficking.

Furthermore, when we talk of the Black Sea basin, we are in fact referring to an area that extends beyond the immediate environment of its waters: in addition to a respectable total of six riparian countries (Romania, Bulgaria, Turkey, Georgia, Russia and Ukraine), Greece, Armenia and Azerbaijan can also be considered to fall within its bounds.

There are obviously wide-ranging differences between the above-mentioned countries, according to a variety of indicators: economic development; governance, democracy and human rights protection; the pace at which reform is taking place in these aspects; access to energy resources (whilst one riparian country can be considered a veritable energy superpower, others are energy deficient and highly dependent on imports; some countries are energy producers, and others are simply countries through which energy passes); and their relationship with the EU. In the latter regard, some are EU Member States, and others are not: one is an EU candidate country (Turkey), others are European Neighbourhood Policy (ENP) partners and another (Russia) is a strategic partner for the EU. Amongst the ENP partner countries, three have shown willingness to achieve closer ties with the EU (Georgia, Moldova and Ukraine), whilst others evidence a certain degree of reticence and appear to favour partnerships with other interlocutors (Armenia, Azerbaijan).[2] Additionally, there are a number of frozen conflicts within the area, such as those in the Republic of Moldova (Transnistria) and Georgia (Abkhazia and South Ossetia) or that between Armenia and Azerbaijan (the Nagorno-Karabakh enclave). To this we must add the serious conflict provoked by Russia's illegal annexation of Crimea in 2014, which has brought about a substantial modification in the strategic landscape not only of the Black Sea Basin itself but also of the outlying area, seen as an example of a broader systemic challenge to the European security architecture.[3]

In addition to the complex political situation described above, it must also be noted that the Black Sea ecosystem is suffering from substantial environmental degradation: as a virtually enclosed inland sea, it is particularly fragile from a physical standpoint, and its vulnerability has regrettably not been sufficiently compensated for by the introduction of appropriate policies to prevent its deterioration. The enormous pressure resulting from numerous human activities such as industrialisation, urbanisation, overfishing or transport (not only of hydrocarbons since the arrival of invasive species in ships' ballast water has also been proved to represent a serious environmental threat) has led to serious problems of pollution, loss of biodiversity, extinction of species and eutrophication, amongst others.[4]

[2]In relation to this policy, currently under review, see *Review of the European Neighbourhood Policy*, JOIN (2015) 50 final, 18.11.2015.

[3]Cf. The European Parliament, *Report on the strategic military situation in the Black Sea Basin following the illegal annexation of Crimea by Russia*, A8-0171/2015, DE 21-5-2015; p. 6.

[4]See, *inter alia*, Adams (1998), pp. 209–217; Postiglione (2007), pp. 489–500; Oral (2013), pp. 787–804, particularly pp. 789 ff.

As is always the case in areas that suffer high volumes of traffic, the fragility of the environment is increased, particularly as far as hydrocarbons are concerned; in this regard, the Black Sea is a much-used corridor along which the latter are transported, mainly from the Caspian Sea, bringing with it the associated risk of accidental spillage.[5] And as if this situation were not in itself cause for concern, the risks may be even greater in future, given the possibility of hydrocarbon exploitation in the Black Sea itself, with offshore oil and gas deposits pending exploration in Romania, Bulgaria and Turkey, which could add new sources of pollution to the existing ones.[6] If this scenario were to occur, the Black Sea could possibly never recover from the consequences of a spill such as that which occurred in the Gulf of Mexico in 2010.

2 Sustainable Development and Blue Energy: From a Universal Strategy to That of the European Union

A suitable starting point for discussion would appear to be the parameters established by the recently published Sustainable Development Goals included in the 2030 Agenda for Sustainable Development, passed by the UN General Assembly in September 2015.[7] Although all 17 goals are interrelated, some of them are more closely linked than others, amongst them those that serve as a basis for this chapter. The goals in question are numbers 7, 8, 9, 13 and 14, namely those that state the need to ensure access to 'affordable, reliable, sustainable and modern energy for all'; promote 'sustained, inclusive and sustainable economic growth, full and productive employment and decent work for all'; build 'resilient infrastructure, promote inclusive and sustainable industrialization and foster innovation'; take 'urgent action to combat climate change and its impacts'; and, last but not least, 'conserve and sustainably use the oceans, seas and marine resources for sustainable development'.

This new framework in turn rests on the foundations provided by previous initiatives in the energy field promoted by the same platform, the United Nations, in recent years. Sustainable development is considered to be a pillar (although this situation, unfortunately, is still more theoretical than real), thus a significant proportion of the 'Sustainable Energy for All' initiative, launched by the UN Secretary-General in 2012 to mobilise action from all sectors of society in support of three interlinked objectives to be achieved by 2030: providing universal access to

[5]See Triantaphyllou (2009), pp. 225–241, particularly p. 229.

[6]*High-level Black Sea Stakeholder Conference* (2014), *loc. cit.* p. 4.

[7]UNGA Resolution 70/1, *Transforming Our World: the 2030 Agenda for Sustainable Development*, A/RES/70/1.

modern energy services, doubling the global rate of improvement in energy efficiency and doubling the share of renewable energy in the global energy mix. Similarly, the objective of sustainable development is one of the main foundations of a number of significant Reports issued by the United Nations Secretary-General, of which we will mention some of the most relevant for our purposes. Thus, the principle of sustainable development to a greater or lesser extent permeates the structure of the following documents: the *United Nations Secretary-General's Report on Marine Renewable Energies*, 2012;[8] the *Climate Change Expert Group's Report on Renewable Energies*, published some months previously;[9] or the report on *new and emerging technologies*.[10] Additionally, the United Nations General Assembly's open-ended informal consultative process on the Oceans and the Law of the Sea (UNICPOLOS), whose mandate is precisely to deal with matters relating to oceans within the context of sustainable development, devoted its 13th meeting, held in 2012, to discussing above all the subject of marine renewable energies, with a focus that can generally considered to be highly positive.[11]

Within the sphere of the European Union, it should be remembered that along with its exclusive competence on conservation on marine biological resources,[12] the EU also has shared competences on other aspects under the common fisheries policy, energy, environment or transport, amongst other fields that can be relevant here.[13] In addition to this, sustainable development is a general and transversal goal,[14] and in line with this the Integrated Maritime Policy is one of the EU's vehicles for promoting the coherent adoption and coordination of decisions aimed at maximising sustainable development, economic growth and cohesion between Member States. Amongst the policies included under this umbrella are two with particular relevance for our case: *blue growth* and *sea basin strategies*, that for the Black Sea being included amongst the latter. In any event, it should also be borne in mind that the Black Sea basin is also a target for other EU policies and initiatives, which will be considered below; a case in point is the European Neighbourhood Policy (ENP), in which sustainable development is seen as a common value that partner States agree to accept.[15]

Remaining for the moment within the general sphere, it should be noted that blue growth has been an ever-present discourse within the European Union in recent years, but particularly since 2012, when the Commission drafted its

[8]UN Doc. A/67/79.

[9]IPCC, *Special Report on Renewable Energy Sources and Climate Change Mitigation* (2011).

[10]"New and emerging technologies: renewable energy for development", UN Doc. E/CN.16/2010/4.

[11]See http://www.un.org/Depts/los/consultative_process/consultative_process.htm.

[12]TFEU, Article 3.

[13]TFEU, Article 4.

[14]TEU, Article 3 and TFEU, Article 11.

[15]See *Joint Consultation Paper. Towards a new European Neighbourhood Policy*, JOIN (2015) 6 final, 4-3-2015, particularly pp. 1–3.

Communication on *Blue Growth: Opportunities for Marine and Maritime Sustainable Growth*. In this document, together with other aspects of the blue economy, blue energy is seen as one of its priority areas for action and one that could aid job creation, basically in coastal regions. The Commission mentions EU industry's position as a world leader in the sector and highlights blue energy's capacity to contribute to 'reductions in carbon emissions outside Europe' through exports, the possibility of exploring 'synergies [. . .] with the offshore conventional energy sector' (e.g. with regard to infrastructure and safety challenges) and the potential to 'secure affordable energy supplies in the EU'.[16]

With this as a starting point, more recent documents have also acknowledged the important role that can be played by marine energy resources, for example the 2013 Communication on *Energy Technologies and Innovation*[17] or the Communication on *Blue Energy: Action Needed to Deliver on the Potential of Ocean Energy in European Seas and Oceans by 2020 and Beyond*, adopted in 2014. The latter includes, in addition to an overview of the current situation and the main opportunities and threats remaining, an 'Action Plan for Ocean Energy', which envisages a two-step approach: a first phase (2014–2016) that includes the setting up of an Ocean Energy Forum to bring stakeholders together in order to develop a shared understanding of the main problems and devise workable solutions, as well as the development of an Ocean Energy Strategic Roadmap, and a second phase (2017–2020) that contemplates the possibility of developing a European Industrial Initiative based on the outcomes of the first stage.[18] A few months later, in its Communication *Innovation in the Blue Economy: Realising the Potential of Our Seas and Oceans for Jobs and Growth*, also dated 2014, the Commission highlights, amongst other aspects, the need to increase knowledge of our seas in order to promote growth in the blue economy, thereby eliminating the hindrances caused by a current lack of information that is holding back innovation in this area; the setting up of a 'sustainable process', through a variety of channels, in order to 'ensure that marine data is easily accessible, interoperable and free of restrictions on use, with a specific target of developing a multi-resolution map of the entire seabed and overlying water column of European waters by 2020'; the creation of an information platform across the whole Horizon 2020 programme in which, in collaboration with Member States, it is intended to include information on nationally funded marine research projects; and the encouragement of 'stakeholders in the blue economy to apply for a Knowledge Alliance and marine Sector Skills Alliance'.[19]

---

[16]COM (2012) 494 final, 13-9-2012, p. 8.

[17]COM (2013) 253 final, 2-5-2013.

[18]COM (2014) 8 final, 20-1-2014, particularly pp. 5–9.

[19]COM (2014) 254 final, 8-5-2014.

3 Sources of Marine Renewable Energies and Adequacy in the Case of the Black Sea

3.1 Kinds of Marine Renewable Energies

Marine renewable energies are a form of renewable energy deriving from the various natural processes that take place in the marine environment. There are four kinds of such energy, namely ocean energy, wind energy from turbines located in offshore areas, geothermal energy derived from submarine geothermal resources and bioenergy derived from marine biomass, particularly ocean-derived algae. In turn, renewable ocean energy comes from six distinct sources, each with different origins and requiring different technologies for conversion but having in common the fact that they are all obtained from the potential, kinetic, thermal and chemical energy of seawater. These six distinct sources are waves, tidal range, tidal currents, ocean currents, ocean thermal energy conversion and, finally, salinity gradients. More specifically, waves, which are generated by the action of wind on water, produce energy that can be harnessed. With regard to tides, their amplitude generates energy through the cyclical rise and fall in the height of the ocean. The same is true of tidal currents, which are generated by horizontal movements of water, their flows resulting from the rise and fall of the tide. Ocean currents, which exist in the open ocean, are another source of energy. Ocean thermal energy conversion, on the other hand, is a technology for taking advantage of the solar energy absorbed by the oceans, based on the temperature difference between the top layers of water and those at a greater depth, which are much colder. However, a minimum temperature difference of 20°C between layers is needed in order to harness this energy, which can therefore only be produced in certain parts of the world, such as equatorial and tropical regions. Finally, salinity gradients arise from the mixing of freshwater and seawater, which takes place at river mouths and releases energy as heat. This energy can be harnessed through a process of inverse electrodialysis, based on the difference in chemical potential between freshwater and seawater, or through an osmotic power process based on the natural tendency of the two types of water to mix together.[20]

The development status of these technologies differs widely, although most of them are still either embryonic or in their infancy, ranging as they do from the conceptual stage to the prototype stage, taking in the pure research and development stage on their way.[21] The IPCC highlights tidal range technology as being the most advanced and in fact as the only form of ocean energy technology (excluding marine wind energy technology) that can currently be considered 'mature'.[22]

---

[20]IPCC, *Special Report on Renewable Energy Sources and Climate Change Mitigation* (2011)..., *loc. cit.*, pp. 503 ff.

[21]Ibid., Chap. 6.3.1.

[22]Ibid.

Although marine energy technologies are still generally at an early stage of development, it has to be said that they could make much swifter progress if investment in them were higher. Prominent amongst the leaders in the development and commercialisation of marine renewable energy technologies are nations such as the United Kingdom, Ireland, the United States, Australia, New Zealand, Finland, Denmark, Belgium, France, Germany and Japan.[23] However, the economic crisis that has been affecting a number of the world's developed countries has had necessarily a negative effect on the flow of investment towards technologies of this kind.

Although forecasts vary widely, depending on who is making the prediction, a prudent approach indicates that any significant deployment of ocean energy technologies is unlikely to occur before 2030, whilst commercial deployments are expected to continue expanding beyond 2050.[24] It remains to be seen, therefore, when these technologies will be able to make a significant contribution to the global energy supply. At the moment, only marine wind energy can be considered to be relatively close to beginning to be competitive with fossil fuels or nuclear energy. However, it must be said that in spite of the incipient status of all marine renewable energies, forecasts of their potential are on the whole clearly optimistic. According to the IPCC, the potential for technically exploitable marine renewable energies, marine wind power excluded, is estimated at some 7400 exajules (EJ) per year.[25] This figure is considered to be more than enough to meet human energy needs not only at present but also well into the future.[26]

3.2 Marine Renewable Energies and the Black Sea

If we take the parameters of sustainable development, and by extension its three constituent dimensions, namely its economic, social and environmental aspects, it is clear that marine renewable energies score very highly in this regard, as the UN Secretary General's 2012 report demonstrates.[27] A similar conclusion was also reached in the UNICPOLOS meeting devoted to marine renewable energies,[28] the

[23]Nevertheless, the list of leading countries in this sector varies according to the source consulted. For example, the countries mentioned in the Report of the UN Secretary-General on marine renewable energies, published in 2012, do not exactly coincide with those that appear in other places, such as specialist websites. See, in any case, the above-mentioned report, UN Doc. A/67/79, dated 4 April 2012, p. 8.

[24]IPCC, *Special Report on Renewable Energy Sources and Climate Change Mitigation* (2011)..., *loc. cit.*, p. 527.

[25]Ibid., p. 501.

[26]Ibid. and UN Doc. A/67/79, pp. 6–7.

[27]UN Doc. A/67/79, pp. 4 ff.

[28]See, for example, 25 *Earth Negotiations Bulletin*, Number 88 (2012), p. 5; and *Report on the work of the United Nations Open-ended Informal Consultative Process on Oceans and the Law of the Sea at its thirteenth meeting*, Doc. A/67/120 (2012).

idea also being supported by doctrinal studies on the subject.[29] Although it is true that certain problems or challenges can always be mentioned, particularly in the economic and environmental spheres,[30] the overall balance is nevertheless clearly favourable since the benefits of sustainable development from all angles are self-evident (job creation, stimulus to the economy, improved access to energy, energy security, reduction of emissions, climate change mitigation, zero risk of hydrocarbon spills and a reduction in the probability of hazardous accidents, to name but a few).

Without prejudice to the above, however, it should be realised that it will never be possible to obtain all of the various kinds of renewable energy in all possible surroundings. We have seen how some kinds of marine energy are dependent on certain particular physical characteristics such as temperature or the existence of currents, amongst others. Taking this into account, the Black Sea has the potential for at least some forms of marine energy, namely marine wind energy, wave energy, tidal barrages and the production of biofuels.[31] Furthermore, it should be noted that the current situation of environmental degradation affecting the Black Sea makes it an ideal space for investing in climate-friendly technologies since they help to reduce emissions and avoid the risk of accidents with serious consequences, unlike, for example, offshore oil rigs.

However, we should also be aware that many aspects of the harnessing and use of energy resources, marine energy included, often require transnational management and inter-State cooperation (e.g. basic issues such as cable laying, data exchange, network connections, etc.) that are not always easy to achieve in a space that has historically been marked not only by the absence of mutual trust but also by rivalries between neighbouring states and even open conflict.

---

[29]Abad Castelos (2014), pp. 221–237; particularly pp. 223–225.

[30]It must be acknowledged that issues can also rise in the social sphere, for example a rejection of the more visible kinds of technology in certain surroundings; see Kerr et al. (2015), pp. 108–115. Above all, however, the main challenges are to be found in the economic sphere, due to the huge costs involved and the massive investments needed, and in the environmental sphere, resulting from other possible negative impacts; see Wright (2014), pp. 23–30. Nevertheless, further research is needed to determine the scope of certain potential problems (e.g. the impact of certain devices on marine fauna and the possible adverse impact of tidal barrages). A more detailed overview is provided in Copping, Battey et al. (2014), pp. 1–11.

[31]*Study to Support the Development of Sea Basin Cooperation in the Mediterranean, Adriatic and Ionian, and Black Sea*, Task 4 Report, *Black Sea—Identification of Elements for Sea Basin Cooperation* (2014), MARE/2012/07-Ref. No 2, pp. 3 ff. also see United States Agency for International Development, *Black Sea Regional Transmission Planning Project: Renewable Energy Compendium Report*, Washington, 2012, pp. 17 ff.

4 The European Union and Blue Energy in the Black Sea

It is essential to bear in mind that the EU's Integrated Maritime Policy (IMP), which first came into being in 2007, in 2009 acquired an international dimension transcending its borders before adopting, in 2012, blue growth as one of its main pillars, at least from the theoretical standpoint. In this sphere, the EU has carried out a strategic assessment of the potential for cooperation in the context of Blue Growth in the various sea basins concerned and has sponsored a series of studies, through DG MARE, to analyse its blue growth potential, examining in detail each of the different development models of its maritime industries, with the aim of drafting specific plans for the future. In this context, the Black Sea has also come under the spotlight in order to explore its current situation and the potential added value that maritime cooperation could bring to the surrounding area, identifying the main maritime players in the region and the aspects that would benefit from a sea-basin approach. This has taken the form of a report, published in 2014 and titled *Black Sea—Identification of Elements for Sea Basin Cooperation*, which lists the most significant initiatives and programmes in the area of maritime cooperation at sea basin level, maps the existing projects and initiatives with a maritime dimension and enumerates the possible sources of funding for blue growth projects in the Black Sea.[32] The report also identifies what are considered to be the priorities,[33] which in the case of sectoral categories include offshore renewable energies, together with offshore oil and gas, as a means of ensuring energy security in the region.[34] Horizontal actions cover four main areas, each with its corresponding sub-categories, namely 'Planning a blue economy' (Maritime Spatial Planning; development of smart infrastructure, etc.), developing knowledge (joint data collection; capacity-building across individuals, institutions and society; sharing maritime culture and heritage), supporting business growth (facilitating access to finance; promoting innovation; development of maritime clusters), and enhancing the environment (preserving, protecting and improving the quality of the coastal and marine environment and heritage; ecosystem monitoring; building resilience to the impacts of climate change).[35]

---

[32]Black Sea—Identification of Elements for Sea Basin Cooperation (2014)... loc. cit., introduction.

[33]Nevertheless, it should be pointed out that the starting point for the study is the acknowledged fact that cooperation between the EU and other Black Sea riparian countries have to date taken place largely on a bilateral basis, which is in contrast to EU initiatives in other geographic regions such as the Baltic, where actions were conceived from the beginning in a regional format and have therefore benefited from a significant institutional presence; cf. *ibid*., p. 8; also see *ibid*., p. 30.

The fact that EU cooperation with Black Sea regions countries is basically *bilateral* has in turn meant that multilateral initiatives have largely been *sectoral*, such as those which will be referred to below (INOGATE, TRACECA and PETrA).

[34]*Black Sea—Identification of Elements for Sea Basin Cooperation* (2014)... *loc. cit*., introduction.

[35]*Ibid.*

Within the same framework, two high-level Black Sea Stakeholder Conferences have been organised, the first in Bucharest (2014) and the second in Sofia (2015).[36] The EU's expressly declared aim in this regard is to promote dialogue between all stakeholders, both public and private, to build their capacity and to support cooperative actions.

Mention should also be made of the publication of another EU study in November 2015, *Project in Support to the Development of Blue Economy and Integrated Maritime Policy in the Black Sea. Concept Paper*.[37] This project concept is currently under discussion with the coastal countries and regional organisations.[38] It is, however, worth noting that its priorities do not include energy issues, the *leitmotiv* of the report being that the development of maritime and coastal tourism should be the central theme.

Furthermore, in 2007 the EU adopted a specific regional initiative, its *Black Sea Synergy*, which lays no claim to being a new policy, but rather a complementary initiative aimed at reinforcing existing ones, since the EU has either adopted or is a partner in various programmes affecting the Black Sea through a number of channels, and thus funded from a variety of sources (and, therefore, with a different status with regard to the various States, depending on their situation). Thus, before looking at *Black Sea Synergy*, it must be noted that the EU's institutions have adopted significant measures regarding the Black Sea in the framework of Turkey's pre-accession process, the ENP[39] and the Strategic Partnership with Russia. Similarly, *Horizon 2020*, the EU Framework Programme for Research and Innovation for 2014–2020, also contains a specific call for the Black Sea region. Although *Black Sea Horizon* does not specifically include energy issues amongst its explicit aims, any renewal energy project would fit perfectly with them, especially given the fact that its seventh and final stated sub-objectives is precisely to 'identify

[36]For documents and minutes of discussions see: "Sustainable development of the blue economy of the Black Sea", *Enhancing marine and maritime cooperation*, Bucharest, Romania, 30 January 2014 (Summary of Presentations and Discussions); and *2nd Black Sea Stakeholders Conference Sofia*, 24th March 2015 Background paper (see http://ec.europa.eu/maritimeaffairs/events/2015/03/events_20150324_01_en.htm).

[37]*Project in support to the development of Blue Economy and Integrated Maritime Policy in the Black Sea* (2015), *Concept paper*, EU, 20 November 2015.

[38]See http://ec.europa.eu/maritimeaffairs/policy/sea_basins/black_sea/index_en.htm.

The European Commission also supports the effort of a number of research institutes and public stakeholders from all Black Sea countries to compile all relevant data and create a digital map of the Black Sea seabed, including geology, habitats and marine life. A first version of the map is expected to be ready in 2016; *ibid.*

[39]The *ENI Cross-Border Cooperation Programme* (CBC) (2014–2020), successor to the *Joint Operational Programme* (2007–2013), lies within the framework of the ENP and is thus financed through its funding instrument, although it should be borne in mind that most of the projects currently envisaged within its framework have no direct connection with the maritime sphere, being related instead with stimulating entrepreneurship and other aspects, etc. See European External Action Service and European Commission—DG for Development and Cooperation—EuropeAid, Programming of the European Neighbourhood Instrument (ENI) 2014–2020; Programming document for EU support to ENI Cross-Border Cooperation (2014–2020).

challenging thematic areas for mutual science, technology and innovation cooperation'.[40] And, finally, various initiatives affecting the Black Sea have been carried out through other cooperative programmes in the energy sphere in which the EU is a partner, such as INOGATE,[41] TRACECA[42] and PETrA,[43] although to date no significant initiatives having to do with blue energy appear to have arisen within them.

Black Sea Synergy, as we have already seen, is a regional initiative that came into being in 2007 with very broad goals that went far beyond maritime, energy, transport or environmental aspects, its cornerstone being the Commission's communication *Black Sea Synergy—A New Regional Cooperation Initiative*. The 'primary task' of this initiative would be 'the development of cooperation within the Black Sea region and also between the region as a whole and the European Union', based on the common interests of the EU and the Black Sea region. The scope of its actions could extend beyond the region itself since many activities are linked to neighbouring regions such as the Caspian Sea, Central Asia and South-Eastern Europe and such cooperation would therefore include 'substantial interregional elements'.[44] The *Synergy* refers to a wide range of cooperation areas, which in turn include other matters such as *democracy, respect for human rights and good governance*; *managing movement and improving security*; *'frozen' conflicts*; *fisheries*; *trade*; *research and education networks*; *science and technology*; *employment and social affairs*; and *regional development*, amongst others.[45]

As far as energy is concerned, although from the very beginning reference was made to the need to 'develop a clearer focus on alternative energy sources',[46] the approach comes from the underlying perspective of the region's strategic importance for EU energy security, in part because it is an energy-producing region but mainly because it is a transport corridor for conventional hydrocarbons. The

[40]See http://ec.europa.eu/maritimeaffairs/policy/sea_basins/black_sea/black-sea-horizon_en.htm.

[41]INOGATE is a regional energy cooperation programme between the European Union, Turkey and various States from the former Soviet Union (Armenia, Azerbaijan, Belarus, Georgia, Kazakhstan, Kyrgyzstan, Moldova, Turkmenistan, Ukraine, Uzbekistan and Tajikistan) that began in 1995 (Russia is not a member, although it enjoys observer status). Although its original focus was the oil and gas pipelines running from the Caucasus to the European Union, in 2004, as a result of the *Baku Initiative*, it widened its goals. This initiative was the outcome of the dialogue on energy cooperation between the EU and INOGATE member countries with a view to incorporating the following areas: enhancing energy security; harmonising legal and institutional frameworks in order to liberalise the energy market between partner countries; developing sustainable energy; and attracting investment towards energy projects of common and regional interest; see http://www.inogate.org/.

[42]This is another international cooperation programme in the field of energy transport: the *Transport Corridor Europe-Caucasus-Asia*, in which the partners are the EU and 14 states from the Eastern Europe, Caucasus and Central Asia region; see http://www.traceca-org.org/en/home.

[43]Black Sea Pan-European Transport Area.

[44]COM (2007) 160 final, 11-4-2007, p. 3.

[45]*Ibid.*, pp. 3 ff.

[46]As well as energy efficiency and energy saving; *ibid.*, p. 5.

Commission's proposal thus contemplates, on the one hand, ongoing improvement of the EU's relations with energy producer, transit and consumer countries, within the framework of a *dialogue on energy security* (with a view to promoting legal and regulatory harmonisation through the Baku Initiative[47]) and, on the other, to increase energy stability by constructing new energy infrastructure and upgrading the existing one.[48] The following year (2008), the European Parliament highlighted the importance of strengthening cooperation between the EU and countries in the region,[49] whilst the Commission proposed, in its *Report on the First Year of Implementation of Black Sea Synergy*, the establishing of 'sectoral partnerships' in the fields of 'transport, environment [and] energy'.[50] In the same vein, the European Parliament made a second appeal to develop EU policies towards the region in a subsequent resolution (2011).

In 2015, the Commission and the High Representative of the European Union for Foreign Affairs and Security Policy adopted a *Joint Staff Working Document* titled *Black Sea Synergy: Review of a Regional Cooperation Initiative*, covering the years 2009–2014.[51] The document provides a review of the initiative and highlights a number of 'lessons learnt' intended to inform the future development of the Synergy, given that the events in Ukraine after the illegal annexation of the Crimean peninsula by the Russian Federation had a significant impact, leading to the suspension of all EU-funded projects in the affected area (with the exception of those in support of civil society) and a reassessment of relations with Russia.[52]

However, the 2015 Report on the review of *Black Sea Synergy* makes no mention whatsoever of any progress regarding renewable energies in general, not to mention marine ones. Reference is made to EU support for certain projects concerning specific hydrocarbon deposits, pipelines and means of transport, as well as to Moldova and Ukraine becoming members of the Energy Community and to the roadmap on energy cooperation with Russia until 2050,[53] which is currently suspended as a result of recent events and will anyway need reviewing in the future. In this regard, it should be pointed out that the weakness of political determination amongst the region's countries, when taken in combination with recent conflicts, significantly increases the complexity of this scenario.[54]

---

[47]Referred to above, in the footnote on INOGATE.

[48]COM (2007) 160 final..., *loc. cit.*, p. 5.

[49]Resolution of 17-6-2008.

[50]COM (2008) 391 final, 19-6-2008.

[51]SWD (2015) 6 final, 20-1-2015.

[52]Renewal of cooperation depends on fulfilment of the 2014 and 2015 Minsk Agreements by Russia; see European Parliament, *Report on the strategic military situation in the Black Sea Basin following the illegal annexation of Crimea by Russia*, A8-0171/2015, 21-5-2015; p. 8.

[53]*Roadmap EU-Russia Energy Cooperation until 2050* (2013); see pp. 21 ff. on the subject of renewable energies in general.

[54]*Report on the strategic military situation in the Black Sea Basin following the illegal annexation of Crimea by Russia ... loc. cit.*, pp. 1, 4–5 and 11.

5 The Main Challenges in the Black Sea Basin that the European Union Can Help to Overcome

The Black Sea needs a regional approach because the challenges it faces, one of them energy, are on the same scale. And energy, in turn, is linked to other aspects that can only be tackled properly from an international perspective and on a regional basis, such as transport and protection of the environment.

This is not the place to examine all the various challenges that the EU can help to overcome since it would go far beyond the scope of the present work, which will only look at some of them, whilst acknowledging that the matter is indeed an extremely complex one. It is important to remember that in addition to the many challenges posed by marine renewable energies themselves (technology costs; transport infrastructure network costs, suitable port installations and specialised vessels; authorisation and licensing procedures; lack of subsidies; possible objections by the general public; technical problems such as connecting to the grid; etc.), there are two other problem areas that have to be considered. Firstly, the Black Sea basin is a particularly complex physical and geographic area in which a variety of different policies come into play, for example the EU's Integrated Maritime Policy, which involves international elements, development cooperation policy, the ENP, Turkey's pre-accession process and certain complementary regional strategies, some of which contain interregional elements. Furthermore, all the above elements come together in a region that includes countries that are EU Member States and others that are not. Secondly, and as if the above were not enough, from a geopolitical standpoint the area is home to a number of major conflicts, some of them 'frozen' and others that have only recently arisen. The area is one in which simply attempting to establish cooperation between certain countries in any field whatsoever is a veritable challenge. Without prejudice to the foregoing, it is necessary here to point out a further set of issues that come into play.

The first point to note is that the *Black Sea Synergy* contains an excessive number of spheres of action: it tries to approach too many issues but, by neglecting to establish priority goals, focuses on none in particular, which amongst other implications could dilute its power.[55] Furthermore, the EU could also make more use of the *Black Sea Economic Cooperation* (BSEC), an organisation created in 1992 for the purpose of promoting cooperation in the regions and that could help to enhance its effectiveness to plan useful projects,[56] particularly when we consider that one of its spheres of cooperation is precisely that of energy. Another interesting factor in this regard is that Russia (in common with other member countries) has

---

[55]A critique of the confusion created by the excessive number of possible areas of cooperation combined with the lack of any hierarchy between them can also be found in Devrim and Grau (2010), pp. 244–251; p. 248.

[56]Its Member States are Albania, Armenia, Azerbaijan, Bulgaria, Georgia, Greece, Moldova, Romania, Russia, Serbia, Turkey and Ukraine; see http://www.bsec-organization.org/Pages/homepage.aspx.

always sought to maintain the organisation's openly non-political nature, rejecting any attempt to include other issues that might refer to territorial disputes or security matters.[57]

Second, there is room for improving the consistency of the EU's actions. If *blue energy* is a key aspect of *blue growth*, which is in turn a key aspect of the EU's Integrated Maritime Policy, why is it not given the same importance across all the EU's policies and strategies? Thus, for example, in the *Energy Union Package* contained in the Commission's 2015 Communication concerning a *Framework Strategy for a Resilient Energy Union with a Forward-Looking Climate Change Policy*, there is no specific section devoted to blue energy: in fact, it does not even deserve a mention,[58] the principal focus being the diversification of supply as far as suppliers and transport routes are concerned in order to guarantee energy security. The most innovative vision of the future to appear in the document would appear to relate more to exploring the full potential of liquefied natural gas (LNG) rather than to renewable energies.[59] In a similar vein, the 2015 concept paper *Project in Support to the Development of Blue Economy and Integrated Maritime Policy in the Black Sea*, referred to earlier, in reality revolves around promoting tourism.[60] Much the same can be said of the most recent revision of the *Black Sea Synergy*, carried out in 2015 through the afore-mentioned *Joint Staff Working Document*, which also makes no mention whatsoever of blue energy and deals only with hydrocarbon deposits or recent and future gas pipeline projects.[61] In light of all the above, the EU could do worse in the future than to turn the spotlight on renewable energies in the various sea basins, of which the Black Sea is one, and thus by extension on blue energy, in order to improve consistency between all its different actions and instruments.

Another challenge that the EU can undoubtedly do much to help overcome is that of spatial planning, in order to plan when and where human activities take place at sea. Maritime spatial planning reduces conflicts, encourages investment, increases coordination not only between administrations in each country but also between countries and protects the environment by helping with the early identification of impact and opportunities for multiple use of space.[62] After the adoption of the *Directive on maritime spatial planning*, Member States are obliged to establish

---

[57] See the press release on the declarations made by the Minister for Foreign Affairs in this regard; "Rusia apuesta por mantener el carácter apolítico de la Organización para la Cooperación Económica del Mar Negro (BSEC)", *Sputnik Mundo*, 10-12-2015 (http://mundo.sputniknews.com/economia/20151210/1054682272/rusia-bsec-apolitico.html).

[58] COM (2015) 80 final, 25-2-2015.

[59] *Ibid.*, pp. 4 ff.

[60] *Project in support to the development of Blue Economy and Integrated Maritime Policy in the Black Sea. Concept paper* (2015)..., *loc. cit.*

[61] SWD (2015)... *loc. cit.*., pp. 4 ff.

[62] Cf. the European Commission's webpage on maritime spatial planning: http://ec.europa.eu/maritimeaffairs/policy/maritime_spatial_planning/index_en.htm (last accessed 20-2-2016).

and implement a procedure for planning activities and uses in their marine waters[63], in which it therefore also becomes necessary to include all possible blue energy projects.[64] However, there is a lack of maritime spatial planning (MSP) in the Black Sea basin as a whole and in the maritime areas adjoining the majority of its riparian states, as highlighted in the 2014 report produced on behalf of the Commission, *Black Sea—Identification of Elements for Sea Basin Cooperation.*[65] It would therefore seem essential for the EU to also promote the adoption of national maritime spatial plans in other Black Sea riparian countries.[66]

The third and final challenge is that of public–private partnerships, which are encouraged in a number of documents relevant to the topic of this chapter, such as the Commission's 2014 Communication on Blue Energy,[67] the *Black Sea Synergy*, the Horizon 2020 programme[68] or the 2014 report produced on behalf of the Commission, Black *Sea—Identification of Elements for Sea Basin Cooperation.* The premise is obviously that companies are a vital element of society, and their contribution to it is indispensable. And this is indeed the case: private sector intervention should clearly represent an obvious advantage,[69] a condition that in this case is fulfilled *ab initio* since the role that plays in matters of energy exploration and exploitation is an irreplaceable one. Nevertheless, there is still a dual challenge to be faced. The first of these is that it is hard to forge certain links. The *Black Sea—Identification of Elements for Sea Basin Cooperation* report highlights this issue when it says that even 'where co-operative platforms exist for the sea basin's key MEAs, they often do not bring together all relevant parties (public, private, academic partners)'[70]. Second, the fact that such initiatives could involve

---

[63]Directive 2014/89/EU of the European Parliament and of the Council, 23-8-2014, establishing a framework for maritime spatial planning; *OJEU* L 257/135, 28-8-2014.

[64]See O'Hagan (2012) and Soininen (2012), pp. 85–118.

[65]*Black Sea—Identification of Elements for Sea Basin Cooperation* (2014) ... *loc. cit.*, introduction.

[66]On their importance in relation to marine renewable energies, see Wright et al. (2016), pp. 126–134; pp. 131 and 132.

[67]COM (2014) 8 final, 20-1-2014, pp. 10 ff. See also: Commission Staff Working Document, Impact Assessment (Accompanying the document Communication from the Commission to the European Parliament, the Council, the European Economic and Social Committee and the Committee of the Regions), *Ocean Energy: Action needed to deliver on the potential of ocean energy by 2020 and beyond*, SWD (2014) 13 final, Brussels, 20-1-2014; pp. 22 ff.

[68]See SWD (2015)... *loc. cit.*, pp. 9 and 10.

[69]This is reflected in the views of experts in the subject, as well as in national plans and regulations, particularly from the development cooperation standpoint, taking into consideration, amongst other requirements, that of compatibility between objectives (e.g. in social, environmental and sustainable development terms), complementarity, the significant nature of the private company's contribution in terms of human and material resources, etc.; see, amongst others, Caplan (2006), pp. 11–14; Dizon-Reyes (2012–2013), pp. 799–819; Vinnyk (2013), pp. 17–36; Tiganescu (2013), pp. 519–526.

[70]*Black Sea—Identification of Elements for Sea Basin Cooperation* (2014) ... *loc. cit.*, introduction.

States with differing degrees of implementation of reforms in areas such as good governance and the fight against corruption implies an additional challenge in that the utmost precautions must be taken in order to ensure that public–private partnerships are structured in the best way possible. In this regard, a key point that should always be kept in mind is that the first priority in energy exploitation is that it should be done in the general interest, including all citizens, and not only in that of the companies concerned.

6 Conclusion

Marine renewable energies, like all renewable energies in general, appear to be the ideal solution from a sustainable development perspective. The range of difficulties that blue energy can help to surmount is enormous. Indeed, as the twenty-first century progresses, there is growing awareness that the energy potential of the seas and oceans may be so vast that it surpasses our current understanding.

The conflicts in the Black Sea basin, whether recent or ‘frozen’, condition a geopolitical scenario in which it is particularly difficult to construct any kind of regional cooperation. Some form of international collaboration, at least sub-regional in scope, will be a necessary precondition for establishing certain projects in the field of marine renewable energies in the area, as well as others relating to them. There is no magic formula for achieving such cooperation, but at the very least the EU should identify all the aspects in which it can help to pave the way. Similarly, it should also strive to achieve maximum coherence between its strategies, thereby maximising their effectiveness. Blue growth and blue energy should play a greater role in the European Union’s projections and initiatives for all the sea basins within its scope, particularly that of the Black Sea.

References

Abad Castelos M (2014) Marine renewable energies: opportunities, law, and management. Ocean Dev Int Law 45:221–237

Adams R (1998) The ecological decline of the black sea. Colo J Int Environ Law Policy 9:209–217

Battey H, Brown-Saracino J, Massaua M, Smith C (2014) An international assessment of the environmental effects of marine energy development. Ocean Coast Manag 99:1–11

Caplan K (2006) Creating space for innovation: understanding enablers for multi-sector partnerships. Partnership matters. Curr Issues Cross Sect Collab 4:11–14

Devrim D, Grau M (2010) El (in)hospitalario mar Negro: la imaginación occidental y la estrategia multilateral en una región en disputa. Quaderns de la Mediterrània 13:244–251

Dizon-Reyes MGN (2012–13) Public-private partnership towards growth & development: is it working? Philipp Law J 87:799–819

Kerr S, Colton J, Wright G (2015) Rights and ownership in sea country: implications of marine renewable energy for indigenous and local communities. Mar Policy 52:108–115

O'Hagan AM (2012) Marine Spatial Planning (MSP) in the European Union and its application to marine renewable energy (WWW Document). Int Energy Agency Ocean Energy Syst. https://www.ocean-energy-systems.org/library/in-depth-articles/document/marine-spatial-planning-in-the-eu-and-its-application-to-marine-renewable-energy/

Oral N (2013) PSSA for the Black Sea. Univ Hawai'i Law Rev 35:787–804

Postiglione A (2007) The Mediterranean and black sea ecosystem under discussion. Env Policy Law 37:489–500

Soininen N (2012) Planning the marine area spatially – a reconciliation of competing interests? Int Law Mak Dipl Rev 12:85–118

Tiganescu AM (2013) Legal aspects of the contract of public-private partnership. Contemp Read Law Soc Just 5:519–526

Triantaphyllou D (2009) The 'security paradoxes' of the Black Sea region. Southeast Eur Black Sea Stud 9:225–241

Vinnyk OM (2013) The public-private partnership agreements: problems of legal regulation. Law Innov Soc 1:17–36

Wright G (2014) Strengthening the role of science in marine governance through environmental impact assessment: a case study of the marine renewable energy industry. Ocean Coast Manag 99:23–30

Wright G et al (2016) Establishing a legal research agenda for ocean energy. Mar Policy 63:126–134

Part II
The National and International Response to Maritime Crimes

Exploring the Ambiguity of Operation Sophia Between Military and Search and Rescue Activities

Giorgia Bevilacqua

1 Introduction

On 17 February 2016, WikiLeaks released the first six monthly Report concerning the performed and planned phases of Eunavfor Med Operation Sophia (Six Monthly Report).[1] This is an ongoing military mission that was recently undertaken by the European Union (EU) to disrupt the business model of migrant smuggling and human trafficking networks in the Mediterranean (hereinafter Operation Sophia).[2]

One of the main challenging elements within the Six Monthly Report is the planned but still pending transition from the phase involving enforcement actions on the high seas to the subsequent phase involving the exercise of the same enforcement actions also in the territorial and internal waters of Libya.[3] The rationale of the entry of naval forces up to the Libyan coastline is to intercept alleged criminals before they depart.[4] And, in effect, the vast majority of

[1]See Eunavfor Med Op Sophia (2016)—Six Monthly Report 22 June–31 December 2015. Available via WikiLeaks https://wikileaks.org/eu-military-refugees/EEAS/EEAS-2016-126.pdf. Accessed 31 May 2016.

[2]The Operation was initially titled 'EUNAVFOR MED' and was subsequently renamed 'EUNAVFOR MED operation SOPHIA' after a baby girl who was given birth to by a woman of Somali origin on a European vessel in the summer months of 2015 after being rescued. See Council Decision (CFSP) 2015/778 of 18 May 2015 on a European Union military operation in the Southern Central Mediterranean, Official Journal (2015) L 122/31 and Council Decision (cfsp) 2015/1926 of 26 October 2015 amending Decision (CFSP) 2015/778 on a European Union military operation in the Southern Central Mediterranean Official Journal (2015) L 281/13.

[3]See the Six Monthly Report, pp. 3 and 18.

[4]On the need to have forces close to the Libyan shore, see Lehmann (2015).

G. Bevilacqua (✉)
Institute for Research on Innovation and Services for Development, Italian National Research Council, Naples, Italy
e-mail: g.bevilacqua@iriss.cnr.it

G. Andreone (ed.), *The Future of the Law of the Sea*,
DOI 10.1007/978-3-319-51274-7_9

undocumented migrants, refugees, and asylum seekers (hereinafter generically referred to as irregular migrants),[5] who attempt to cross the Mediterranean sea, depart from Libya,[6] where the volatile government situation and the consequent inability to control the territory are contributing to facilitate the development of the two distinct criminal phenomena of migrant smuggling[7] and human trafficking.[8] These two different phenomena in the hands of transnational criminal networks share the common feature of exploiting the migratory movements for personal gains and disregard for human life. In practice, what smugglers and traffickers do is to offer a transportation service to hundreds of thousands of persons who crossed many borders already, mainly in African and Middle Eastern countries, and take advantage of the fact that the only concern of these persons is to flee from their countries of origin by whatever means possible, accepting the risk of sinking, and, indeed, many do sink.[9]

In light of this complex and tragic scenario, this chapter aims to explore the ambiguity of Operation Sophia, focusing on two sensitive and interrelated aspects: the use of coercive powers against suspected smugglers and traffickers and the rescue of migrants carried out by naval forces at sea. To this end, we will first introduce the operational and legal background in which the EU operates within the

[5]According to the United Nations High Commissioner for Refugees (UNHCR), in 2015 over 80 percent of the irregular migrants came from the world's top 10 refugee-producing countries, including Syria, Afghanistan, Eritrea, and Iraq. Available via UNHCR http://data.unhcr.org/mediterranean/regional.php. Accessed 31 May 2016. On the different definitions of 'irregular immigrants', see Trevisanut (2012), pp. 1–22.

[6]The Six Monthly Report, p. 6.

[7]Pursuant to Art. 3 of the Protocol against the Smuggling of Migrants by Land, Sea and Air, the Smuggling of migrants is: 'the procurement, in order to obtain, directly or indirectly, a financial or other material benefit, of the illegal entry of a person into a State Party of which the person is not a national or a permanent resident'. See the Protocol against the Smuggling of Migrants, supplementing the United Nations Convention against Transnational Organized Crime, New York, 15 November 2000, in United Nations Treaty Series, Vol. 2241, Doc. A/55/383, p. 507.

[8]Pursuant to Art. 3 of the Protocol to Prevent, Suppress and Punish Trafficking in Persons, especially Women and Children, the Trafficking in persons is: 'the recruitment, transportation, transfer, harboring or receipt of persons, by means of the threat or use of force or other forms of coercion, of abduction, of fraud, of deception, of the abuse of power or of a position of vulnerability or of the giving or receiving of payments or benefits to achieve the consent of a person having control over another person, for the purpose of exploitation. Exploitation shall include, at a minimum, the exploitation of the prostitution of others or other forms of sexual exploitation, forced labour or services, slavery or practices similar to slavery, servitude or the removal of organs'. See the Protocol to Prevent, Suppress and Punish Trafficking in Persons, especially Women and Children, supplementing the United Nations Convention against Transnational organized Crime, New York, 15 November 2000, in United Nations Treaty Series, Vol. 2237, Doc. A/55/383, p. 319.

[9]By way of example, in October 2013 an overcrowded boat carrying asylum seekers from Eritrea, Somalia and Ghana capsized within sight of Italy's shores. Despite the vessel's stated capacity of 35 passengers, it carried around 500 souls on board that night. For 360 of them, dreams of a better life away from poverty and war died in the depths of the Mediterranean, see BBC: http://www.bbc.com/news/world-europe-24440908. Accessed 31 May 2016.

area of immigration at sea, assessing, in particular, the main relevant characteristics both of the civilian (Sect. 2.1) and military operations (Sect. 2.2) recently adopted in the Central Mediterranean Route. As the Operation Sophia's mandate stipulates that the activation of its crucial phase in the territorial waters of Libya requires an authorization of the United Nations Security Council (Security Council) or the consent of Libya, we will then assess whether and, eventually, under which conditions military assets are allowed to exercise coercive powers against vessels suspected of being used for migrant smuggling and human trafficking.[10] The analysis will consider the different legal regimes that may apply in the different jurisdictional marine areas and, specifically, on the high seas (Sect. 3.1) and in the territorial sea (Sect. 3.2). Furthermore, since naval forces may be and frequently are called upon saving human lives at sea,[11] we will verify the content and the legal framework of the positive obligation to render assistance to people in distress at sea (Sect. 4), as well as the complex risks raised by search and rescue (SAR) interventions and disembarkation procedures, especially when these activities are undertaken in cooperation with third countries (Sect. 5). The conclusions contain some remarks on the EU military mission and, more in general, on the role played by the EU and its Member States in the Mediterranean Sea in order to manage the phenomenon of irregular immigration by sea (Sect. 6).

2 The Engagement of the EU in SAR Activities: A Missed Opportunity

For the purpose of managing the Mediterranean migration crisis, a wide number of different routes have been undertaken at national and European levels. While in the past immigration control programs were implemented unilaterally and exclusively by the most affected coastal States, in the last decade an increasing role has been played by Europe. The following illustrates the relevant background of Frontex Joint Operations (Sect. 2.1) and of the new EU Operation Sophia (Sect. 2.2).

[10]The mandate of Operation Sophia refers to 'human smuggling or trafficking', whereas the established terminology in international law for these two criminal phenomena is 'smuggling of migrants' and 'trafficking of persons'. See Art. 3 of the Protocol against the Smuggling of Migrants and Art. 3 of the Protocol against the Trafficking in Persons, cit.

[11]On recent salvage operations carried out by military forces, see EEAS: http://www.eeas.europa.eu/csdp/missions-and-operations/eunavfor-med/news/index_en.htm. Accessed 31 May 2016.

2.1 Frontex Joint Operations

Since 2005, the EU has been dealing with migration by sea through the European Agency for the Management of Operational Cooperation at the External Borders of Member States of the European Union (Frontex).[12] One of its most important competences is to plan, coordinate, implement, and evaluate Member States' border control and surveillance activities through the so-called Joint Operations at the EU external borders (sea, land, and air). In the past years, as a result of these operations, hundreds of irregular migrants, while attempting to cross the European external maritime border, rather than being saved, have been forced to return to the State from which they departed or were presumed to have departed.[13]

The main focus of Frontex Joint Operations remained essentially the same over the years, but some steps forward toward an increasing engagement in SAR activities have been slowly undertaken. Rather important in this context is the adoption of EU Regulation No. 656/2014 (Sea Border Regulation),[14] which is the result of a laborious series of negotiations and institutional issues.[15] It replaced Council Decision 2010/252/EU,[16] which was entirely annulled by the European Court of Justice in September 2012. The provisions laid down in the challenged Decision contained essential elements of the surveillance of the sea external borders of the Member States and constituted "a major development" in the Schengen Borders Code system.[17] Moreover, according to the Court of Luxembourg, these

[12]Frontex was established by Council Regulation (EC) 2007/2004 establishing a European Agency for the Management of Operational Cooperation at the External Borders of the Member States of the European Union (EU). This Regulation was later amended by the Regulation (EC) No 863/2007 of 11 July 2007 establishing a mechanism for the creation of Rapid Border Intervention Teams and amending Council Regulation (EC) No 2007/2004 as regards that mechanism and regulating the tasks and powers of guest officers. It was then amended by the Regulation (EU) No 1168/2011 of 25 October 2011 amending Council Regulation (EC) No 2007/2004 establishing a European Agency for the Management of Operational Cooperation at the External Borders of the Member States of the EU.

[13]On the practice of 'push-backs' in the Mediterranean, see the European Court of Human Rights, *Hirsi Jamaa and others v. Italy*, Judgment of 23 February 2012; See Borelli and Stanford (2014), pp. 29–69.

[14]See Regulation (EU) No. 656/2014 of the European Parliament and of the council of 15 May 2014 establishing rules for the surveillance of the external sea borders in the context of operational cooperation coordinated by the European Agency for the Management of Operational Cooperation at the External Borders of the Member States of the European Union, in Official Journal of the European Union L 189/93.

[15]For further considerations on the negotiations and institutional conflicts within the EU, see den Heijer (2016), pp. 53–71.

[16]See Council (2010), Decision 2010/252/EU of 26 April 2010 supplementing the Schengen Borders Code as regards the surveillance of the sea external borders in the context of operational cooperation coordinated by Frontex at the external borders of the Member States of the EU, *OJ* L 111/20, 04.05.2010.

[17]The Schengen Borders Code applies to any person crossing the external borders of all EU countries, except those of the United Kingdom and Ireland, and the internal borders of the

provisions entailed political choices and questions over fundamental rights. For these reasons, their adoption must fall "within the responsibilities of the European Union legislature" and require the ordinary legislative procedure and not the comitology procedure.[18] During the subsequent legislative process that finally brought to the adoption of the Sea Border Regulation, the Member States have accepted the EU's involvement in SAR activities, but only as far as it remains limited to Frontex Joint Operations.[19]

The first Joint Operation to which the Sea Border Regulation was applied is the Operation Triton. It was actually launched in October 2014 in order to solve the struggle between Italy and the EU over a follow-up to the Italian SAR operation *Mare Nostrum*. And, even though the operational area and main activities of Triton were initially very limited in scope, in the aftermath of two grave shipwrecks in April 2015, Frontex adopted a new operational plan.[20] The latter expanded its mission with an increased budget, additional assets and an extended operational area from 30 up to 138 nautical miles south of Lampedusa, almost reaching the extent that had been covered earlier by *Mare Nostrum*.[21] The expansion of this Frontex' operation could be seen as an implicit admission of guilt by the EU for its deadly policy of retreat and for its passive role toward the frequent drownings. The Mediterranean Sea has been defined as a firm and fatal dividing border between "North" and "South"; according to the International Organization for Migration, since the year 2000, close to 25,000 migrants have perished in the Mediterranean, making it the world's deadliest border.[22] In practice, however, Triton remained first and foremost a border control operation, whereas rescue activities continue to be incidental and a secondary task of this primary mission.

Schengen Area, a border-free area comprising 22 EU countries, along with Iceland, Liechtenstein, Norway and Switzerland.

[18]See Court of Justice of the European Union (Grand Chamber) (2012), Judgment of 5 September 2012, Case C-355/10, *Parliament v Council,* par. 65–85. For a comment on the case, see Andreone (2014).

[19]See Carrera and den Hertog (2016), pp. 1–20.

[20]For practicalities on Frontex support to SAR activities, see http://ec.europa.eu/dgs/home-affairs/what-we-do/policies/european-agenda-migration/background-information/docs/frontex_triton_factsheet_en.pdf.

[21]On the expiration of Frontex Joint Operation Triton, see Frontex: http://frontex.europa.eu/news/frontex-expands-its-joint-operation-triton-udpbHP. Accessed 31 May 2016.

[22]See International Organization for Migration (IOM), Migration Trends Across the Mediterranean: Connecting the Dots, June 2015, p. 1. Available via IOM file:///C:/Users/win/Downloads/Altai_Migration_trends_accross_the_Mediterranean.pdf. Accessed 31 May 2016.

2.2 The Novel EU Engagement in Operation Sophia

The adoption of the second edition of the Triton operation was accompanied by the deployment of an EU novel undertaking at sea. Having regard to Art. 42 of the EU Treaty and to the proposal from the High Representative of the Union for Foreign Affairs and Security Policy, on 18 May 2015 the Council adopted Decision No. 2015/778 (Council Decision) approving the Crisis Management Concept for a Common Security and Defense Policy operation.[23] This is part of the EU's comprehensive approach to migration and, as anticipated in the introduction, constitutes the military response designated to tackle the business model of migrant smuggling and human trafficking in the Southern Central Mediterranean.

The Council Decision is a non-legislative act adopted on an intergovernmental basis by the Council and represents the legal instrument that sets out the purpose of the mission, its mandate, and other practical information on how Operation Sophia shall be handled.[24] With specific respect to its mandate, the mission aims to identify, capture, and dispose vessels and assets used or suspected of being used by smugglers and traffickers. The mission is intended to be conducted in three sequential phases, and the Political and Security Committee has the power to decide on the transition between the different phases, subject to the assessment of the Council.[25] Whereas the first phase, which was completed from 22 June to 7 October 2015, was focused on the surveillance and assessment of existing smuggling and trafficking networks,[26] the two subsequent phases involve the exercise of real enforcement actions against the boats carrying irregular migrants. More specifically, pursuant to Art. 2 of the Council Decision, Phase 2 consists of two distinct subphases: (1) a phase that includes "boarding, search, seizure and diversion" of suspected boats on the high seas (Phase 2—High Seas)[27] and (2) a phase that includes "boarding, search, seizure and diversion" of suspected boats in the territorial and internal waters of Libya (Phase 2—Libyan Territorial Sea).[28] Finally, the third phase, would enable Eunavfor Med forces to "take all necessary measures" against suspected vessels, "including through disposing of them or rendering them inoperable," in the territory of the coastal State concerned, i.e., in Libyan territorial and internal waters, in its ports, and in its coastal areas.[29]

[23] See Council Decision (CFSP) 2015/778 of 18 May 2015 on a European Union military Operation in the Southern Central Mediterranean, cit.

[24] For details on the legal framework governing the adoption of Operation Sophia, see Butler and Ratcovich (2016), pp. 235–259.

[25] See Council Decision (CFSP) 2015/972 of 22 June 2015 launching the European Union military operation in the southern Central Mediterranean, Official Journal (2015) L 157/51.

[26] See Art. 2.2(a) of the Council Decision.

[27] See Art. 2.2(b)(i) of the Council Decision.

[28] See Art. 2.2(b)(ii) of the Council Decision.

[29] See Art. 2.2(c) of the Council Decision.

It is noteworthy that the Operation has obtained an extensive support from EU Member States. Since the EU cannot rely on any military personnel or weaponry on its own, the Operation is reliant upon 22 contributing Member States.[30] The Operation's Headquarter has been established in Rome, and the Rear Admiral of the Italian Navy has been appointed as the Operation Commander.

Within the field of immigration at sea, whereas the EU remains focused on the protection of the security of internal and external borders, the target of EU operations has—at least formally—changed. Indeed, whereas previous Frontex push-back operations were addressed to irregular migrants as such, the new EU naval operation in the Mediterranean Sea[31] is addressed to migrant smugglers and human traffickers. Meanwhile, however, the full engagement of the EU in a real salvage mission remains a missed opportunity.

3 Enforcement Jurisdiction Against Vessels Suspected of Migrant Smuggling and Human Trafficking and the Difficult Transition to Phase 2 into the Libyan Territorial Sea

Following the political guidance provided by the defense and foreign affair ministers at their informal meetings in September 2015, the EU Council established that the conditions for Phase 2 of the Operation Sophia were met but only insofar as actions in international waters are concerned.[32] In what follows, we will take a closer look at the legal framework applicable to Phase 2—High Seas and Phase 2—Libyan Territorial Sea of the Operation. For these subphases, the Council Decision stipulates different legal conditions that require distinct considerations and assessments. Notably, when conducting "boarding, search, seizure and diversion" of suspected vessels on the high seas, naval forces shall act in accordance with the conditions provided for by applicable international law, including the 1982 United

[30]The Member States participating to the mission as contributing states are: Belgium, Bulgaria, Cyprus, Czech Republic, Spain, Estonia, Finland, France, Germany, Great Britain, Greece, Hungary, Italy, Latvia, Lithuania, Luxemburg, Malta, the Netherlands, Poland, Rumania, Slovenia Sweden.

[31]Distant from European shores, the first EU naval mission 'EUNAVFOR' was launched in the Gulf of Aden off the Eastern coast of Africa in 2008 in order to combat piracy and armed robbery at sea. See the EU Council, Joint Action 2008/851/CFSP of 10 November 2008 on a European Union military operation to contribute to the deterrence, prevention and repression of acts of piracy and armed robbery off the Somali coast, 10 Nov. 2008, Official Journal (2008) L301/33. For doctrine, see Geiss and Petrig (2011), p. 18.

[32]On 28 September 2015, the Political and Security Committee adopted Decision (CFSP) 2015/1772 concerning the transition by EUNAVFOR MED operation SOPHIA to the second phase of the operation, as provided for in point (b)(i) of Article 2(2) of Decision (CFSP) 2015/778 which also approved adapted Rules of Engagement for that phase of the operation, OJ L 258, 3.10.2015.

Nations Convention on the Law of the Sea (UNCLOS)[33] and the Protocol against the Smuggling of Migrants.[34] Differently, when operating in territorial or internal waters, foreign naval forces shall act in accordance with any applicable Security Council Resolution or consent by the coastal State concerned, *i.e.*, Libya.

With the aim of assessing whether and, eventually, under which conditions military assets are allowed to exercise enforcement jurisdiction in the different marine zones against vessels suspected of being engaged in migrant smuggling and human trafficking, the following section identifies and analyzes the legal framework that is applicable both on the high seas (Sect. 3.1) and in Libyan territorial and internal waters (Sect. 3.2).

3.1 On the High Seas

When sailing at sea, ships are subject to the jurisdiction of their flag State, whose exercise differs according to the maritime zone in which the ship is sailing.[35] As a ship sails away from a State's coastline, the extent of jurisdiction shifts in favor of the flag State, until it becomes exclusive on the high seas. Conversely, as the ship approaches a State's coastline, the balance shifts in favor of the coastal State. In this regard, the basic legal framework is set by the UNCLOS. Its provisions stipulate the current division in maritime zones and codify States' jurisdiction, including the jurisdiction to enforce.[36] In particular, as far as the high seas is concerned, the UNCLOS codifies the following relevant customary principles of international law of the sea: the principle of the freedom of navigation and the principle of the exclusive jurisdiction of the flag State. On the high seas, by virtue of these principles, ships are free to navigate[37] and are subject to the exclusive jurisdiction of the flag State,[38] and no claims of sovereignty can be validly put forward by any State.[39]

In exceptional circumstances, however, pursuant to Art. 110 UNCLOS, "a warship which encounters on the high seas a foreign ship" may exercise the boarding of a suspected foreign private ship.[40] Excluding the relevance of the

[33]The United Nations Convention on the Law of the Sea (UNCLOS) was signed on 10 December 1982, in Montego Bay, entered into force on 16 November 1994 and was ratified by 165 States as of 19 July 2013, in United Nations Treaty Series, Vol. 1833, p. 3.

[34]For references on the Protocol against the Smuggling of Migrants, cit.

[35]See Ringbom (2015), pp. 1–454.

[36]On the provisions of international law of the sea applicable to immigration in the different marine jurisdictional zones, see Scovazzi (2014), p. 216.

[37]See Art. 87 of the UNCLOS (*Freedom of the high seas*).

[38]See Art. 92 of the UNCLOS (*Status of ships*).

[39]See Art. 89 of the UNCLOS (*Invalidity of claims of sovereignty over the high seas*).

[40]See Art. 110 of the UNCLOS (*Right of visit*).

case in which the ship is engaged in piracy,[41] the first case of the list laid down in Art. 110 UNCLOS that deserves attention is the case of "slave trade."[42] In theory, since trafficking in persons is often referred to as a modern form of slavery,[43] this norm may be used as a specific legal basis for exercising the right to visit against vessels suspected of this crime. In practice, however, this legal basis may appear quite weak, at least because the exception concerning ships engaged in slavery trade has never been used for previous cases of human trafficking yet. Moreover, when operating at sea, it might be hard to identify and distinguish quickly the cases of migrant smuggling from the case of human trafficking. Accordingly, it would be safer to use a stronger legal basis applicable to both criminal phenomena.

In effect, a stronger legal basis may be found in Art. 110(d) UNCLOS, which considers the case in which there is reasonable grounds for suspecting "that a ship is without nationality."[44] Since the boats used for irregular migration in the Mediterranean Sea are very often non-registered small vessels without any flag, the exception concerning the absence of nationality may represent a very useful instrument for exercising the boarding of the suspected vessels. Furthermore, we believe that government vessels may also be entitled to act beyond the right of visit. Article 110 UNCLOS does not explicitly allow the exercise of additional coercive powers, but at the same time it does not prohibit them.[45]

The approach taken by the UNCLOS, which is mainly based on the freedom of navigation on the high seas and on the exclusive jurisdiction of the flag State, is not substantially changed by the more recent Protocol against the Smuggling of Migrants,[46] which is also recalled in the Council Decision.[47] Specifically, under Art. 8(7) of this Protocol, on the high seas the boarding of the suspected ships can take place only after having received authorization by the flag State, unless the ship suspected of migrant smuggling is without nationality. In addition, Art. 8(7) stipulates that "[i]f evidence confirming the suspicion is found, [the] State Party shall take appropriate measures." Such "appropriate measures" may be interpreted as those necessary to exercise further enforcement powers in order to act beyond the boarding of the suspected vessel,[48] provided that such powers are exercised "in accordance with relevant domestic and international law."[49]

---

[41]See Art. 110(a) UNCLOS.

[42]See Art. 110(b) UNCLOS.

[43]See Siller (2016), pp. 405–427; Scarpa (2008).

[44]See Art. 110(d) UNCLOS.

[45]See Papanicolopulu (2016), pp. 2–22.

[46]See Scovazzi (2014), cit.

[47]The Protocol against the Smuggling of Migrants, cit.

[48]See Papanicolopulu (2016) cit.

[49]For a thorough analysis on the criteria for a lawful exercise of the use of force at sea in accordance with human rights law, see International Tribunal for the Law of the Sea, *The M/V "Saiga" (No. 2) Case (Saint Vincent and the Grenadines v. Guinea)*, Judgment of 1 July 1999D. For doctrine, see Guilfoyle (2009), p. 268.

Differently, the 2000 Protocol to Prevent, Suppress and Punish Trafficking in Persons, especially Women and Children, supplementing the United Nations Convention against Transnational Organized Crime,[50] is silent on enforcement measures that can be exercised at sea.

The analysis of the above-recalled international legal framework clarifies that naval forces that encounter on the high seas a suspected private foreign ship are justified in boarding it if the flag State of the ship suspected of migrant smuggling or human trafficking has given its express authorization, if there is reasonable grounds for suspecting that the ship is without nationality, or when the suspect is of trafficking in human beings, that is, engaged in slave trade.[51]

3.2 *In Territorial Waters*

The launch of the Operation in Libyan Territorial Sea would be extremely important as criminals would be intercepted before they depart and, in turn, before they jeopardize the lives of hundreds of persons. However, while the transition to Phase 2—High Seas was not problematic, conversely the transition to the subsequent Phase 2—Libyan Territorial Sea is not immune from challenges as the Council Decision requires either a Security Council mandate or the consent of Libya.

3.2.1 The UN Security Council Resolution

With respect to the first alternative condition required by the Operation Sophia's mandate, according to the powers granted by Articles 39 and 42 of the United Nations Charter (UN Charter), the Security Council may authorize the use of force, whenever it determines "the existence of any threat to the peace, breach of the peace, or aggression."[52] The notion of obtaining the support of the United Nations, in the absence of authorization by the State concerned, is in line with a consolidated State practice of past decades, according to which UN Member States attempted to

[50] The Protocol to Prevent, Suppress and Punish Trafficking in Persons, especially Women and Children, cit.

[51] Even if the analysis in the text is restricted to the international legal framework, it seems worth noting that with respect to enforcement jurisdiction on the high seas some issues may be raised by the domestic legal systems which may restrict the jurisdiction against migrant smugglers to the territorial sea. For doctrine, see Andreone (2011), pp. 183–188. With respect to the Italian system, more recently, the Supreme Court has affirmed that against ships without nationality encountered on the high seas, coercive powers can be exercised on the basis of a valid reason, such as art. 8 of the Protocol against Smuggling of Migrants. See Corte di Cassazione, judgment of 23 May 2014, No. 36052.

[52] See Articles 39 and 42 of the UN Charter. For doctrine on the powers of the UN Security Council, see Conforti, Focarelli (2015).

legitimize unilateral interventions through the label of Security Council resolutions adopted under Chapter VII of the UN Charter.[53]

On 11 May 2015, the High Representative of the Union for Foreign Affairs and Security Policy has officially informed the Security Council of the need for the EU to work with its support in order to manage the Mediterranean migration crisis.[54] The UN bodies, however, have often stressed the importance of focusing the European action on saving lives when dealing with migration rather than on military actions.[55] The negotiations between the EU and the UN ended with a strange compromise: with 14 votes in favor and one abstention,[56] acting under Chapter VII of the UN Charter, the Security Council has adopted a Resolution to maintain international peace and security, condemning, in particular, "all acts of migrant smuggling and human trafficking into, through and from the Libyan territory and off the coast of Libya, which undermine further the process of stabilization of Libya and endanger the lives of thousands of people."[57]

In the view of the EU institutions, the Resolution "reinforces the authority to take measures against the smuggling of migrants and human trafficking from the territory of Libya and off its coast."[58] Conversely, we would minimize the relevance of the Resolution since it is hard to identify the legal rationale behind its adoption. As requested by Russia at the UN Security Council's preparatory meetings,[59] the mandate of the Resolution covers only the high seas off the coasts of Libya, rather than the Libyan territorial sea.[60] As seen above (Sect. 3.1), in this marine zone, naval forces can already act against stateless ships in accordance with the applicable provisions both of the UNCLOS and the Protocol against the Smuggling of Migrants.[61] Moreover, even though the Resolution may, in theory,

[53]On the unilateral use of force, see Picone (2015), pp. 3–32.

[54]See the Council Decision, Rec. 4.

[55]See, for instance, the speech of the UN Secretary-General, Ban Ki-Moon, at the European Parliament in plenary session on 27 May 2015. Available via European Parliament News http://www.europarl.europa.eu/news/en/news-room/20150526STO59634/Ban-Ki-moon-on-migration-%E2%80%9CSaving-lives-should-be-the-top-priority%E2%80%9D. Accessed 31 May 2016.

[56]The abstention was of Venezuela.

[57]See Resolution 2240 (2015), adopted by the Security Council at its 7531st meeting, on 9 October 2015, Doc. S/RES/2240 (2015), para. 1.

[58]On the perspective of the EU, see Political and Security Committee Decision (CFSP) 2016/118 of 20 January 2016, concerning the implementation by Eunavfor Med Operation Sophia of United Nations Security Council Resolution 2240 (2015) (Eunavfor Med Operation SOPHIA/1/2016, Rec. 3; A similar approach on the Resolution 2240 (2015) is used by the Operation Commander in the Six Monthly Report, p. 10, cit.

[59]On the express request of Russia, see the Security Council meeting records adopted at the 7531st meeting on 9 October 2015, Doc. S/PV.7531 and the Press Release of the Security Council Resolution 2240 (2015), including Statements after action. Available via United Nations Meetings Coverage and Press Releases: http://www.un.org/press/en/2015/sc12072.doc.htm. Accessed 31 May 2016.

[60]See Resolution 2240 (2015), cit., para. 7.

[61]See the Protocol against the Smuggling of Migrants, cit.

extend the possibility to intervene to suspected vessels that fly the flag of a third State, in practice, according to the current *modus operandi* of suspected criminals who act in the Mediterranean sea, the vast majority of the vessels used by them are stateless vessels.[62] Furthermore, the Resolution does not appear to be a blanket mandate authorizing the use of force as resolutions adopted under Chapter VII usually are. Indeed, if there is confirmation that the vessels are being used for migrant smuggling or human trafficking, "all measures commensurate to the specific circumstances" can be used. This is not an authorization to use "all necessary measures" in confronting migrant smugglers and human traffickers, which was the wording originally adopted in the initial draft of the Resolution circulated by the United Kingdom. The unusual expression ultimately adopted in the final version of Resolution 2240 (2015) is the result of the amendments wanted by some Security Council Member States concerned that the Resolution could mean, as it namely means in the language generally used by the Security council, a blanket mandate to the use of force.[63]

Allegedly, notwithstanding the adoption of the Security Council Resolution, the essential parts of Operation Sophia's mandate risks to remain unaccomplished, unless Libya decides to authorize the international fight against migrant smugglers and human traffickers.

3.2.2 The Consent of Libya

The second alternative route identified by the Council Decision consists of the consent of Libya. In this respect, the legal analysis revolves around the principle of territorial sovereignty. The boarding, search, seizure, and diversion activities envisioned by the Operation's mandate are, indeed, enforcement measures that may involve the use of coercive powers.[64] The exercise of such powers by foreign authorities may interfere with the fundamental principle of territorial sovereignty. This principle of general international law is codified also by the UNCLOS. The latter specifically provides that States have full sovereignty within their territorial waters, which may extend up to 12 nautical miles from the baselines.[65] In this marine area, the coastal State enjoys the exclusive right to exercise coercive powers just like on the territory of the mainland. Accordingly, the principle of territorial

[62]On the question of jurisdiction over suspected vessels, see Papastavridis (2016b).

[63]See Vote on a Resolution on Human Trafficking and Migrant Smuggling in the Mediterranean. In What's in Blue. Insights on the Work of the UN Security Council, 8 October 2015. Available via What's in Blue http://www.whatsinblue.org/2015/10/vote-on-a-resolution-on-human-trafficking-and-migrant-smuggling-in-the-mediterranean.php. Accessed 31 May 2016.

[64]See Art. 2 of the Council Decision.

[65]See Articles 2–4 of the UNCLOS.

sovereignty does not allow for other States to participate in this exercise, unless expressly authorized.[66]

The EU has been seeking to obtain the consent to enter Libyan territorial sea, but up to now Libyan authorities did not appear as willing to provide such authorization. The international relationships are clearly complicated by the fact that since 2014, Libya has been facing a grave political crisis and, as a consequence of such crisis, Libyan authorities lack the capacity to effectively control their territory, as well as their territorial sea.

The current political scenario may probably change in light of the recent UN-led formation of a government of national unity, which, according to the Security Council Resolution 2259 (2015), is the sole legitimate Government of Libya.[67] Accordingly, only this government would be considered by the International Community as the authority legitimate to authorize the entry into Libyan territorial waters for the purpose of disrupting the business model of migrant smuggling and human trafficking. This legal option may find support in a precedent case and, specifically, in the context of the international response to maritime piracy in Somalia. Over 20 years, Somalia has lacked any functioning institutions or any form of political control of the territory, to the extent of being defined by the literature as the *locus classicus* of a failed State.[68] In 2004, after nearly 2 years of insidious negotiations and numerous international attempts to deal with the problem of the crisis of sovereignty, a provisional transition federal government (TFG) was formed. The latter was a highly precarious body held together artificially by the UN and, at the same time, the sole Somali government recognized by the International Community. Likewise, in Libya, the political instability of Somalia contributed to the fervent and atrocious resurgence of a violent crime, i.e., maritime piracy. In order to address this phenomenon, the TFG (replaced in 2012 by the new Somali Authorities) explicitly asked for international assistance.[69] In light of this request and thus in light of the TFG's consent, a number of Security Council resolutions were adopted to ensure the implementation of the rules of international law concerning coercive powers against piracy on the high seas also in the territorial sea and even on the mainland of Somalia.

---

[66]For a thorough analysis on the relevance of the State's consent, see Wippman (1996), p. 209; with specific reference to the coastal State's consent in the context of piracy in Somalia, see Treves (2009), p. 406; Tancredi (2008), p. 937.

[67]See Resolution 2259 (2015), adopted by the Security Council at its 7598th meeting, on 23 December 2015, Doc. S/RES/2259 (2015), para. 3.

[68]For a reconstruction of the Somali crisis, for doctrine see Gordon (1997), pp. 903–974; Kreijen (2004), p. 65 ff.; for case-law see European Court of Human Rights, *Sufi and Elmi vs. United Kingdon*, Apps. No. 8319/07 and 11449/07, concerning the appeal by two Somali citizens at risk of inhumane treatment if returned to Mogadishu.

[69]According to paragraph 9 of Resolution 1816 (2008) 'the authorization set out in paragraph 7' has been provided only following receipt of the letter from the Permanent Representative of the Somalia Republic to the United Nations to the President of the Security Council dated 27 February 2008 conveying the consent of the TFG'. Similar formulations, referring to further letters conveying the consent of the TFG, are in Security Council Resolutions 1846 (2008) and 1851 (2008).

It is probably bearing in mind the Somali precedent that the Operational Commander states in the above recalled Six Monthly Report on Operation Sophia that "In order to move to phase 2 in Libyan territorial waters, we need firstly an invitation from the GNA [Government of National Accord], as the sole legitimate Government of Libya under UNSCR 2259 (2015), and secondly a UN Security Council Resolution to provide the necessary legal mandate to operate".[70]

4 Between the Need and the Obligation to Turn into a Rescue Scheme

At the current state of play, following Operation Sophia's entry into Phase 2—High Seas, criminals suspected of migrant smuggling and human trafficking appear to enjoy less freedom of maneuver. The geographical limits of the Operation forced them to act in the Libyan territorial sea where, however, they may act uncontrolled as national authorities are not capable to prevent and repress their criminal actions. And while alleged smugglers and traffickers are remaining within the national borders, smuggled and trafficked persons are left on unsafe boats with limited food, water, and, above all, insufficient fuel to reach anywhere further than 30–50 nautical miles from the Libyan coast.[71]

By consequence, Operation Sophia is often turned into a salvage mission. According to the Six Monthly Report, as of 31 December 2015, military forces have completed the rescue of more than 8336 persons, initiated both by detection of boats in distress by military assets or by request from the Rescue Coordination Centre in Rome.[72] Also in 2016, the Operation is often turned into a rescue mission that saved hundreds of migrants at sea while they were attempting the Central Mediterranean route to Europe.[73]

The relevance of the duty to render assistance is expressly recalled both in the Council Decision and Resolution 2240 (2015). Specifically, Recital (6) of the Council Decision[74] stipulates that the Operation shall be conducted in accordance with international law and, in particular, with the relevant provisions of the UNCLOS, the Safety of Life at Sea Convention (SOLAS Convention),[75] and the

[70]See the Six Monthly Report, section on Legal mandate—UNSCR and Libyan Invitation, p. 19.

[71]Ibid, Section on Tactics, Techniques and Procedures (TTP) Evolution, p. 7. For a description of the conducts of smugglers and traffickers, see also Cataldi (2015), pp. 1498–1502; in particular, the author examines the decision of the Italian Supreme Court, *Criminal proceedings against Radouan Hai Hammouda,* No. 3345, 23 January 2015.

[72]Ibid, Section on Phase II.A (High Seas) Activities, p. 11.

[73]On more recent news, see EU External Action: http://www.eeas.europa.eu/csdp/missions-and-operations/eunavfor-med/news/index_en.htm (5/16).

[74]See Council Decision, cit.

[75]The International Convention for the Safety of Life at Sea, adopted 1 November 1974, in force 25 May 1980; 1184 *UNTS* No. 1861.

Maritime SAR Convention,[76] which include the obligation to assist persons in distress at sea. More intensively, the Security Council affirms "the necessity to put an end to the recent proliferation of, and endangerment of lives by, the smuggling of migrants and trafficking of persons in the Mediterranean Sea off the coast of Libya"[77] and urges Member States and regional organizations to render "assistance to migrants and victims of human trafficking recovered at sea, in accordance with international law".[78]

In light of the above, even though Operation Sophia was created as a military mission, its naval assets, when encountering people boats in distress at sea, have a positive obligation to rescue them. Several international treaties include provisions relating to a duty to render assistance at sea. The prevalence of treaties has been ascribed in part to the quite old but still well-known sinking of the *Titanic*, which raised demand for international cooperation in safety matters. The most widely applicable treaty rule, however, is Art. 98 of the UNCLOS:

> Every State shall require the master of a ship flying its flag, in so far as he can do so without serious danger to the ship, the crew or the passengers:
> (a) to render assistance to *any person* found at sea in danger of being lost;
> (b) to *proceed with all possible speed to the rescue of persons in distress*, if informed of their need of assistance, in so far as such action may reasonably be expected of him;
> (c) after a collision, to render assistance to the other ship, its crew and its passengers and, where possible, to inform the other ship of the name of his own ship, its port of registry and the nearest port at which it will call (emphasis added).[79]

According to this provision, every flag State must require the master of a vessel, whether a State or private vessel flying the State's flag, to proceed with all possible speed to the rescue of persons in peril when informed of their need for assistance or whenever there is reasonable grounds for retaining that they are threatened by a grave and imminent danger. This obligation is applicable to all vessels, and assistance must be provided to any person regardless of the nationality or status of such a person or the circumstances in which that person is found. Although this article is located in the UNCLOS section on the high seas, the duty to rescue applies in all maritime zones.[80] It is closely connected with the principle of safety of life at sea, which is the only real limit to the freedom of navigation on the high seas. Accordingly, and because of its repetition in treaty and domestic law and in State

[76]The International Convention on Maritime Search and Rescue, adopted 27 April 1979, in force 22 June 1985; 1405 *UNTS* No. 23489.

[77]See Resolution 2240 (2015), cit. preamble.

[78]IbId, para. 3.

[79]See Art. 98 UNCLOS.

[80]Among others, on the content of the duty to rescue at sea, see Pallis (2002), pp. 329–364; Trevisanut (2010), p. 523.

practice, the duty to render assistance is generally recognized as a principle of customary law binding on all States.[81]

5 The Search for Clarity on the Place of Safety

Undertaking rescue operations does not exhaust the duty to render assistance. In this context, it is important to emphasize that this duty is only fully met if the rescued persons can disembark in a place of safety. In other words, following their rescue, survivors shall be conducted to a safe place. This complementary aspect seems implicit in the logic of any rescue attempt, which is to save lives. In practice, however, the prompt identification of a place of safety where irregular migrants rescued at sea can be disembarked is often quite problematic. The main challenge derives from the fact that while the obligation of States to rescue people at sea is, as seen above, a long-established rule of international customary law codified in a number of treaties, for disembarkation purposes a comparable binding provision does not exist in the law of the sea.

During the general revision of the International Maritime Organization's (IMO) SAR system, the IMO's Maritime Safety Committee (MSC) faced the problem of where rescued persons can and should be disembarked, without distinction based on their legal status, nationality, or place where they were found.[82] The MSC adopted two Resolutions that entered into force on 1 July 2006 and amended the abovementioned SOLAS and SAR Conventions. In particular, following the amendments to these Conventions, people rescued at sea must be promptly conducted to a "place of safety," which is defined as follows:

> [. . .] a place where the survivors' safety of life is no longer threatened and where their basic human needs (such as food, shelter and medical needs) can be met. Furthermore, it is a place from which transportation arrangements can be made for the survivors' next or final destination.[83]

According to the renewed SAR regime, making available a place of safety is the responsibility of the State in whose SAR zone the survivors are rescued.[84] This rule, however, does not mean that this State or the intervening State is automatically obliged to disembark recovered people on their own territories.[85] In theory, survivors can also be disembarked in a third country that is willing to receive them. In

---

[81]Ibid.

[82]See Tondini (2012), p. 59; Trevisanut (2010), cit.

[83]IMO-Maritime Safety Committee 'Guidelines on the treatment of persons rescued at sea' (MSC Guidelines) art 6.12 Resolution MSC.167(78) (20 May 2004) www.unhcr.org/refworld/docid/432acb464.html.

[84]Ibid Art. 2.5.

[85]On the obligation to render assistance at sea and on the responsibility for failing to save lives, see Papastavridis (2016a), pp. 31–47.

practice, if compared to "ordinary" situations of distress at sea, the fact that the people in question are mainly represented by a large amount of undocumented migrants, refugees, and/or asylum seekers does not facilitate the disembarkation procedures. The combination of "irregular migrants" on the one hand and "distress" on the other hand tends to generate questionable dynamics of burden avoidance and burden shifting between States.[86]

While the amendments to the SAR and SOLAS Conventions decided by the IMO were aimed to guarantee assistance to rescued persons and, at the same time, to minimize negative consequences for the rescue vessel, it is clear that in practice the disembarkation burdens rest primarily upon the warship's flag State, with the SAR coordinating State concurring. When the latter is unable or unwilling to find a proper place of disembarkation, it is the warship's flag State that in the end must find an appropriate solution to the stalemate.[87]

In the current context of the Mediterranean migration crisis, the international practice and discussions around the follow-up of rescue operations at sea show how the SAR regime is under pressure due to the fact that the State accepting disembarkation is bound to assume responsibility of asylum seekers and to manage the presence of other migrants, with an irregular status as far as entry and stay are concerned. As Operation Sophia's assets are incidentally engaged in rescue operations, the legal issues relating to the prompt identification of a place of safety to disembark survivors rescued at sea assume renewed relevance. The key legal question we wish to focus on hereinafter concerns the criteria that can and should be applied to establish disembarkation procedures in the case of SAR interventions carried out by Eunavfor Med naval forces on the high seas and, eventually, in the Libyan territorial sea.

5.1 On the High Seas

In the Council Decision, the above-recalled Recital (6) stipulates that survivors must be delivered to "[. . .] a place of safety, and to that end the vessels assigned to Eunavfor Med will be ready and equipped to perform the related duties under the coordination of the competent Rescue Coordination Centre."

The generic rule entailed by the Council Decision is in line with the generic wording used in the Six Monthly Report in the section on "Cooperation with EU organizations and Agencies."[88] In this section, the Operation Commander clarifies to have with Frontex "a general agreement and specific operational Procedures". Such cooperation with Frontex led to the adoption of the operational coordination structures that have been formalized by means of an exchange of letters on 14 July 2015

[86]For an analysis of the disputes of Mediterranean States with regard to SAR operations, see Trevisanut (2010) cit.

[87]See Tondini (2012), p. 62, cit.

[88]The Six Monthly Report, p. 14.

and the subsequent adoption of Standard Operating Procedures (SOPs) on 30 September 2015. The cooperation with this Agency shall cover, among other aspects, "the procedures for disembarkation in a place of safety." Furthermore, in the following section on "Cooperation with the Italian Authorities,"[89] the Operation Commander adds that "EUNAVFOR Med must comply with the EEAS (2015) 885 guidance to follow the TRITON Operation Plan for the disembarkation of persons rescued at sea." In addition, likewise generic is the official news reporting on Eunavfor Med rescue interventions. It refers to disembarkation either in Sicily ports or in an undefined place of safety.[90]

However, as far as we may understand from both the Council Decision and the Six Monthly Report—the EEAS (2015) 885 guidance is not publicly available—we presume that by virtue of this cooperation with Frontex and Italy, Eunavfor Med's operational plan and SOPs are in compliance with Triton's operational plan. The latter, in turn, must comply with the above-recalled Sea Border Regulation, which governs Frontex joint operations and includes specific modalities for the disembarkation of persons (intercepted or) rescued in a maritime operation.[91] On this basis, we deem that the Sea Border Regulation applies also to disembarkation procedures undertaken within Operation Sophia's interventions on the high seas.

For a number of reasons, our interpretation would be well desirable. First, this Regulation expressly contemplates the specific case in which border surveillance operations at sea turn into rescue interventions. For these cases, recalling the content of the duty to rescue, provided for by Art. 98 UNCLOS, the Regulation stipulates as follows:

> [...] every State must require the master of a vessel flying its flag, in so far as he can do so without serious danger to the vessel, the crew or the passengers, to render assistance without delay to any person found at sea in danger of being lost and to proceed with all possible speed to the rescue of persons in distress. Such assistance should be provided regardless of the nationality or status of the persons to be assisted or of the circumstances in which they are found. The shipmaster and crew should not face criminal penalties for the sole reason of having rescued persons in distress at sea and brought them to a place of safety.[92]

Second, when people are rescued, pursuant to Art. 10(c) of the Sea Border Regulation, the responsibility of the operational decision shall be shifted to the host State, which—both for Triton and Eunavfor Med operations—is Italy. This decision must be adopted in accordance both with the principle of *non-refoulement*[93] and the specific disembarkation modalities that would prevent

[89] The Six Monthly Report, p. 15.

[90] On more recent news, see EU External Action http://www.eeas.europa.eu/csdp/missions-and-operations/eunavfor-med/news/index_en.htm. Accessed 31 May 2016.

[91] See Regulation 656/2014, Art. 10.

[92] See Regulation 656/2014, Rec. 14.

[93] On the broad interpretation of the principle of non-refoulement, see the European Court of Human Rights, *Hirsi Jamaa and others v. Italy*, cit. For a comment, see Liguori (2012), p. 415. For doctrine on the implementation of the principle of non refoulement at sea, see Trevisanut (2008), pp. 205–246.

naval units from transferring people (intercepted or) rescued at sea in a place where, *inter alia*, there is a serious risk that they would be subjected to the death penalty, torture, persecution, or other inhuman or degrading treatment or punishment or where their life or freedom would be threatened on account of their race, religion, nationality, sexual orientation, membership of a particular social group, or political opinion.[94]

Third, pursuant to Frontex' Annual Report on the implementation of the Sea Border Regulation, with respect to 2014, since handover and disembarkation in third countries were not foreseen in Triton's operational plan, "[a]ll the migrants intercepted or rescued were disembarked in Italy."[95]

At the time of writing, the subsequent Frontex Annual Report concerning the implementation period of 2015 has not been published yet. However, from the cross-checked analysis of the puzzle of documents above recalled, we would conclude assuming that irregular migrants rescued on the high seas during SAR operations carried out within the Operation Sophia are disembarked in Italy, which may be considered a safe place.

5.2 *In the Libyan Territorial Sea*

Our interpretation is less optimistic if Phase 2—Libyan Territorial Sea begins. In this respect, while the Council Decision does not contain any specific information regarding how EU military forces shall act with irregular migrants saved by Sophia's units when the Operation patrols will begin to operate within Libyan territorial waters, from the Six Monthly Report emerges the idea that such forces shall interact and cooperate with the Libyan Navy and Coast Guard. In more detail, the Report clarifies that, if requested and if the operational mandate is amended, Eunavfor Med will provide Libyan forces operating at sea with capability and capacity building. Initially, this interaction in Libyan territorial waters would include Libyan "cooperation in tackling the irregular migration issue," with the expectation that at a later point in time "Libyan authorities could take the lead in patrolling and securing their Territorial Waters, with support being provided by EUNAVFOR Med."[96]

The illustrated scenario seems to be inspired by the bilateral agreements concluded by Spain with a number of States in Northwest Africa, including Morocco, Senegal, Mauritania, and the Cape Verde, which stipulate joint sea patrols of the Spanish Coast Guard and the border authorities of the partner States and which

[94]Ibid, Regulation 656/2014, Rec. 13.

[95]See Frontex' Annual Report on the implementation on the EU Regulation 656/2014 of the European Parliament and of the Council of 15 May 2014 establishing rules for the surveillance of the external sea borders. P. 7.

[96]See the Report, section Capacity and Capability Building, p. 20.

intend to prevent people boats from exiting the territorial sea of the latter States.[97] These agreements have been highly instrumental in closing the Atlantic migration route to the Canary Islands and continental Spain and have raised questions, among the other aspects, about the sharing of responsibilities between the various States involved in relation to the handover of intercepted migrants.[98]

Bearing in mind the issues raised by these arrangements, we fear that similar arrangements may be concluded with third States that may not be qualified as a safe place in accordance with the SOLAS and SAR Conventions and/or as a safe third country in accordance with the customary principle of nonrefoulement.[99] Our fear, in particular, concerns the possible cooperation with Libya. In this regard, Frontex's past experience with the Triton Operation proves that Libyan authorities are not ready to cooperate. According to the abovementioned Frontex Annual Report, the Italian Rescue Coordination Center received several distress calls from people boats in Libyan SAR zone and even though the Triton operational area was extended up to the SAR regions of Italy and Malta, every attempt to communicate with Libyan SAR authorities was always denied.[100]

This may be probably explained once again by the above-recalled political crisis that has existed in Libya since 2014. Due to this crisis, in many parts of the State, Libyan authorities lack the capacity to effectively control the territory. They are basically omitting to adopt measures capable to prevent and repress on their territory a wide range of serious violent threats such as the rising trend of terrorist groups in Libya proclaiming allegiance to the Islamic State in Iraq and the Levant (also known as Da'esh) and the continued presence of other Al-Qaida-linked terrorist groups and individuals operating in Libya.[101] Also, the area of Tripoli and its ports, from which most smugglers and traffickers depart, are subject to dangerous militias. Additional concerns regarding stability in Libya and in the Region derive from the uncontrolled proliferation of unsecured arms and ammunition.[102]

Our conclusion is accordingly that the legality of SAR activities in Libyan territorial waters will depend on how rescued migrants will be processed and

[97] The agreements themselves are confidential. See for an extensive analysis Garcia Andrade (2010), pp. 311–346.

[98] Ibid.

[99] This principle of non-refoulement appears in several central instruments of international law. See, among others, Art. 33(1) of the Convention Relating to the Status of Refugees (opened for signature 28 July 1951, entered into force 22 April 1954) and Art. 3(1) of the Convention against Torture and Other Cruel, Inhuman or Degrading Treatment or Punishment (opened for signature 4 February 1985, entered into force 26 June 1987) 1465 UNTS 85 (CAT).

[100] See Frontex Annual Report, p. 7.

[101] On the use of force against terrorism and other violent activities of non-state actors in acquiescent States, see Tancredi (2007), p. 969 ff.

[102] On these grounds, the Security Council has affirmed that 'the situation in Libya constitutes a threat to international peace and security'. See *Security Council Resolutions 2213 (2015) and 2238 (2015),* cit.

where they will be disembarked. EU Member States operating within Operation Sophia would necessarily be exercising effective control over migrants when operating unilaterally or jointly with Libyan forces within Libyan territorial waters, and EU Member States would therefore be bound by SAR regime and the non-*refoulement* obligations. Moreover, this time it is hard to imagine that the Sea Border Regulation will be further geographically extended. Its provisions, indeed, are remarkably silent on SAR interventions in the territorial sea of third States, even though in earlier policy documents the intention was to clarify the scope of Member State powers also in respect of operations in the territorial sea of third countries.[103]

6 Concluding Remarks

When Frontex was created, the core of its mandate was described rather broadly as rendering "more effective the application of existing and future Community measures relating to the management of external borders."[104] Over the years, in the wake of frequent embarrassing shipwrecks, the EU approach vis-à-vis the phenomenon of immigration by sea has been slowly evolving toward the respect of migrants' fundamental rights and the inclusion of SAR activities in Frontex join operations.

With Operation Sophia, the idea was to adopt a new strategy. In addition to Frontex joint operations—which nevertheless remain primarily focused on border management—the EU is now committed to a military mission, having the specific goal of disrupting migrant smuggling and human trafficking routes and capabilities. In effect, these transnational organizations play a crucial role in the current escalation of migratory movements toward Europe. But, even if illegally, the fact is that they represent the sole route available to escape hunger, civil wars, and other unimaginable situations in their countries of origin.

The extraordinary situation in the Mediterranean led also the Security Council to intervene and, specifically, to make use of the powers granted by Chapter VII of the UN Charter in order to authorize Member States and international organizations to use enforcement jurisdiction against irregular migration on the high seas.[105] With regard to this maritime zone, however, States are already provided with sufficient coercive powers by virtue of the applicable international legal framework, including Art. 110 UNCLOS and Art. 8 of the Protocol against the Smuggling of Migrants. Therefore, what is currently really needed to activate what is deemed to be the crucial phase of Operation Sophia is the consent of an effective Libyan government.

[103]See COM(2006), 733 final, par. 34. For doctrine, see M. den Heijer, "How the Frontex Sea Borders Regulation avoids the hot potatoes", cit.

[104]See Reg. 2007/2004, Art. 1(2).

[105]See Resolution 2240 (2015).

The recent UN-led formation of a government of national unity may slowly go in this direction. In fact, based on the readiness expressed by the President of the Presidency Council of the Libyan Government of National Accord to cooperate with the EU, the latter went ahead with the militarization of the waters off the coast of Libya.[106] Last June, the Operation Sophia's mandate was extended until 27 July 2017 with two additional assignments: the assistance in the development of the capacities and in the training of the Libyan Coast Guard and Navy in law enforcement tasks and the implementation of the UN arms embargo[107] on the high seas off the coast of Libya. Meanwhile, however, while both the EU and the UN remain focused on fighting criminal networks, it persists the primary humanitarian need of saving migrants' lives at sea. This old problem affected and affects many coastal States in the world. Already in the mid-1970s, for instance, many boat people fled from the communist regime in Vietnam. By the end of the 1980s, the number of people fleeing Vietnam was increasing and, in parallel, the willingness of host States in the region to offer protection and of third countries outside the region to offer resettlement was declining.[108] In the context of the Mediterranean migration crisis, the need to render assistance to migrants in danger at sea is unfortunately more evident by frequent mass drownings. This is an urgent need and also a positive obligation binding on all States.

Against this factual and legal backdrop, both Frontex and Eunavfor Med operations, even if initially and primarily focused on different objectives, have incidentally turned into rescue operations. This reaction is certainly appreciated under a humanitarian perspective. But, facing the increasing number of deaths at sea, we hardly understand the reasons why SAR interventions are still adopted incidentally, rather than officially and on a regular basis. The rescue of irregular migrants at sea, indeed, requires the adoption and the implementation of an appropriate legal regime. From the chapter arises the importance of relying on clear and transparent criteria to govern operational decisions concerning the disembarkation of rescued migrants in a safe place. The crossed interpretation of a number of EU legal and operational documents on Operation Sophia led us to argue that to SAR interventions on the high seas applies the Sea Border Regulation. Should our interpretation be correct, each operational decision to establish the place of disembarkation would

[106]See the Council Decision (CFSP) 2016/993 of 20 June 2016 amending Decision (CFSP) 2015/778 on a European Union military operation in the Southern Central Mediterranean, OJ L 162/18 of 21 June 2016.

[107]On 14 June 2016, the UN Security Council adopted the Resolution 2292 (2016) on the arms embargo on Libya, expressing in particular concern that the situation in Libya is exacerbated by the smuggling of illegal arms and related materiel. Pursuant to this Resolution, Member States are authorized 'to inspect [...] on the high seas off the coast of Libya, vessels bound to or from Libya which they have reasonable grounds to believe are carrying arms or related materiel to or from Libya [...]'. With previous resolutions, the Security Council has imposed, modified and reaffirmed the arms embargo in Libya. See resolutions 1970 (2011), 1973 (2011), 2009 (2011), 2040 (2012), 2095 (2013), 2144 (2014), 2174 (2014), 2213 (2015), 2214 (2015) and 2278 (2016).

[108]Coppens and Somers (2010), pp. 377–403.

be taken by the Italian Rescue Coordination Center. The latter shall act and decide in compliance with the migrants' fundamental rights and the disembarkation procedures laid down in the Regulation itself. Nevertheless, we stress that this is only a presumption, a desirable presumption that would need to be confirmed by official sources.

Lack of clarity concerns also the possible activation of Phase 2—Libyan Territorial Sea. After a number of mistakes, however, we cannot forget the negative effects of the patrol agreements between Spain and a number of States in Northwest Africa, the difficult experience of Triton with Libyan authorities, and, above all, the grave political situation in Libya. Against these factors, the intention to cooperate with Libyan enforcement authorities is induced to skepticism in terms of possible violations of the legal framework on the identification of a safe place in accordance with the UNCLOS, SOLAS, and SAR Conventions and/or a safe third country in accordance with the principle of nonrefoulement.

An extensive application of the Sea Border Regulation to every operation of the EU and its Member States must be more intensively considered together with an extensive cooperation with more stable and reliable States.

References

Andreone G (2011) Immigrazione clandestina, zona contigua e Cassazione italiana: il mistero si infittisce. Diritti Umani e Diritto Internazionale 5(1):183–188

Andreone G (2014) Chronique de la Jurisprudence. Annuaire du Droit de la Mer 2013, 63

Borelli S, Stanford B (2014) Troubled waters in the *Mare Nostrum*: interception and push-backs of migrants in the Mediterranean and the European Convention on Human Rights. Rev Int Law Polit 10:29–69

Butler G, Ratcovich M (2016) Operation Sophia in uncharted waters: European and international law challenges for the EU Naval Mission in the Mediterranean Sea. Nord J Int Law 85:235–259

Carrera S, den Hertog L (2016) A European border and coast guard: what's in a name? Liberty Secur Europe 88:1–20

Cataldi G (2015) Giurisdizione e intervento in alto mare su navi impegnate nel traffico di migranti. Giurisprudenza italiana 167(6):1498–1502

Conforti B, Focarelli C (2015) Nazioni Unite. Cedam Wolters Kluwer, Milan

Coppens J, Somers E (2010) Towards new rules on disembarkation of persons rescued at sea? Int J Mar Coastal Law 25:377–403

Court of Justice of the European Union, Judgment of 5 September 2012, Case C-355/10, *Parliament v Council,* par. 65–85

den Heijer M (2016) Frontex and the shifting approaches to boat migration in the European Union; a legal analysis. In: Zaiotti R (ed) Externalizing migration management Europe, North America and the spread of 'remote control' practices. Routledge, London, pp 53–71

Corte di Cassazione, judgment of 23 May 2014, No. 36052

European Court of Human Rights, *Hirsi Jamaa and others v. Italy*, Judgment of 23 February 2012

European Court of Human Rights, *Sufi and Elmi vs. United Kingdon*, Apps. No. 8319/07 and 11449/07, concerning the appeal by two Somali citizens at risk of inhumane treatment if returned to Mogadishu

Garcia Andrade P (2010) Extraterritorial strategies to tackle irregular immigration by sea: a Spanish perspective. In: Ryan B, Mitsilegas V (eds) Extraterritorial immigration control. Legal challenges. Martinus Nijhoff, Leiden, pp 311–346

Geiss R, Petrig A (2011) Piracy and armed robbery at sea, the legal framework of counter-piracy operations in Somalia and the Gulf of Aden. Oxford University Press, Oxford

Gordon R (1997) Saving failed states: sometimes a neocolonialist notion. Am Univ Int Law Rev 12(6):903–974

Guilfoyle G (2009) Shipping interdiction and the law of the Sea. Cambridge University Press, New York

International Tribunal for the Law of the Sea, *The M/V "Saiga" (No. 2) Case (Saint Vincent and the Grenadines v. Guinea)*, Judgment of 1 July 1999D

Italian Supreme Court, *Criminal proceedings against Radouan Hai Hammouda,* No. 3345, 23 January 2015

Kreijen G (2004) State failure, sovereignty and effectiveness: legal lessons from the decolonization of sub-Saharan Africa. Brill/Academic, Leiden/Boston

Lehmann J (2015) the use of force against people smugglers: conflicts with refugee law and human rights law. In: EJIL:Talk! Blog of the European Journal of International Law. Available via EJIL:Talk! http://www.ejiltalk.org/the-use-of-force-against-people-smugglers-conflicts-with-refugee-law-and-human-rights-law/. Accessed 31 May 2016

Liguori A (2012) La Corte europea dei diritti dell'uomo condanna l'Italia per i respingimenti verso la Libia del 2009: il caso Hirsi. Rivista di diritto internazionale 95(2):415–443

Pallis M (2002) Obligations of states towards asylum seekers at sea; interactions and conflicts between legal regimes. Int J Refugee Law 14(2/3):329–364

Papanicolopulu I (2016) Immigrazione irregolare via mare ed esercizio della giurisdizione: il contesto normativo internazionale e la recente prassi italiana. In: Antonucci A, Papanicolopulu I, Scovazzi T (eds) L'immigrazione irregolare via mare nella giurisprudenza italiana e nell'esperienza europea, 1st edn. G Giappichelli Editore, Torino, pp 2–22

Papastavridis ED (2016a) The forgotten responsibilities of the flag state to save people at sea. In: Nikodinovska Krstevska A, Tushevska Gavrilovikj B (eds) Migration at sea. International law perspectives and regional approaches, 1st edn. Giannini, Napoli, pp 31–47

Papastavridis ED (2016b) EUNAVFOR MED Operation Sophia and the question of jurisdiction over transnational organized crime at sea In: Questions of International Law. Available via QIL: http://www.qil-qdi.org/eunavfor-med-operation-sophia-question-jurisdiction-transnational-organized-crime-sea/

Picone P (2015) Gli obblighi *erga omnes* tra passato e futuro. In: Interesse collettivo e obblighi *erga omnes* nel diritto internazionale contemporaneo. In: Questions of International Law. Available via QIL: http://www.qil-qdi.org/wp-content/uploads/2015/07/Picone_Gli-obblighi-erga-omnes-tra-passato-e-futuro.pdf. Accessed 31 May 2016

Ringbom H (2015) Jurisdiction over ships: post-UNCLOS developments in the law of the sea. Brill, Leiden

Scarpa S (2008) Trafficking in human beings: modern slavery. Oxford Scholarship Online

Scovazzi T (2014) Human rights and immigration at sea. In: Rubio-Marín R (ed) Human rights and immigration, 1st edn. Oxford University Press, New York, pp 212–260

Siller N (2016) 'Modern Slavery' does international law distinguish between slavery, enslavement and trafficking? J Int Crim Justice 14:405–427

Tancredi A (2007) Il problema della legittima difesa nei confronti di milizie non statali alla luce dell'ultima crisi tra Israele e Libano. Rivista di diritto internazionale 90(4):969–1007

Tancredi A (2008) Di pirati e stati "falliti": il Consiglio di sicurezza autorizza il ricorso alla forza nelle acque territoriali della Somalia. Rivista di diritto internazionale 91(4):937–966

Tondini M (2012) The legality of intercepting boatpeople under search and rescue and border control operations. J Int Marit Law 18:59–74

Treves T (2009) Piracy, law of the sea, and use of force: developments off the coast of Somalia. Eur J Int Law 20(2):399–414

Trevisanut S (2008) The principle of non refoulement at sea and the effectiveness of asylum protection. Max Planck Yearbook United Nations Law 12:205–246

Trevisanut S (2010) Search and rescue operations in the Mediterranean: factor of cooperation or conflict? Int J Mar Coastal Law *25*:523–542

Trevisanut S (2012) Immigrazione irregolare via mare. Diritto internazionale e diritto dell'Unione Europea, Jovene Editore, Napoli

Wippman D (1996–1997) Military intervention, regional organizations, and host-state consent. Duke J Comp Int Law 7:209–239

Private Maritime Security Contractors and Use of Lethal Force in Maritime Domain

Jasenko Marin, Mišo Mudrić, and Robert Mikac

1 Introduction

The following chapter analyzes the recently emerged phenomenon of the use of force at sea by private actors (private maritime security companies). The utilization of force in the maritime domain by private actors—a novelty in the modern common international experience[1]—proved to be a considerable challenge to international law and good practice. After initial antagonism, the international community has gradually accepted the notion of utilizing private maritime security companies. Following the plight of individual coastal states (most notably, the United States (US)) and certain parts of the shipping industry, more and more nongovernmental actors became reliant on the private maritime security services within a short period of time, forcing the International Maritime Organization (IMO) to reassess its position with regard to the use of force at sea by private actors.[2] This made it necessary to adjust the international approach as, up to that point, it was in principle prohibited for private actors to carry arms—the notable exception (recognized by the relevant international maritime law and law of the sea

[1]For a general background with regard to the standard private security industry regulation, see: Chesterman and Lehnardt (2007), Mikac (2013), Chapter III; Tonkin (2011), Percy (2006) and Jäger and Kümmel (2007).

[2]For more information, especially with regard to the soft-law bottom-up approach to the regulation of private maritime security companies, see: Mudrić (2015a), pp. 53–55.

J. Marin (✉) • M. Mudrić
Department for Maritime and Transport Law, Faculty of Law, University of Zagreb, Zagreb, Croatia
e-mail: jasenko.marin@pravo.hr; miso.mudric@pravo.hr

R. Mikac
Department of International Politics and Diplomacy, Faculty of Political Sciences, University of Zagreb, Zagreb, Croatia
e-mail: robert.mikac@fpzg.hr

G. Andreone (ed.), *The Future of the Law of the Sea*,
DOI 10.1007/978-3-319-51274-7_10

conventions and relevant domestic law maritime codes and acts) allows the Masters of Vessels and First Officers to make use of personal firearms that have to be kept under lock and key at all times. The security of people and goods at sea—falling at the same time within the sphere of individual, national, and also common interests—thus became reinforced by an unprecedented enforcement methodology.

The present chapter particularly focuses on the issue of lethal force utilized by privately contracted armed security personnel (Contractors)[3] as governed by the standard security services on board vessel contract form, the GUARDCON—Contract for the Employment of Security Guards on Vessels (GUARDCON).[4] The contractual provisions, such as the example of GUARDCON, are often accompanied by the standards of conduct and separate rules on the use of force, developed either by an individual private maritime security company or by a professional body or association (usually consisting of private maritime security companies and/or other interested stakeholders). Several guidelines and recommendations, drafted particularly for the maritime domain (i.e., the IMO Guidance, the Baltic and International Maritime Council (BIMCO) Guidance, the 100 Series Rules, the IAMSP Rules on Use of Force, the ISO/PAS 28007:2012, and the ASIS/ANSI PSC. 4) will be analyzed and compared in order to assess the extent to which, inter alia, the issue of the use of force is harmonized on a global level. Due to the fact that none of these documents are mandatory and legally binding, they are necessarily accompanied and intertwined with domestic regulation or nonbinding recommendations, as well as general domestic criminal and civil laws, especially in connection to the issue of self-defense. A number of jurisdictions will, therefore, be analyzed in order to examine the noted interconnection.

2 The Shifting Nature of Maritime Security

The core actors in the maritime-related privatization of violence encompass the full spectrum of shipping-related stakeholders, including nongovernmental actors concerned with the issue of human rights at sea—a separate drive that reviews the phenomenon of the use of force at sea by private actors as a possible insecurity factor, leading to an increased level of violence and increased insecurity of people and goods at sea. Thus, the conjunction of individual, national, and common interests is interrupted by a conflict of interest within its core—in order to achieve better security, one forgoes principles that have ensured security thus far. This dilemma is ever so problematic, especially when considering that the primary goals of any sound ocean security governance are to promote peaceful use of the seas and oceans (as stipulated by the 1982 Convention on the Law of the Sea and other

[3]Often used abbreviation: PCASP.

[4]Baltic and International Maritime Council, see: BIMCO (2012a). For a legal analysis of GUARDCON, see: Mudrić (2015a, b).

related international instruments). It, therefore, continues to be true that the *privati-militarization* of sea in the name of peace remains a dubious concept, despite the argument that the utilization of violence by private actors only serves to promote the noted principle of peace.

The initial conflict of interest has slowed down the bottom-up lawmaking process. This is especially true for the period when the majority of states have continued to adhere to the state's monopoly on violence principle at sea,[5] despite the fact that this principle has long been breached with regard to the activities of private military and security companies on land, including such companies working directly for government institutions in and outside of conflict (war) zones. But when the shipping and insurance driving forces ushered in the private contractors in the maritime domain, the states, already accustomed to the utilization of private security industry on land, soon adjusted to the emerging subbranch of the global private security industry. Initially—keeping in mind that the predominant maritime-related interest of the international community as a whole continues to be the security of people and goods at sea—the introduction of firearms held by private entities into that realm was considered as a violation of the common interest (leading to the possible spread of and an increase in violence). This perception has, however, gradually transformed into a compromise solution whereby the security services are to be offered by professional entities trained and properly equipped to utilize violent means when no other means are available to thwart the realization of danger to people and goods at sea in accordance with the best industry practice, such as the example of the *Montreux Document* (not relevant for the maritime sector).[6]

The role of the state was, therefore, diminished to a certain extent, partially due to the private contractual nature of private maritime security services (as opposed to such instances when states contract private actors to do their bidding), even though states persist in their determination to remain the *watchful dog* through carefully drafted legislation. The impact of this effort is, however, dubious due to the practical difficulties of control and supervision enforcement. At the same time, the role of the private sector in the protection of seas and oceans continues to increase—the scope of utilization echoing beyond any precedent.

Nevertheless, the compromise solution does not disentangle the continuous dilemma as to whether the introduction of private actors' right to utilize force at sea equals a step closer to or a step further away from the general security of people and goods at sea, posing a serious challenge to the law of the sea and ocean security governance. The latter is especially true when considering the appearance of private armed flotillas that today offer security services to individual vessels while tomorrow they may potentially be utilized for a broader (private) purpose, exempt from

[5]For more information, see: Petrig and Geiβ (2011), Tondini (2012), Williams (2014), Coito (2013) Berube and Cullen (2012), Andreone et al. (2013), Kraska (2011) and Mejia et al. (2013).

[6]For more information on the Montreux Document, see: International Committee of the Red Cross (2009) and Buckland and Burdzy (2013).

any form of organized, institutional control. In the context of the emerging field of Maritime Security Studies,[7] private armed flotillas may alternatively function under a hidden organized, institutional control, as one means of hybrid maritime warfare, somewhat similar to when the *Letter of Marque* granted certain formal authority to the private actors of the past.

At the same time, when states ratify international instruments, they are granted certain rights but are, at the same time, bound to accept certain obligations as well. The security of people and goods at sea is a general obligation present in various international conventions, including the 1982 Convention on the Law of the Sea. The emergence of maritime crimes and piracy poses such security threats that must be tackled swiftly and without delay, and the lack of proper domestic tools necessary to combat such occurrences is a poor excuse for nonperformance. An analogy with the private salvage industry serves to demonstrate that the need for such an industry is brought about by the inability of the majority of coastal states to train, equip, and maintain a domestic salvage capacity, thus being heavily reliant on the private salvage industry to meet their needs and fulfill their obligations.

It remains to be seen whether the utilization of private actors in the context of providing maritime security services may potentially become a standard mechanism in case a certain set of conditions is met, thus entangling the private sector into the regular network mechanism of domestic maritime security. It should be noted that the sanctioning of private actors' right to utilize force was just one aspect of the international response to the scourge of piracy, and it took a while to formulate, following a careful evaluation. In the context of the recent Mediterranean crisis and the increased human trafficking by sea, whereas certain stakeholders argue for the employment of private actors in, among other things, combating organized crime, state navies—European Union in particular—have long initiated joint efforts to fight the smugglers. The effort has been reinforced with the relevant United Nations Security Council's resolutions, enabling further steps to be undertaken, much similar to what was witnessed in Somalia several years ago (which has, among other factors, enabled the overt in-land combat activities as conducted by the *Joint Special Operations Command* and other militaries that have by far most contributed to the elimination of the pirate outposts in the region). During the height of the piracy threat, similar state navies' activities were initiated in the Somalian waters and the Gulf of Aden (still active today), providing safe passage to the vessels that can afford to wait, and patrolling over the High Risk Area, representing predominantly moral support to all the endangered seamen. At the same time, the newly developed piracy tribunals were kept busy in prosecuting captured pirates—courtesy of the previously mentioned navies—whereas a small number of cases appearing before the regional coastal states' courts have tackled the issue of private maritime security companies offering their services without any proper documentation or licenses. Quite recently, one case has appeared before the International

[7]For more general information and background, see: Bueger (2015), pp. 159–164, and Kraska and Pedrozo (2013).

Tribunal for the Law of the Sea that deals with the issue of alleged unlawful use of force at sea (*Enrica Lexie*).

3 Use of Force, Regulation, and Consequences: General Overview

On 22 October 2014, the Washington jury found four former Blackwater (US private military and security company) private security guards guilty of several charges—murder in the first degree, manslaughter, and attempted manslaughter. This is the result of a 7-year investigation and proceedings following the all-out shoot-out at Nisour Square, Baghdad, Iraq, on 16 September 2007.[8] The first instance decision (pending appeal), based on the jury's verdict, represents a rare example where the employees of a private military and security company have been found accountable and responsible for conduct exhibited during the performance of a contracted security service that has caused bodily harm and death to innocent bystanders. It also represents a rare occasion where the issue of the use of force by private actors has been legally scrutinized.

On 15 February 2012, two Italian marines, providing government-authorized Vessel Protection Detail (VPD)[9] service to the Italian tanker *Enrica Lexie*, allegedly shot and killed two Indian fishermen. The proceedings are still under way.[10]

The former is an example of (questionably) excessive use of force when providing private military and security service on land and the latter of (questionably) excessive use of force when providing (private) maritime security services. Both examples represent what is often referred to as a "hot potato" or an "elephant in the room"—an issue very difficult to tackle, assess, and regulate. The law, in general, allows the use of force (the right of an individual to use force as a defensive measure), including the lethal kind, only when absolutely necessary and only when it serves to prevent equal or greater harm. This is generally accepted as lawful use of force. What amounts to excessive use of force or unlawful use of force is something difficult to ascertain through clear regulation and very much depends on the circumstances of each particular case. Any conclusion reached by a tribunal thus easily falls into the arena of criticism.

Incidents at sea with deadly consequences are being reported at an increasing rate. Beginning with as early as March 2010,[11] unfortunate events involving the death of innocent fishermen are becoming a harsh reality,[12] placing a growing shadow on the provision of private maritime security services in general.

[8]See: Los Angeles Times, October 22, 2014.

[9]For more general information on VPDs, see: Guilfoyle (2013), p. 221 et seq.

[10]See: The Hindu, December 16, 2014.

[11]See: Symmons (2012), 36—the "*MZ Alinezaan*" case.

[12]See, for example: Katz (2012).

Simultaneously, a growing number of reports indicate possible excessive behavior of individual armed security guards.[13]

With regard to the maritime domain, IMO and the private maritime security services industry have produced a number of legally nonbinding rules with the purpose of providing guidance to states, private maritime security service providers, and clients with regard to, inter alia, the use of (lethal) force.[14] An increasing number of states have also adopted guidelines and/or ordinances with regard to the provision of private maritime security services. A number of such legal documents will be examined in this chapter in order to ascertain the extent to which the issue of the use of force has been regulated and/or considered. Only a small number of available documents are legally binding in nature—most are offered on a voluntary basis, with the private maritime security industry, which claims to be a professional industry, being under increased pressure to willingly incorporate such recommendations into their standard operation procedures. The courts will have a final say, but the fact that only a fraction of reported incidents concerning the provision of private security services on land and none in the maritime domain have reached the courts speaks of the sensitivity and difficulty of the issue at hand.

Several attempts were made in the past to produce an international convention regulating the activity of private military and security companies in general,[15] but no such project came to life, likely due to the fact that the use of private security providers, although present throughout the ages (especially in the few recent decades),[16] contradicts the fundamental monopoly of the state over the use of force and, as such, creates ambiguous feelings towards a formal recognition of their indispensable status in the modern world. However, one should not neglect the fact that, very often, states directly contract private military and security companies' services, thus effectively authorizing such companies to utilize violent measures (derived from the state's monopoly on violence) when necessary and appropriate. In other words, depending on the nature of a particular operation, an activity performed by the engaged private actor may constitute an activity so inherent to the notion of state sovereignty and state performance that any formal line of differentiation—of where the state's functions end and private actor's separate operations begin—is easily blurred in practice. Any proper international instrument that would regulate the conduct of private security companies would have to take that into consideration and provide for an equal level of measures in case of a breach of an obligation, against both the private actors and government institutions that employed them, and possibly other relevant states (in line with the *Montreux* principles). This is perhaps another reason for ambiguity with regard to the existence of such an instrument, which requires further consideration.

---

[13]See: Dutton (2013), p. 111 et seq.—the "*Avocet*" case.

[14]See generally: Mudrić (2011), p. 165.

[15]For more general information, with the focus on maritime domain, see: Dutton (2013), p. 140. Especially see: Scheimer (2009).

[16]For more information, see: Mikac (2013), Chapter II.

3.1 Prelude to Precedent

In the much-discussed and still ongoing case *Enrica Lexie*, the use of force utilized by a six-strong Italian military protection team led to the undesired consequence of two Indian fishermen being shot dead.

Although most of the facts are continuously being disputed by the involved parties,[17] awaiting the final court determination, the early sources indicated that (a) the incident occurred during the day; (b) the fishing trawler had at one point allegedly maneuvered toward the tanker (in accordance with some accounts, this consisted of a sharp move towards the tanker, in order to pass by the stern—difficult to be confirmed due to the fact that seven out of nine fishermen present on the trawler were asleep at the time of the incident, and the two in charge of navigation were killed); (c) the vessel made no attempt to avoid the approach; (d) the Team Leader failed to consult the Master of Vessel or any of the crew with regard to the steps to be taken (in accordance with the Master's and First Officer's testimony[18]); (e) warning shots fired (the Italian marines claim that they utilized warning light signals prior to the shooting—an action with a dubious effect during the day) resulted in lethal consequences, thus failing to fulfill their purpose; (f) the fishermen were not armed, nor did they exhibit any hostile intentions (the inspection of the tanker failed to produce any evidence of the tanker being fired upon, which was argued by the Italian marines); (g) the tanker failed to immediately report the incident to the proper authorities and proceeded with its voyage; and (h) the Indian Coastal Police alerted the Indian Coastal Guard, who contacted the tanker, which only then altered its course and steered back to the port of Kochi.[19]

The questions pending before the Indian Supreme Court[20] are whether the Italian marines acted in accordance with the Rules of Engagement (it is relevant to note that the service provided by the Italian marines is derived from the state's monopoly over the use of force[21]) and whether their actions entail criminal and/or civil responsibility and liability. The decision to be reached by the Indian court will make a significant impact on the provision of private maritime security services and the issue of rules on the use of force as it will provide a much-awaited court determination with regard to the use of force in general and the use of lethal force in particular.

[17]For more information on the Indian position and the decisions reached by the Kerala High Court, see: Gandhi (2013), pp. 3–5. See, however, the formal position of the parties as submitted to the International Tribunal for the Law of the Sea portal, The "Enrica Lexie" Incident (Italy v. India), Case No. 24, available at: https://www.itlos.org/en/cases/list-of-cases/case-no-24/.

[18]See, for example: Reuters, 10 June 2013.

[19]See: Eboli and Pierini (2012), p. 4.

[20]Subject to the International Tribunal for the Law of the Sea Order 24 of August 2015, whereby both parties are ordered to suspend all court proceedings which might produce a negative effect with regard to the dispute submitted to the arbitral tribunal, see: International Tribunal for the Law of the Sea (2015).

[21]For more information, see: Petrig (2013), pp. 669–670.

3.2 *Use of Force*

3.2.1 Right to Self-Defense

In accordance with the generally established principle of self-defense, a person causing damage (actor) has a legally valid defense when utilizing reasonable measures to protect himself or a third person, provided that the injured party has endangered the protected interests.[22] The actor must demonstrate the necessity of the defensive measures and ensure that the defensive measures are proportionate to the perceived threat.[23] If that is the case, it will be deemed that any damage so caused is a legally relevant damage and that it is for the sole accountability and responsibility of the other (injured) person endangering the protected interests to suffer the consequences (liability) of the legally relevant damage.[24]

In the context of the current examination, a good example is the incident that occurred on 15 October 2014 in the Gulf of Aden, where a motor yacht reported an approach by a skiff with four people on board, up to a distance of 500 meters, followed by warning shots fired by the Contractors after the weapons and ladders were sighted.[25] The skiff followed for a while but decided not to pursue. In another good example that occurred in the Gulf of Aden, three pirate skiffs made two attempts to approach the vessel.[26] The first approach was deterred through the use of nonviolent evasion measures, whereas during the second approach it was necessary to utilize indirect force (first, a so-called *parachute flare*—on three separate occasions, followed by the second measure in the form of indirect warning shots) in order to *persuade* the skiffs to move away. Such examples represent a model scenario on how force is to be used, under what circumstances, and up to what extent of severity.

The actor may, however, be held liable (for damage) if the defensive measures are unreasonably excessive or disproportional to the actual threat (excessive self-defense) or if there was no actual threat (putative self-defense), as examined in further text.

[22]Compare: von Bar and Clive (2010), VI. – 5:202, 3665.

[23]Compare: *The M/V "SAIGA" (No. 2) Case (Saint Vincent and the Grenadines v. Guinea)*, International Tribunal for the Law of Sea, ITLOS Rep. 10, § 155. For more information on the case, see: International Tribunal for the Law of Sea portal, available at: https://www.itlos.org/cases/list-of-cases/case-no-2/case-no-2-merits/.

[24]Von Bar and Clive (2010), VI. – 5:202: Self-defense, benevolent intervention, and necessity.

[25]Maritime Security Centre—Horn of Africa (MSCHOA).

[26]Aburgus Risk Management, October 2014c.

3.2.2 Relevant Maritime-Related National Legislation and Guidelines

A growing number of jurisdictions have enacted laws, ordinances, and recommendations with respect to the provision of private maritime security services. Such legal documents usually contain several important considerations with regard to the right to use force and the limitations of such use.

The US *Port Security Advisory (3-09) Guidance on Self-Defense or Defense of Others by U.S. Flagged Commercial Vessels Operating in High Risk Waters* (PSA), by definition, includes the use of deadly force within the scope of the right to self-defense (Rule 2(a)),[27] whereby the Master of Vessels retains final authority with respect to the decision to utilize force (Rule 3(a)).[28] The PSA defines imminent danger as a situation when "... an attacker manifests apparent intent to cause great bodily harm or death..." to others, provided that the attacker possesses adequate means (i.e., weapons, climbing gear, etc.) and acts when the opportunity so permits.[29]

The United Kingdom's (UK's) *Interim Guidance to UK Flagged Shipping on the Use of Armed Guards to Defend against the Threat of Piracy in Exceptional Circumstances*,[30] recognizing the Master's overall authority (Rule 5.1), sets a number of rules to be adhered to when considering the use of deadly force. The use of force must be proportionate and reasonable (Rules 5.6, 8.9, and 8.10), kept to the minimum necessary level, and can only be leveled up at a gradual basis (Rule 8.3). The Interim Guidance allows for preemptive strikes provided that an attack is imminent (Rule 8.12), keeping in mind that a mere sighting of a possible threat is not to be considered as an imminent danger (Rule 8.13).

A similar provision is available in the Croatian Ordinance on the requirements for legal persons providing the services of boarding armed escort on Croatian-flagged vessels,[31] whereby the Team Leader must sign a statement recognizing the Master's final authority over the use of firearms that are to be utilized at a minimum possible capacity sufficient to thwart an attack (Article 9(2)), subject to the general Croatian criminal law provisions.

---

[27]U.S. Department of Homeland Security, United States Coast Guard, *Port Security Advisory (3-09), Guidance on Self-Defense or Defense of Others by U.S. Flagged Commercial Vessels Operating in High Risk Waters*. For a critical assessment of Port Security Advisory, see: Patrick (2014), pp. 350–355.

[28]PSA 3(a)., subject to: Title 33—Navigation and Navigable Waters, Chapter 6—General Duties of Ship Officers and Owners after Collision or other Accidents, 33 U.S.C. § 383—Resistance of pirates by merchant vessels.

[29]PSA, ibid.

[30]UK Department for Transport, *Interim Guidance to UK Flagged Shipping on the Use of Armed Guards to Defend Against the Threat of Piracy in Exceptional Circumstances*, Version 1.2, Updated May 2013.

[31]*Pravilnik o uvjetima koje moraju ispunjavati pravne osobe koje pružaju usluge ukrcaja osoba za naoružanu pratnju na brodove hrvatske državne pripadnosti*, Narodne novine (Official Gazette), broj (No.) 123/12.

The requirement of adherence to the general criminal law provisions on self-defense is restated throughout the Italian *Regulation on the employment of contractors on board Italian-flagged ships sailing in international waters under piracy risk*,[32] whereby lawful and proportionate use of force is limited to self-defense.

The Indian *Guidelines on Deployment of Armed Security Guards on Merchant Ships*[33] refer to the necessity of taking all reasonable steps in order to avoid the use of force that may only be utilized if necessary and in a proportionate manner (Rule 6.9(2)), and only provided that there is an imminent danger (Rule 6.9(3)).[34]

The *Norwegian Provisional Guidelines—Use of Armed Guards on Board Norwegian Ships*[35] allow the use of force only when the threat is "direct, immediate, significant and otherwise unavoidable" and only to the extent that the utilization of force is "necessary, justifiable and proportionate" (Rule 10(2)). Irrespective of the fact that the Master of Vessel retains final authority, the responsibility for the decision to use force resides on the individual making such decision (Rule 10(7)). The Norwegian Guidelines mention a very specific determination of a distance of 2000 meters as viable to assess an actual threat (Rule 10(3)). Another relevant Norwegian-originated set of guidelines is available under the auspices of the Norwegian Shipowners' Mutual War Risks Insurance Association—*The Guidance on the Selection of Private Security Companies*.[36] The *Guidance* stipulates that when a threat has been reasonably identified and classified as a hostile intent or a hostile act (Team Leader and Master of Vessel deciding in conjunction), the Team Leader is to assume tactical command, with the possibility of using firearms as a last resort, leading towards, with a gradual progression, the use of deadly force as an ultimate resort (Rule 2).

3.2.3 Excessive Self-Defense and Putative Self-Defense

Determination of excessive self-defense is applicable for such occurrences where the measures used to thwart the attack are excessive when compared with the level of danger threatened.[37] Putative self-defense denotes a situation when a measure of self-defense is utilized under the impression that there is an attack whereas, in reality, no attack is taking place.[38]

The *Nisour Square* incident provides a perspective for consideration. In response to the call from the Blackwater protection detail providing personal escort and

[32]*Regulation on the employment of contractors on board Italian flagged ships sailing in international waters under piracy risk*, 29th March 2013, Italian Official Gazette.

[33]Indian Ministry of Shipping (2011).

[34]See: *Darshan Singh v. State of Punjab & another*, Criminal Appeal No. 1057 of 2002.

[35]Norwegian Mar. Directorate (2011).

[36]Den Norske Krigsforsikring for Skib—Gjensidig Forening, 2011.

[37]Von Bar and Clive (2010), VI. – 5:202, 3667.

[38]Id.

protection to the USAID director—who was under attack—and upon request from the US State Department's regional security officer,[39] the second Blackwater team left the *Green Zone* in an attempt to provide support and security on the route to be taken by the first team in order to reach the safety of *Green Zone*. The second detail reasoned that *Nisour Square* would make a good place for staging an ambush to the advancing first team. What follows has been heavily disputed in the proceedings and media coverage.[40]

At one point, a Kia motor vehicle had entered the Nisour Square's motorcade circle. The Blackwater team, as alleged by the defendants,[41] made several attempts to signal the vehicle to stop in its tracks. As the vehicle was unresponsive and continued moving towards the armored column, and due to the fact that the Blackwater team, as well as any other security force in the theater of operations, was aware of the danger that such a vehicle could easily represent, the Blackwater team and the local police officers decided to stop the car, first by hand gestures and vocal commands and, after the vehicle remained unresponsive, by the use of force. The force was direct and mortal. What followed was complete chaos, resulting in many dead and wounded.[42]

The Blackwater company was operating at the time in accordance with its own rules of engagement but was additionally bound, by the virtue of Worldwide Personal Protective Service contract,[43] by the US Embassy's Escalation of Force policy[44] that requires a gradual and proportional use of force, taking into consideration that deadly force may have to be utilized immediately, depending on the circumstances. The Contractors further claimed that soon after the vehicle was repeatedly shot at, they came under enemy fire, and that they continued to act in self-defense.[45] One could easily imagine that, being in the theater of operations such as the Baghdad, a *better safe than sorry* policy is a policy often utilized by all engaged forces, armed or otherwise. In this sense, excessive or putative self-defense is just one step away from becoming a victim, making it extremely difficult to tell the difference, let alone devise a set of rules defining any points of differentiation in advance.

It has often been reported that, especially in the High Risk Areas, fishermen sometimes carry weapons to protect themselves from potential pirate attacks.[46] A mere sighting of weapons on board may not constitute a sufficient reason for immediate use of lethal force, unless the Contractors have a strong reason to believe

[39]As noted by Erik Prince, see: Prince (2013), Chapter 13.

[40]See: *United States v. Slough*, Criminal No. 2008-0360 (D.C. 2014), District Court, District of Columbia, Filed: May 23rd, 2014.

[41]See: Prince (2013), Chapter 13.

[42]See: Tarzwell (2009), pp. 181–183.

[43]For more information, see: Cheadle (2009), p. 689 et seq.

[44]See: U.S. Department of State (2006). Also see: U.S. Department of State (2009).

[45]See: Prince (2013), Chapter 13.

[46]See: Symmons (2012), p. 29. Also see: Murdoch (2011), p. 40.

that the threat is imminent or present. At the same time, practical experience has shown that assailants are capable of concealing weapons and striking when unexpected, placing security guards or armed forces in a difficult position of telling the difference between an approaching vessel with a hidden agenda (simply acting as a fisherman) and an innocent passer-by (an actual fisherman).[47] In addition, should they opt for the use of force, the question remains as to which rules they are to base their decision on whether the use of force is in order.[48]

3.2.4 Assessment of Reasonable Conduct

In order to understand how the utilization of self-defense in the maritime domain is determined and understood by the courts, it is necessary to point to several comparative examples of self-defense regulation applicable to all self-defense manifestations, including ones in the maritime domain. Various domestic law regulations and case law, both with regard to criminal and civil laws, in principle, presuppose similar elements and standards necessary for establishing the right to claim the self-defense exemption from or exclusion of liability. However, several important differences in the evaluation of excessive and putative self-defense conduct are to be noted.

French law and Belgian law require an objective assessment based on the conduct of a reasonable person or a reasonable professional under the same or similar circumstances.[49] English and Welsh case law recognizes the principle of proportionality in comparing the act of self-defense with the act of attack but at the same time acknowledges the extreme circumstances where the balance of proportionality may be difficult to assess by the actor, stipulating the doctrine of reasonable vs. unreasonable mistakes.[50] In accordance with Spanish law, in order to claim the self-defense exception, the actor must prove the existence of an unlawful and unprovoked attack where necessary and proportionate force was utilized in order to thwart that attack.[51] Italian law follows the same principles but stipulates that in case of putative self-defense, it is likely that the principle of contributory negligence will be employed.[52] Portuguese law further stipulates that in case of excessive self-defense applied out of fear, the actions will have been justified,[53] leaving

---

[47]A good example being the attack on the *USS Cole* in 2000. For more information, see: Mikac (2013), p. 119.

[48]For more practical examples and general consideration, see: Neri (2012), p. 83.

[49]Compare: von Bar and Clive (2010), VI. – 5:202, 3669.

[50]Compare: Murphy and Witting (2012), pp. 334–335. Also see: *Cross v. Kirkby* (2000) Times, 5th April (CA)—where it was deemed relevant what was the defendant's genuine apprehension of the circumstances, and whether the critical moment increased the defendant's anguish. Also see: *Palmer v. R* [1971] AC 814.

[51]Compare: von Bar and Clive (2010), VI. – 5:202, at 3669–3670.

[52]Compare: ibid., at 3670.

[53]Compare: ibid., at 3672.

no possibility for the actor's liability. Contrary to all the above, Bulgarian law does not provide any grounds for exclusion from civil liability, even if excessive self-defense was applied due to fear or fright.[54] German law, inter alia, requires the attack to be unlawful and imminent,[55] with an additional rule stipulating that a professional person must be reasonably prepared for such circumstances and ready to use proportional force,[56] whereas Dutch law, inter alia, requires the attack to be present.[57]

3.3 *Standards, Guidelines, and Recommendations*

Parallel to IMO's publication of guidelines with regard to the provision of private maritime security services, a number of interested stakeholders, particularly on the side of shipping and related industries, have endeavored to issue similar guidance and recommendations, either to enhance the security of people and goods in relation to the provision of security services or to increase the popularity of such services. The emerging *soft law*, forged in the dwellings of corporate interests, soon began to develop into *hard law*—the so-called *bottom-up law-making*—with an increasing number of coastal and shipping states enacting laws and ordinances with regard to the provision of private maritime security services on board vessels flying their flags.

3.3.1 IMO Interim Guidance

In 2012, IMO issued the *Interim Guidance to private maritime security companies providing contracted armed security personnel on board ships in the High Risk Area*.[58] With regard to the use of force, recognizing the Master's overall authority (Clause 5.6(1)), the Guidance stipulates (Clause 5.15(2)) that it is necessary to undertake all reasonable steps in order to avoid the use of force, but should the use of force be deemed necessary, it should be conducted in a gradual manner, applying only such measures that are necessary and reasonable in the given circumstances (Clause 5.15(3)).[59]

[54]Compare: ibid., at 3671.

[55]Compare: ibid., at 3671.

[56]Compare: ibid., at 3668.

[57]Compare: ibid., at 3673.

[58]IMO (2012a).

[59]The noted recommendations are, in essence, repeated in the: IMO (2012b).

3.3.2 BIMCO Guidance and GUARDCON

The BIMCO issued the *Guidance on Rules for the Use of Force (RUF) by Privately Contracted Armed Security Personnel (PCASP) in Defence of a Merchant Vessel* (MV) (BIMCO Guidance).[60] The BIMCO Guidance should be read in conjunction with GUARDCON provisions relevant for the relationship between the Master of Vessel and Team Leader (Clause 5 BIMCO Guidance).

In accordance with Clause 8(b) GUARDCON, when under an "actual, perceived or threatened act of piracy," the Contractors' Team Leader has the right to invoke the Rules for the Use of Force, and the responsibility for and potential liability arising out of the discharge of weapons resides with the Contractors. Whereas, in accordance with Clause 8(d) GUARDCON, the Master of Vessel can order a cease-fire, each Contractor retains the right of self-defense (the use of lethal force included). This right is based on the Contractors' main duty (Clause 3 (b) GUARDCON) to provide protection and defense of the vessel ("using all reasonable skill and care"—Clause 6(a)), in accordance to which the Contractors do not guarantee the safety of Vessel (Clause 9) but, instead, promise to act to the best of their abilities.

In accordance with Clause 3(d), Clause 4(c) and Clause 4(d) BIMCO Guidance, the use of (lethal) force should be utilized only when essential and strictly necessary, ensuring that the measures undertaken are proportional and appropriate to the circumstances, and utilized at a minimum necessary level. Clause 4(f) BIMCO Guidance stipulates that all reasonable steps should be taken to avoid the use of lethal force. Furthermore, Clause 7 BIMCO Guidance defines the scope of graduated and proportional defense, requiring (Clause 7(a)(iii)) the use of nonviolent means first (except when circumstances necessitate immediate use of force)—such as the show of weapons—followed by (Clause 7(f)) a discharge of weapons in a graduated flow (warning shots, disabling fire, and, finally, deliberate direct fire). Clause 7(g) BIMCO Guidance requires certain exemplary conditions to be met before being allowed to use lethal force, such as the fact that the attack is ongoing despite the show of weapons and warning shots, with a clear and visible intention on behalf of the attacker to board the vessel, at the same time demonstrating the use of weapons.

3.3.3 The 100 Series Rules™

Another legally nonbinding set of rules with regard to the use of force, enjoying support from many relevant stakeholders, is the *100 Series Rules: An International Model Set of Maritime Rules for the Use of Force (RUF)* (100 Series Rules).[61] The

[60]BIMCO (2012b).

[61]Globus Intelligence Ltd (2013).

100 Series Rules follows (Rule 100) the GUARDCON stipulation with regard to the relationship between the Master of Vessel and Team Leader.

The 100 Series Rules stipulate (Clause 17) that force should be used when necessary and/or reasonably required as a deterrent when an imminent threat to life is present and commensurate to the threat posed. Rule 101 stipulates the use of nonviolent means when there is a reasonable belief that a potential attack is due. Rule 102 stipulates the use of, *inter alia*, warning shots to thwart the attack. Finally, Rule 103 stipulates the right to use lethal force in case of an imminent attack, which is defined as a manifest, instant, and overwhelming occurrence.

The GUARDCON, BIMCO Guidance, and 100 Series Rules are interlinked with the *ISO/PAS 28007:2012 Ships and marine technology—Guidelines for Private Maritime Security Companies (PMSC) providing privately contracted armed security personnel (PCASP) on board ships (and pro forma contract)*[62] (ISO/PAS 28007:2012) due to the fact that the organizations publishing or supporting the noted documents constitute the same or similar stakeholders, and often relate to each other. In addition, the "pro forma contract" mentioned in the title of ISO/PAS 28007:2012—the quality standard with regard to the conduct of private maritime security companies—refers directly to the GUARDCON.

3.3.4 IAMSP-2011-01-UOF-001 v2.0

The International Association of Maritime Security Professionals (IAMSP) has issued the *Use of Force, IAMSP-2011-01-UOF-001 v2.0* (IAMSP Rules).[63] The IAMSP Rules are aligned with the *Quality Assurance and Security Management for Private Security Companies Operating at Sea—Guidance ANSI/ASIS PSC. 4 – 2013*,[64] the quality standard that sets similar but more detailed provisions with regard to the conduct of private maritime security companies, when compared to the ISO/PAS 28007:2012.[65]

The IAMSP Rules place an emphasis on risk assessment (Appendix E IAMSP Rules), stipulating (Clause 63 IAMSP Rules) that the choice of use of force depends on the feasibility study. Clause 64 IAMSP Rules further stipulates that the use of force depends on the nature of attack, the potential for escalation, the attacker's intention, and other possibilities (such as the evasion of attack by other means). Clause 66 IAMSP Rules enhances the previous stipulation by requiring the Contractors to ensure due care and undertake all reasonable steps prior to the utilization of lethal force. Clause 68 IAMSP Rules stipulates that the final choice of whether to use lethal force is subject to the reasonability test, in accordance to which the escalation of force must equal the perceived threat in order for the use of lethal force

[62]International Organization for Standardization (2012).

[63]International Association of Maritime Security Professionals (2011).

[64]American National Standards Institute, Inc. ASIS International (2013).

[65]For more information on, see: Mudrić (2015b), pp. 61–62.

to be allowed. The responsibility arising out of the use of lethal force (Clause 93 IAMSP Rules) resides on the person who, in accordance with the ship's log, authorized the use of lethal force. The IAMSP Rules provide a detailed reference with regard to the chain events leading to the use of lethal force (Appendix B IAMSP Rules). In order to establish the right to use lethal force, the Contractors must demonstrate, if possible, the following elements/steps: (a) suspicious vessel, (b) sighting of weapons and boarding equipment, (c) potential use of nonviolent means,[66] (d) warning shots, and (e) direct fire. With regard to the escalation of force, the IAMSP Rules detail (Appendix C IAMSP Rules) that in case the attacker has fired upon the vessel or directed weapons upon the vessel, and provided that the vessel cannot evade the attack, the minimum use of (lethal) force is allowed in order to stop the attack, with an additional clarification that the use of (lethal) force is allowed to escalate in case of a continued attack, until the attack is broken off.

3.4 *Open Issues*

3.4.1 Present and Imminent Attack

Clause 6(a)(iii) GUARDCON refers to "monitoring suspicious vessels or craft during the Transit" as one of the main Contractors' duties. It is, however, unclear to what extent a suspicious vessel constitutes a threat that would justify a series of measures to be undertaken, including possible use of force. Under what circumstances are the Contractors expected to conclude that a suspicious vessel represents a clear possibility of an imminent attack, and under what circumstances is such a vessel to be considered to constitute a clear possibility of a present attack? In addition, does an approach of a suspicious vessel constitute the necessary elements required for legitimate use of lethal force?

In a recently reported case occurring in the Straits of Hormuz, after a sighting of a high-speed skiff approaching the tanker, the Master of Vessel decided to take a series of evasive measures as preemptive action. When the skiff approached the tanker, the crew sighted three armed persons equipped with boarding hooks and ropes. The tanker continued with the nonlethal evasive measures and successfully thwarted the attack.[67] This is a good example of how nonlethal and nonforceful measures can successfully be utilized to thwart a clearly imminent attack. Thus, and as many previously examined guidelines suggest, the (lethal) force should only be used when the attack is imminent and when no other means are available to prevent the occurrence of the attack. From the few available undisputed facts in connection to the *Enrica Lexie* case, it is difficult to ascertain whether the Team Leader made

[66] A detailed description of the available means, with or without the presence of non-armed or armed guards on the vessels, is available in: Industry Stakeholders (2011). The document was endorsed by the IMO, and published as: IMO (2011).

[67] Aburgus Risk Management, October 2014c.

any sort of suggestions to the Master of Vessel with regard the use of other means prior to resorting to deadly force.

3.4.2 Reasonability and Proportionality vs. Extreme Circumstances

A pirate skiff may follow the vessel and scout the premises. Is the use of lethal force necessary in order to protect the vessel under such circumstances? In addition, what constitutes a reasonable measure as opposed to unreasonable measures? Is the rule of proportionality a guiding principle to be used for all circumstances, or can the existence of extreme circumstances (such as the occasion of exchange of fire, including heavy weaponry) negate the rule of proportionality and exempt the actor from liability?

In accordance with the report of the International Chamber of Commerce's (ICC's) International Maritime Bureau with regard to the attack on the bulk carrier *Golden Ice*, which occurred on 9 December 2013,[68] following an armed attack of a pirate skiff,[69] the Master of Vessel undertook a series of actions in accordance with the Best Management Practices (BMP, consisting of the following activities: alarm, fire hoses activated, evasive maneuvers, and, finally, the crew's retreat to the protection of citadel).[70] Irrespective of the fact that the armed security guards had made their presence visible,[71] the pirates continued with the aggression. Following a warning flare (general discharge of weapons) and a warning shot (discharge of weapons in the vicinity of the pirate skiff),[72] the pirates nevertheless pursued the attack and reengaged the vessel with weapons' fire. Finally, the armed security guards engaged the pirates directly, resulting in the pirate's withdrawal. Similarly, in a recent incident that occurred in the Gulf of Oman, two skiffs approached a vessel at full speed, with visible ladders on board the skiffs. The pirates opened fire on the vessel, and the Contractors responded, resulting in the withdrawal of attackers.[73] Such examples indicate circumstances under which the Contractors are ready (and able) to determine that the use of (lethal) force is the only available means to thwart the attack.

A pirate skiff may attempt to board the vessel by the use of boarding ladders, supported by the use of guns. Is the use of lethal force unavoidable under such circumstances? Is a professional maritime security service provider expected to use proportional measures to thwart the attack without resorting to ultimate deadly force?

---

[68]ICC International Maritime Bureau (2013), p. 25.

[69]Weapons were sighted. For an example where boarding ladders where sighted, sufficing as a proof of hostile intentions, see the report on an attack on a chemical tanker, available in: ICC International Maritime Bureau (2014), p. 28.

[70]Industry Stakeholders (2011).

[71]For examples where this proved to be a sufficient deterrence, see: ICC International Maritime Bureau (2014)—the "*Gulf Pearl*" case.

[72]For examples where the warning shots have proved to be sufficient, see: ICC International Maritime Bureau (2014)—the "*MSC Jasmine*" case, "*Alba Star*" case and "*Island Splendor*" case.

[73]Aburgus Risk Management (2014b), Issue 29.

In July 2014 in Nigeria, a group of armed pirates attacked an oil rig platform, and a firefight ensued between the attackers and policemen guarding the rig, resulting in the death of five attackers.[74] In the attack on the tanker *SP Brussels*, several people (both pirates and crew) lost their lives, following a firefight between armed guards and pirates and the subsequent withdrawal of armed guards and most of the crew to the safety of the citadel.[75] These cases, as is unfortunately very often the case in some pirate-infested regions such as the West Africa theater of operations, indicate that sometimes the use of force will likely represent the first and, simultaneously, the last means of ensuring security on board a vessel. A stringent rule, stipulating the necessity of following each and every step in accordance with the gradual increase of force severity, may very well hinder or disable the Contractor's ability to provide a successful protection service.

3.4.3 Conclusion

It is unlikely that any new set of recommendations and guidelines will offer a critically different approach toward the use of force by private entities engaged in providing private (maritime) security services. A written rule can only provide so much—the rest is left to the professional service providers, consumers, third party interests, and, finally, tribunals. The *Nisour Square* incident adjudication and the upcoming decision by the Indian Supreme Court (or arbitration decision) in the *Enrica Lexie* case have and certainly will shed more light on the overall accountability and liability of private military and (maritime) security companies. They have and will provide legal precedents and bases for possible further adjudication and will produce a significant impact on the industry. It is, nevertheless, quite likely that a number of such cases will remain at an all-time low due to the fact that the facts and circumstances surrounding occurrences when the force is utilized by private security providers, devoid of objective, neutral, and third-party oversight and control authority, very often remain blurred and one-sided, making it almost impossible to claim otherwise.

References

Journals and Articles

Bueger C (2015) What is maritime security? Mar Policy 53:159–164

Cheadle SP (2009) Private military contractor liability under the worldwide personal protective service II contract. Public Contract Law J 38(3):689–707

[74] Aburgus Risk Management (2014a), Issue 28.

[75] See: Gallagher (2014).

Coito JC (2013) Pirates vs. private security: commercial shipping, the Montreux document, and the battle for the Gulf of Aden. Calif Law Rev 101(1):173–226

Dutton YM (2013) Gunslingers on the high seas: a call for regulation. Duke J Comp Int Law 24:107–160

Eboli V, Pierini JP (2012) The 'Enrica Lexie' case and the limits of the extraterritorial jurisdiction of India. I quaderni europei 39:2–27

Gandhi M (2013) The Enrica Lexie Incident: seeing beyond the grey areas of international law. Indian J Int Law 53:1–26

Neri K (2012) The use of force by military vessel protection detachments. Military Law Law War Rev 51(73):74–93

Mudrić M (2011) Armed guards on vessels: insurance and liability. Poredbeno pomorsko pravo 50 (165):217–268

Mudrić M (2015a) The Guardcon contract, knock-for-knock clauses, DCFR and unfair terms (Part I). JIML 21(1):51–62

Mudrić M (2015b) The Guardcon contract, knock-for-knock clauses, DCFR and unfair terms (Part I). JIML 21(2):115–132

Patrick S (2014) Mahard Blackwater's New Battlefield: toward a regulatory regime in the United States for privately armed contractors operating at sea. Vand J Transnatl Law 47:331–369

Petrig A (2013) The use of force and firearms by private maritime security companies against suspected pirates. Int Comp Law Q 62(3):667–701

Scheimer M (2009) Separating private military companies from illegal mercenaries in international law: proposing an international convention for legitimate military and security support the reflects customary international law. Am Univ Int Law Rev 24(3):609–646

Symmons CR (2012) Embarking vessel protection detachments and private armed guards on board commercial vessels: international legal consequences and problems under the law of the sea. Military Law Law War Rev 51:21–53

Tarzwell A (2009) In search of accountability: attributing the conduct of private security contractors to the united states under the doctrine of state responsibility. Oregon Rev Int Law 11:179–204

Tondini M (2012) Some legal and non-legal reflections on the use of armed protection teams on board merchant vessels: an introduction to the topic. Military Law Law War Rev 51(1):7–20

Books and Chapters

Andreone G, Bevilacqua G, Cataldi G, Cinelli C (2013) Insecurity at sea: piracy and other risks to navigation. Giannini Editore, Napoli

von Bar C, Clive EM (eds) (2010) Principles, definitions and model rules of European private law: draft common frame of reference. Oxford University Press, Oxford

Berube C, Cullen P (eds) (2012) Maritime private security: market responses to piracy, terrorism and waterborne security risks in the 21st century. Routledge, London

Buckland BS, Burdzy A (2013) Progress and opportunities, five years on: challenges and recommendations for Montreux document endorsing states. DCAF: A Centre for Security, Development and the Rule of Law, Geneva

Chesterman S, Lehnardt C (2007) From mercenaries to market: the rise and regulation of private military companies. Oxford University Press, Oxford

Guilfoyle D (2013) Modern piracy: legal challenges and responses. Edward Elgar, London

Jäger T, Kümmel G (2007) Private military security companies: chances, problems, pitfalls and prospects. VS Verlag für Sozialwissenschaften, Wiesbaden

Kraska J (2011) Contemporary maritime piracy: international law, strategy, and diplomacy at sea. Praeger, Santa Barbara

Kraska J, Pedrozo R (2013) International maritime security law. Martiuns Nijhoff, Leiden
Mejia MQ, Kojima C, Sawyer M (2013) Piracy at sea. Springer, Heidelberg
Mikac R (2013) Suvremena sigurnost i privatne sigurnosne kompanije: Privatizacija sigurnosti i posljedice. Jesenki i Turk, Zagreb
Murdoch A (2011) Recent legal issues and problems relating to acts of piracy off Somalia. In: Symmons CR (ed) Selected contemporary issues in the law of the sea. Nijhoff, The Hague
Murphy J, Witting C (2012) Street on torts. Oxford University Press, Oxford
Percy S (2006) Regulating the private security industry. Routledge, London
Petrig A, Geiβ R (2011) Piracy and armed robbery at sea: the legal framework for counter-piracy operations in Somalia and the Gulf of Aden. Oxford University Press, Oxford
Prince E (2013) Civilian warriors: the inside story of blackwater and the unsung heroes of the war on terror. Penguin Publishing Group, New York
Tonkin H (2011) State control over private military and security companies in armed conflict. Cambridge University Press, Cambridge
Williams SO (2014) Assessing state jurisdiction and industry regulation over private maritime security: an international and comparative regulatory review. The Artic University of Norway, Narvik

Cases

Cross v. Kirkby (2000) Times, 5th April (CA)
Darshan Singh v. State of Punjab & another, Criminal Appeal No. 1057 of 2002
Palmer v. R [1971] AC 814
The "Enrica Lexie" Incident (Italy v. India) Case No. 24, International Tribunal for the Law of the Sea
The "Enrica Lexie" Incident (Italy v. India), International Tribunal for the Law of the Sea Order 24 of August 2015, International Tribunal for the Law of the Sea*The M/V "SAIGA" (No. 2) Case (Saint Vincent and the Grenadines v. Guinea)*, Case No. 2, International Tribunal for the Law of Sea, ITLOS Rep. 10, § 155*United States v. Slough*, Criminal No. 2008-0360 (D.C. 2014), District Court, District of Columbia, Filed: May 23rd, 2014

Legal Documents

Pravilnik o uvjetima koje moraju ispunjavati pravne osobe koje pružaju usluge ukrcaja osoba za naoružanu pratnju na brodove hrvatske državne pripadnosti, Narodne novine, broj 123/12
Regulation on the employment of contractors on board Italian flagged ships sailing in international waters under piracy risk, 29th March 2013, Italian Official Gazette
Title 33 – Navigation and Navigable Waters, Chapter 6 – General Duties of Ship Officers and Owners after Collision or other Accidents, 33 U.S.C. § 383 - Resistance of pirates by merchant vessels
UK Department for Transport, *Interim Guidance to UK Flagged Shipping on the Use of Armed Guards to Defend Against the Threat of Piracy in Exceptional Circumstances*, Version 1.2, Updated May 2013
U.S. Department of Homeland Security, United States Coast Guard, *Port Security Advisory (3-09), Guidance on Self-Defense or Defense of Others by U.S. Flagged Commercial Vessels Operating in High Risk Waters*

Other

Gallagher AJ (2014) Marine Market Report: Marine Hull & Machinery and War Risks Market Update – September 2014. http://www.ajginternational.com/media/95376/ajg-marine-market-report_sept-2014.pdf. Accessed 1 Dec 2015

Aburgus Risk Management (2014a) Piracy Situation Risk Assessment, August 2014. Issue 28. http://www.aburgus.com/wp-content/uploads/2014/08/Risk-Assessment-August-14.pdf. Accessed 1 Dec 2015

Aburgus Risk Management (2014b) Piracy Situation Risk Assessment, September 2014. Issue 29, http://www.aburgus.com/wp-content/uploads/2014/09/Risk-Assessment-Sep-14.pdf. Accessed 1 Dec 2015

Aburgus Risk Management (2014c) Piracy Situation Risk Assessment, October 2014. Issue 30, http://www.aburgus.com/wp-content/uploads/2014/10/Risk-Assessment-Oct-14.pdf. Accessed 1 Dec 2015

American National Standards Institute, Inc. ASIS International (2013) Quality Assurance and Security Management for Private Security Companies Operating at Sea - Guidance ANSI/ ASIS PSC.4-2013

BIMCO (2012a) GUARDCON Standard Contract for the Employment of Security Guards on Vessels (2012)

BIMCO (2012b) Guidance on Rules for the Use of Force (RUF) by Privately Contracted Armed Security Personnel (PCASP) in Defence of a Merchant Vessel (MV)

Globus Intelligence Ltd (2013) The 100 Series Rules: An International Model Set of Maritime Rules for the Use of Force (RUF)

Hindu, December 16, 2014. http://www.thehindu.com/news/resources/timeline-the-italian-marines-case/article4538162.ece. Accessed 1 Dec 2015

Indian Ministry of Shipping (2011) Guidelines on Deployment of Armed Security Guards on Merchant Ships. F.No. SR-13020/6/2009-MG(pt.)

Industry Stakeholders (2011) BMP4 - Best Management Practices for Protection against Somalia Based Piracy, Suggested Planning and Operational Practices for Ship Operators and Masters of Ships Transiting the High Risk Area. Version 4

International Association of Maritime Security Professionals (2011) Use of Force. IAMSP-2011-01-UOF-001 v2.0

International Chamber of Commerce, International Maritime Bureau (2013) Piracy and Armed Robbery against Ships: Report for the Period 1 January – 31 December 2013

International Chamber of Commerce, International Maritime Bureau (2014) Piracy and Armed Robbery against Ships: Report for the Period 1 January – 31 December 2014

International Committee of the Red Cross (2009) The Montreux Document: On pertinent international legal obligations and good practices for States related to operations of private military and security companies during armed conflict. Swiss Confederation Federal Department of Foreign Affairs, Geneva

International Maritime Organization (2012a) Interim Guidance to private maritime security companies providing contracted armed security personnel on board ships in the High Risk Area. MSC.1/Circ.1443

International Maritime Organization (2011) Piracy and Armed Robbery against Ships in Waters off Somalia, Best Management Practices for Protection against Somalia Based Piracy. MSC.1/ Circ.1339

International Maritime Organization (2012b), Revised Interim Guidance to Shipowners, Ship Operators and Shipmasters on the Use of Privately Contracted Armed Security Personnel on board Ships in the High Risk Area. MSC.1/Circ.1405/Rev.2

International Organization for Standardization (2012) The ISO/PAS 28007:2012 Ships and marine technology - Guidelines for Private Maritime Security Companies (PMSC) providing privately contracted armed security personnel (PCASP) on board ships (and pro forma contract)

International Tribunal for the Law of the Sea (2015) The "Enrica Lexie" Incident (Italy v. India), Provisional Measures. ITLOS/Press 237

Katz A (2012) Fighting Piracy Goes Awry with Killings of Fishermen. Bloomberg Business, September 17

Los Angeles Times, October 22, 2014. http://www.latimes.com/nation/la-na-iraq-blackwater-verdict-20140903-story.html?track=rss#page=1. Accessed 1 Dec 2015

Maritime Security Centre – Horn of Africa (MSCHOA). http://www.mschoa.org/on-shore/home. Accessed 1 Dec 2015

Den Norske Krigsforsikring for Skib - Gjensidig Forening (The Norwegian Shipowners' Mutual War Risks Insurance Association) (2011) Guidance on the Selection of Private Security Companies (PSC)

Norwegian Mar. Directorate (2011) Provisional Guidelines – Use of Armed Guards on Board Norwegian Ships

Reuters, 10 June 2013, Murder trial of Italian marines in India navigates murky waters

U.S. Department of State (2006) Foreign Operations and Related Programs Appropriations for Fiscal Year 2006: Hearings before a Subcommittee of the Committee on Appropriations. U.S. Senate, Mission Firearms Policy: American Embassy – Baghdad

U.S. Department of State (2009) Private Security Contracting in Iraq and Afghanistan: Statement of Ambassador Richard J. Griffin

United Nations Authorized Embargos and Maritime Interdiction: A Special Focus on Somalia

Magne Frostad

1 Introduction

Somewhat less known than the UN Security Council (UNSC) resolutions on Somali piracy are its resolutions on a weapons embargo of that state—an embargo established due to the anarchy ruling in Somalia after the fall of Dictator Siad Barre.[1] Resolutions have been adopted on this topic since 1992[2] without much effect as no proper enforcement regime has been authorized. This flow of weapons and ammunition has not only helped to arm the factions that fought for supremacy in Somalia but has most certainly also helped the organizers of piracy to equip the tactical teams that until recently brought larger vessels into Somali ports in wait for ransom of ship and crew.

In 2014, the UNSC finally authorized enforcement measures for the upholding of this embargo, as well as for curbing the illegal trade in charcoal from Somalia.[3] The latter trade was targeted as it tends to finance the activities of the rebel-terrorists in southern Somalia known as Al-Shabaab. One of the three international maritime forces in the region—the Combined Maritime Forces[4]—has to some extent applied

[1]In the preamble, the UNSC states that it is "[g]ravely alarmed at the rapid deterioration of the situation in Somalia and the heavy loss of human life and widespread material damage resulting from the conflict in the country and aware of its consequences on the stability and peace in the region"; UNSC Res. 733 of 23 January 1992 (italics in original omitted).

[2]UNSC Res. 733 of 23 January 1992.

[3]UNSC Res. 2182 of 24 October 2014.

[4]The others are NATO's Operation Ocean Shield (http://www.nato.int/cps/en/natohq/topics_48815.htm), and the European Union's Operation Atalanta (http://eunavfor.eu/).

M. Frostad (✉)
Faculty of Law, K.G. Jebsen Centre for the Law of the Sea, University of Tromsø, The Arctic University of Norway, Tromsø, Norway
e-mail: magne.frostad@uit.no

G. Andreone (ed.), *The Future of the Law of the Sea*,
DOI 10.1007/978-3-319-51274-7_11

the charcoal authorization successfully, and its efforts have been welcomed by the UNSC.[5] On the other hand, until recently, no interdiction had been undertaken in order to uphold the weapons embargo. This chapter will, among other things, seek to shed light on why that is so.

A quick terminological note might here be in order: Whereas "embargo" is used in this chapter as a reference to the order that closes a territory in relation to the goods covered by that order, maritime interdiction operations is an operational term of art that also covers a multitude of other maritime operations.[6] The enforcement of an embargo is nevertheless usually referred to as interdiction activities.[7] Moreover, these maritime interdiction operations are different from the *jus in bello* concept of naval blockade, whereby naval vessels and aircraft patrol the coastline of an enemy state in order to stop any in- and outgoing traffic from her ports.[8]

In the following, the chapter will commence by looking at the history of UN authorized embargos where enforcement has been contemplated through, typically, visitation of vessels and confiscation of goods violating the embargo. An issue is here the right to innocent passage in the territorial seas of the states actually enforcing the embargo, and the regime established by UNSC Res. 1973 (2011) will for this purpose be considered in some detail. The chapter will then move on to the regime adopted for Somalia, with a focus on the enforcement authority granted in UNSC Res. 2182 (2014), especially in relation to weapons etc., before some of the challenges resulting from this latter authorization are considered.

2 The History of UN Security Council Authorized Maritime Interdictions

2.1 Material Scope

To the knowledge of this author, the UN General Assembly has never sought to authorize maritime interdiction, whereas it has on some occasions exercised a role in relation to peacekeeping and the coordination of collective self-defense. On the other hand, the UNSC has set up such operations on a few occasions, and weapons embargos are the most often used form of economic sanctions authorized by the UNSC.[9]

---

[5]UNSC Res. 2244 of 23 October 2015, para. 19. There is no reference to this operation on the homepage of the Combined Maritime Forces (https://combinedmaritimeforces.com/).

[6]von Heinegg (2010), p. 375.

[7]McLaughlin (2009), p. 137 (under reference to Soons).

[8]Doswald-Beck (1995), pp. 26–27 and 176–180. The concept was in use in the Vietnam war, Iran-Iraq war and during the 2006 Israel-Lebanon conflict; see McLaughlin (2009), p. 125, n. 2. A special instance of blockade is Israel's blockade of the Gaza strip, see Klein (2010), p. 294.

[9]Fleck (2013), p. 70 (with further references).

The UNSC first applied this measure to the unilateral declaration of independence by the white minority of Southern Rhodesia in 1966. As a consequence of the Portuguese colonial port of Beira being essential for the supply of oil to Southern Rhodesia, the UNSC mandated in Res. 221 (1966) the British "to prevent, by the use of force if necessary, the arrival at Beira of vessels reasonably believed to carrying oil destined for Southern Rhodesia, and empowers the United Kingdom to arrest and detain the tanker known as the *Joanna V* upon her departure from Beira in the event of her oil cargo is discharged there."[10]

No authorizations were then made until the end of the Cold War when the UNSC agreed on a firm response toward Iraq for her invasion of Kuwait *inter alia* with Res. 665 (1990) authorizing maritime interdiction activities. In para. 1 thereof, the UNSC "[c]alls upon those Member States co-operating with the Government of Kuwait which are deploying maritime forces to the area to use such measures commensurate to the specific circumstances as may be necessary under the authority of the Security Council to halt all inward and outward maritime shipping in order to inspect and verify their cargoes and destinations and to ensure strict implementation of the provisions related to such shipping laid down in resolution 661 (1990)."[11] This formulation was used as a model until the Libyan crisis in 2011.[12] During the 1990s, such authorizations were made in relation to the Federal Republic of Yugoslavia (Serbia and Montenegro),[13] Haiti,[14] and Sierra Leone.[15]

A somewhat atypical authorization was issued in 2006 when the UN Interim Force in Lebanon was granted the power to *assist* the Government of Lebanon, at

---

[10]UNSC Res. 221 of 9 April 1966, para. 5 (italics in original). The resolution *inter alia* calls upon flag states to "to ensure the diversion of their vessels reasonably believed to be carrying oil destined for Southern Rhodesia which may be en route for Beira" (para. 4). For more on the Beira patrol, as well as the argument that the UNSC acted under Art. 41, see McLaughlin (2009), pp. 134–135. In the current author's view, the better view is held by Frowein whom argues that this resolution, expressly authorizing the use of force, should instead be seen as authorized by Art. 42; see McLaughlin (2009), p. 135, n. 43.

[11]Italics in original not indicated. UNSR Res. 661of 6 August 1990 established economic sanctions in relations to "all commodities and products originating in Iraq or Kuwait" (para. 3 (a)), hindered similar items from being exported to Iraq and Kuwait ((c)), and stopped the transfer of financial and economic resources (para. 4). For more on this operation, see McLaughlin (2009), pp. 135–138. Dinstein holds that "[i]n practical terms, Iraq was subjected in consequence to a blockade, although Resolution 665 avoided that expression", see Dinstein (2012), p. 320.

[12]Fink (2011), p. 239.

[13]UNSC Res. 787 of 16 November 1992, para. 12, and UNSC Res. 820 of 17 April 1993, paras. 28 and 29. Thus, even vessels merely in transit through the territorial sea of the Federal Republic of Yugoslavia (Serbia and Montenegro) were covered, thereby suspending innocent passage. See McLaughlin (2009), p. 143. On the naval interdictions activities in relation to this set of authorizations, see id., pp. 138–145.

[14]UNSC Res. 875 of 16 October 1993, para. 1, and UNSC Res. 917 of 6 May 1994, paras. 9 and 10.

[15]UNSC Res. 1132 of 8 October 1997, paras. 6 and 8. This authorization was terminated in UNSC Res. 1940 of 29 September 2010, para. 1.

the latter's *request*, in preventing the entry into Lebanon without the government's consent of arms or related materiel.[16]

Another atypical situation, in that it limited the otherwise typically authorized enforcement powers significantly, was the UNSC's response to North Korea in Res. 1874 (2009). This resolution is enacted explicitly under Art. 41 and limits states to "inspect, in accordance with their national authorities and legislation, and consistent with international law, all cargo to and from the DPRK, in their territory."[17] Moreover, it calls upon states "to inspect vessels, with the consent of the flag State, on the high seas, if they have information that provides reasonable grounds to believe that the cargo of such vessels contains items the supply, sale, transfer, or export of which is prohibited."[18] The limited right of enforcement is underscored by para. 13, which "[c]alls upon all States to cooperate with inspections pursuant to paragraphs 11 and 12, and, if the flag State does not consent to inspection on the high seas, decides that the flag State shall direct the vessel to proceed to an appropriate and convenient port for the required inspection by the local authorities pursuant to paragraph 11." Klein correctly holds that the adopted approach would suffice as long as the flag state actually instructs the ship to sail for a specified port and undergo the relevant inspection.[19] If the flag state abstains from doing just that, it might fall out of favors with the UNSC, but the resolution, however, does not include a right to use force against a vessel whose flag state does not consent to its inspection. In later resolutions, states are obliged to deny vessels refusing such inspection entry into their ports unless it is for the purpose of inspection of the said vessel, in case of emergency, or it is returning to its port of origination.[20]

Another case of maritime interdiction not exactly of the embargo kind is here the authorization granted in relation to migrant smuggling and human trafficking from Libya in UNSC Res. 2240 (2015), where vessels suspected on reasonable grounds for such may be inspected on the high seas off the coast off Libya and seized when confirmed as such after inspection.[21]

In February 2011, the UNSC responded to the situation in Libya by authorizing *inter alia* a weapons embargo under the explicit reference to UN Charter Art. 41.[22] The UN member states are in para. 9 obligated to "take the necessary measures to prevent the direct or indirect supply, sale or transfer to the Libyan Arab Jamahiriya, from or through their territories or by their nationals, or using their flag vessels or aircraft" of certain items, assistances, and personnel.

These items, assistances, and personnel are identified as "arms and related materiel of all types, including weapons and ammunition, military vehicles and

[16]UNSC Res. 1701 of 11 August 2006, paras. 12 and 14. On this resolution, see Fink (2013), p. 81.

[17]Para. 11.

[18]Para. 12.

[19]Klein (2010), p. 279.

[20]See e.g. UNSC Res. 2094 of 7 March 2013, para. 17.

[21]UNSC Res. 2240 of 9 October 2015, paras. 7, 8 and 10.

[22]UNSC Res. 1970 of 26 February 2011, paras. 9–14.

equipment, paramilitary equipment, and spare parts for the aforementioned, and technical assistance, training, financial or other assistance, related to military activities or the provision, maintenance or use of any arms and related materiel, including the provision of armed mercenary personnel whether or not originating in their territories."[23] To this list, a range of exceptions is included in the same paragraph. Additionally, para. 10 establishes a prohibition on Libyan export of arms and related materiel.

Notice should here be made of the flexible formulation of "arms and related material of all types," which might, importantly, also cover so-called dual use objects.[24] Moreover, the inclusion of personnel in the embargo—armed mercenaries—was a novelty,[25] and this brought about the question of how to deal with potentially detained personnel.[26]

Paragraph 11 moreover authorizes "all States, in particular States neighbouring the Libyan Arab Jamahiriya, to inspect, in accordance with their national authorities and legislation and consistent with international law, in particular the law of the sea and relevant international civil aviation agreements, all cargo to and from the Libyan Arab Jamahiriya, in their territory, including seaports and airports, if the State concerned has information that provides reasonable grounds to believe the cargo contains items the supply, sale, transfer, or export of which is prohibited by paragraphs 9 or 10 of this resolution for the purpose of ensuring strict implementation of those provisions."

This later authorization is limited to "their territory" and would thus apply to the territorial sea of a member state, seemingly hindering innocent passage. Admittedly, the authorization does not specifically mention territorial seas, whereas it explicitly mentions seaports. This can be understood either as merely highlighting the most important part of the maritime area where sanctions could be enforced, *i.e.*, not to the exclusion of, e.g., internal waters and the territorial sea, or as indicating that only port state authority was contemplated and no restrictions on innocent passage intended. As the authorization does require the state to undertake its enforcement "in accordance with their national authorities and legislation and consistent with international law, in particular the law of the sea [...]," it would seem that the answer to whether the embargo could be enforced in a coastal state's territorial sea would depend on the reach of costal state enforcement jurisdiction under the 1982 UN Convention on the Law of the Sea (UNCLOS).[27]

---

[23]Ibid, para 9.

[24]Fink (2011), pp. 252–253.

[25]Ibid., p. 254.

[26]Ibid., p. 256.

[27]1833 UNTS 3.

2.2 *Innocent Passage*

The issue of innocent passage is regulated by UNCLOS in Arts. 18–26 and their customary international law counterparts. A starting point would here be the requirement in Art. 19 para. 1 of "[s]uch passage shall take place in conformity with this Convention and with other rules of international law."[28] These "other rules of international law" may *inter alia* be mandatory resolutions of the UNSC. However, Res. 1970 (2011) does not grant enforcement powers in relation to the weapons embargo beyond that which was already recognized by international law. Moreover, the said embargo operations are not explicitly covered by the exceptions to innocent passage in Art. 19 para. 2. Arguably, para. 2 is not a closed list of what may be considered "prejudicial to the peace, good order or security of the coastal State" due to the flexible wording of *litra* l ("any other activity not having a direct bearing on passage"),[29] but it might be seen as odd if the relevant nonmentioned situations where just slight variations of those listed. Thus, as regards the exceptions explicitly mentioned in para. 2, an antithetical approach could be held as preferable, and this would still leave room for other scenarios that are different from those identified in *litras* a to k. However, the very wording of *litra* l is so wide in itself ("any other activity") that it would be hard to restrict its interpretation in such a way. On the other hand, it would seem strange if the mere transport of such goods would constitute an activity "not having a direct bearing on passage."

However, no general exemption from the laws and regulations of the coastal state exists for vessels in innocent passage,[30] whereas UNCLOS Art. 21 (1) (h) allows the coastal state to adopt laws and regulations regarding "the prevention of infringement of the customs, fiscal, immigration or sanitary laws and regulations of the coastal State." One approach could then be to argue that customs laws provide the required authority as—it would have to be argued—a state can make these applicable also to transshipment through its territory, if not otherwise obligated. However, such a restriction would seriously undermine the very concept of innocent passage, and this cannot have been the intention of the states negotiating UNCLOS. Tanaka moreover argues that the violation of the laws of the coastal state does not *ipso facto* deprive a passage of its innocent character, unless such violations fall within the scope of Art. 19.[31] What does then Art. 19 allow for in such a setting? Actually, the reference in Art. 19 (2)(g) to "the loading or unloading of any

---

[28] Klein (2010), p. 278. For the view that such transport is in violation of Arts. 17, 19 and 301 due to its non-peaceful purpose, see Song (2007), pp. 116–117.

[29] Tanaka (2015), p. 88. This would then be in contradistinction to the 1989 USA-USSR Joint Statement on the Uniform Interpretation of Rules of International Law Governing Innocent Passage, para. 3 (http://cil.nus.edu.sg/rp/il/pdf/1989%20USA-USSR%20Joint%20Statement%20on%20the%20Uniform%20Interpretation%20of%20Rules%20of%20International%20Law-pdf.pdf).

[30] Rothwell and Stephens (2016), p. 457.

[31] Tanaka (2015), p. 89.

commodity [..] contrary to the customs [...] laws and regulations of the coastal State" would seem to limit customs powers to situations where the cargo is physically transferred off the vessel, and that is wholly different from a vessel merely sailing through the territorial sea of a state.

Be that as it may, as Art. 27 refers to "should not" as opposed to "shall not," the latter being used in Arts. 21 para. 2 and 24 para. 1, the coastal state actually has *criminal* jurisdiction over ships within its territorial sea also beyond the situations mentioned in para. 1 of Art. 27, although this authority should generally not be exercised for purposes outside of those mentioned in Art. 27.[32] Although it might be argued that the possible reach of the laws of the coastal state should be seen in relation to Art. 19, Guilefoyld correctly points out in relation to the transport of weapons of mass destruction (WMD) that the obligations flowing from UNSC Res. 1540 means that "[i]n the context of WMD shipments passing through the territorial sea, it is exactly that capacity to act which becomes an *obligation* to act under UNSCR 1540."[33] Moreover, a transport of WMD through the territorial sea is seen by him, for the purpose of Art. 27 para. 5, as being "continuous acts, breaches of the prohibition occurring as much within territorial waters as without."[34] The WMD regime under Res. 1540 (2004)[35] and the embargo regime under, e.g., Res. 1973 (2011) are both passed under Chapter VII of the UN Charter and oblige states to take the necessary measures to stop the relevant activity.[36] It might thus be argued

[32]Guilfoyle (2009), p. 242, and Tanaka (2015), p. 96. For the view that counter-proliferation and counter-terrorism operations are covered by Art. 27, para. 1, *litras* a or b, see von Heinegg (2010), pp. 378–379.

[33]Guilfoyle (2009), p. 242. This would provide a legal basis for the interdictions considered by the Proliferation Security Initiative to be consistent with international law: "To take appropriate actions to (1) stop and/or search in their internal waters, *territorial seas, or contiguous zones* (when declared) vessels that are reasonably suspected of carrying such cargoes [WMD, their delivery systems, or related materials] to or from states or non-state actors of proliferation concern and to seize such cargoes that are identified; and (2) to enforce conditions on vessels entering or leaving their ports, internal waters or territorial seas that are reasonably suspected of carrying such cargoes, such as requiring that such vessels be subject to boarding, search, and seizure of such cargoes prior to entry": U.S. Department of State, Statement of Interdiction Principles, 2003, No. 4, chapeau and litra d, (http://www.state.gov/t/isn/c27726.htm) (italics by author). Without this resolution, Tanaka would be right in pointing out that this self-authorization is not necessarily in accordance with international law, as Art. 19(2) does not mention WMD, and since it is debatable whether the transport of such through the territorial sea of a state is "prejudicial to the peace, good order or security of the coastal State" under Art. 19(1): Tanaka (2015), p. 393.

[34]Guilfoyle (2009), p. 243. See also von Heinegg (2010), p. 379.

[35]As continued until 2021 under UNSC Res. 1977 of 20 April 2011, para. 2.

[36]UNSC Res. 1540 of 28 April 2004 explicitly obliges states in para. 2 to "adopt and enforce appropriate effective laws which prohibit any non-State actor" to e.g. manufacture WMD, whereas UNSC Res. 1970 para. 9 more generally orders that states "shall immediately take the necessary measures to prevent the direct or indirect supply, sale or transfer" to Libya of the relevant items. Presumably, such measures would by necessity include the rules necessary in order to implement this obligation in domestic law, and this limits the difference between the approaches used. Admittedly, the obligation under Res. 1540 would seem to also cover criminal sanctions, whereas the normal embargo authorization might possibly be limited to non-criminal sanctions.

that coastal states are obliged to use their right to legislate under Art. 27 to hinder "innocent" passage of the items covered by an embargo and that the breaching of such a UN resolution constitutes a continuous act by the vessel undertaking "innocent" passage.

2.3 Other Aspects of the Geographical Application of the Libyan Embargo

As regards the waters beyond the territorial sea, Fink correctly points out that the reference in Res. 1970 (2011) to "their territory" did not mandate actual enforcement at sea.[37] Basically, states were thus limited to checking vessels going to their ports, or flying their own flag, unless the foreign flag state consented to such control. The states could nevertheless monitor the effectiveness of the measures through information gathering, and they could support other actors who actually had jurisdiction over potential embargo breakers in their response toward these.[38]

However, this situation changed with the authorization granted in Res. 1973 (2011) para. 13, where the UNSC "[c]alls upon all Member States, in particular States of the region, acting nationally or through regional organisations or arrangements, in order to ensure strict implementation of the arms embargo established by paragraphs 9 and 10 of resolution 1970 (2011), to inspect in their territory, including seaports and airports, and on the high seas, vessels and aircraft bound to or from the Libyan Arab Jamahiriya, if the State concerned has information that provides reasonable grounds to believe that the cargo contains items the supply, sale, transfer or export of which is prohibited by paragraphs 9 or 10 of resolution 1970 (2011) as modified by this resolution, including the provision of armed mercenary personnel, *calls upon* all flag States of such vessels and aircraft to cooperate with such inspections and authorises Member States to use all measures commensurate to the specific circumstances to carry out such inspections."[39]

Thus, this resolution allows a state to also inspect foreign flagged vessels on the high seas. This is due to the general reference to "vessels" where no exception is made for those flying a foreign flag, the following reference to flag states being called upon to cooperate, and the added "all measures" authorization. With this expansion of the regime under Res. 1970 (2011) to the high seas and including the vessels of other flag states, the reference to enforcement in the territory of the state would, if not earlier, at least now include a right of the coastal state to act against vessels in innocent passage in its own territorial sea: if it can enforce the embargo where it otherwise has no sovereignty, it must even more be able to do so where it actually holds such an entitlement.

[37]Fink (2011), p. 242.

[38]Ibid., p. 243 with further references.

[39]Italics in original.

Similarly, the reference above to the high seas does not exclude the enforcement of such powers in the exclusive economic zone (EEZ) of all other states, including the Libyan EEZ, as the high seas freedoms largely apply even in the EEZ, and consequently NATO also enforced the embargo in the Libyan EEZ.[40]

As there are no phrases in this resolution on Libya that geographically limit the part of the high seas to which this authorization applies, as would arguably have been the case if the formulation "to halt all inward and outward shipping" had been used, the embargo could have been enforced also on the most remote waters.[41] However, the operation was in practice contained to the central Mediterranean Sea.[42] The embargo was nevertheless not enforced in the Libyan territorial sea, or for that matter seemingly in the territorial sea of other states,[43] as the former was understood to fall outside of the latter's authorization as it only referred to "the high seas."[44]

Presumably, the territorial sea limitation in Res. 1973 (2011) could have been bypassed by the authorization granted in para. 4 to take all necessary measures to protect civilians and civilian-populated areas under threat of attack. If that is so, at the very least, such an operation would have to be closely linked to that part of the mandate, i.e., not being seen as an embargo operation.[45] However, although the civilian protection authorization does not have an explicit geographical limitation as such, it nevertheless refers to civilian-populated areas, and these will almost by necessity be on land. Thus, enforcement in the territorial sea of Libya would seemingly have to relate to the halting of troops, weapons, or ammunition intended for use, preferably in the close future, against such civilians and civilian-populated areas.

A different take on the legal situation would be to consider the law of international armed conflicts to be applicable to those states upholding the weapons embargo against Libya that were also involved in the use of armed force for the protection of civilians.[46] As their territories, including territorial seas, would then be covered by the area of that international armed conflict, these coastal states would be in a position to enforce visitation rights over neutral vessels under the law of neutrality, which go far beyond the rights of coastal states in relation to vessels exercising innocent passage in peacetime. Of course, such rights would also apply to the high seas.[47]

The authorization furthermore requires "reasonable grounds" for believing that the embargo is being broken, and this would seem to be similar to the standard used

[40]Fink (2011), p. 247.

[41]Ibid., p. 248.

[42]Ibid., p. 249.

[43]That is the territorial sea of another state than the flag state undertaking the interdiction.

[44]Fink (2011), p. 246.

[45]Ibid., p. 248.

[46]Aaron and Nauta (2013) pp. 363–364.

[47]Doswald-Beck (1995) pp. 8 and 80–82.

in UNCLOS Art. 110 regarding visitations.[48] This threshold links back to the Beira Patrol resolutions as a blanket authority to halt every vessel would seem to have been granted by the post Beira resolutions.[49] The requirement of reasonable grounds differentiates such embargos from a naval blockade.[50] No difference seems here to be made between state-owned vessels and merchant vessels, but as the first category largely enjoys sovereign immunity, clear language to that effect should be expected, before interdiction of state vessels is considered authorized.

The Libyan embargo was loosened in Res. 2009 (2011),[51] whereas NATO ceased its maritime embargo operation after military operations were terminated in Res. 2016 (2011),[52] even though the economic sanctions continued.[53] In relation to illicitly exported crude oil from Libya, a rather narrowly drafted authorization was provided by the UNSC in Res. 2146 (2014).

2.4 The Authorization to Use Force

In the resolution authorizing the Beira patrol, use of force is explicitly referred to ("by the use of force if necessary"), and the resolution is passed under reference to the "situation constitute[ing] a threat to the peace," although no explicit reference is made to Arts. 41 or 42.[54] The other abovementioned resolutions went from vague references to Chapter VII through a reference to other resolutions,[55] via a reference to the situation in the relevant state as constituting a threat to international peace and security,[56] to explicit mentioning of Chapter VII.[57] Thus, it must be decided through a rather contextual analysis whether the UNSC has authorized an embargo under Art. 41 or Art. 42.[58]

McLaughlin raises the question of whether authorizations to implement economic sanctions should rather be seen as uses of force different from military

48 Fink (2011), p. 250.

49 Id.

50 Ibid., pp. 250–251.

51 UNSC Res. 2009 of 16 September 2011, para. 13.

52 UNSC Res. 2016 of 27 October 2011, para. 5.

53 Fink (2011), p. 241.

54 Fink nevertheless holds that this resolution was passed under reliance on Art. 41. See Fink (2011), p. 244.

55 UNSC Res. 665 of 25 August 1990.

56 UNSC Res. 1701 of 11 August 2006.

57 UNSC Res. 787 of 16 November 1992, para. 12, UNSC Res. 820 of 17 April 1993 (text between paras. 9 and 10), UNSC Res. 875 of 16 October 1993, UNSC Res. 917 of 6 May 1994, UNSC Res. 1132 of 8 October 1997.

58 For a similar view, see Fink (2013), p. 83.

sanctions under Art. 42,[59] and he also holds that Art. 41 and Art. 42 should be considered as a continuum making it less important with a clear-cut separation.[60] However, with the development of the expanded notion of Art. 42, where also situations below the threshold of full-scale warfare are covered, it is submitted that the better view would be to consider maritime interdiction activities authorized to use force as having been granted under Art. 42.[61] Correspondingly, the lack of a direct authorization to use force through the traditional formulations used for that purpose would arguably leave naval vessels intent on enforcing a UNSC-established embargo without the right to use any force in so doing.[62]

Res. 1970 (2011) is nevertheless clear as it refers explicitly to Art. 41, whereas Res. 1973 (2011) merely applies an unspecified reference to Chapter VII as such. The reference therein to "all necessary measure" (paras. 4 and 8) and "all measures commensurate to the specific circumstances" (para. 13) nevertheless shows that Art. 42 is the basis for Res 1973 (2011).[63] The latter formulation is a slight rephrasing of the formulation used in relation to Iraq (Res. 665 (1990)), which instead used these words in para. 1: "[U]se such measures commensurate to the specific circumstances as may be necessary." The chosen wording might be seen as a combination of the usual phrasing for such operations and the wording traditionally used for enforcement operations ("all necessary means/measures").[64]

3 Somalia

3.1 General Aspects

Although the counterpiracy resolutions of the UNSC are the Somalia-related resolutions which have generated the most comments and discussion,[65] resolutions are passed in relation to Somalia at least on two other interlinked topics as well: foreign troops authorized to help the new government regain control over Somali territory, and the weapons and charcoal embargo.

The weapons embargo is established in Res. 733 (1992) "for purposes of establishing peace and stability in Somalia"[66] and constitutes "a general and complete embargo on all deliveries on weapons and military equipment to

[59]McLaughlin (2009), p. 132. For seemingly the same view, see Klein (2010), p. 280.

[60]McLaughlin (2009), p. 133.

[61]Fink (2013), pp. 84–87.

[62]In this direction, see ibid., p. 92.

[63]For a similar view, see Fink (2011), p. 251.

[64]Id.

[65]The current resolution is UNSC Res. 2246 of 10 November 2015.

[66]UNSC Res. 733 of 23 January 1992, para. 5.

Somalia."[67] Res. 1425 (2002) expands the embargo to also cover the financing of all acquisitions and deliveries of weapons and military equipment,[68] as well as "direct or indirect supply to Somalia of technical advice, financial and other assistance, and training related to military activities."[69]

Not surprisingly, the UNSC has made a number of exceptions from the embargo:[70] the foreign counterpiracy presence,[71] the presence of the African Union,[72] UN personnel,[73] and the European Union Training Mission in Somalia.[74] Similarly excepted are the security forces of the Somali federal authorities, although some categories of weapons must be approved beforehand by the committee established to oversee that embargo,[75] and where the equipment etc. will solely be used in the development of institutions of the Somalia security sector.[76]

In Res. 751 (1992), the UNSC also established the abovementioned committee,[77] which was merged with the committee on Eritrea following Res. 1907 (2009).[78] Furthermore, the UNSC requested the UN Secretary General (UNSG) to establish a panel of experts in Res. 1425 (2002) to *inter alia* investigate violations of the embargo, and through the continuation of the mandate of this panel,[79] the UNSG was requested to establish a Monitoring Group in Res. 1519 (21) to undertake largely the same activities.[80] The mandate of the Monitoring Group is

---

[67]Id.

[68]UNSC Res. 1425 of 22 July 2002, para. 1.

[69]Ibid., para. 2.

[70]The newest list is found in UNSC Res. 2244 of 23 October 2015, paras. 1–10.

[71]UNSC Res. 2184 of 12 November 2014, paras. 15.

[72]UNSC Res. 1744 of 21 February 2007, para. 6, and UNSC Res. 1772 of 20 August 2007, paras. 11. Including the strategic partners of this force: UNSC Res. 2093 of 6 March 2013, para. 36, and UNSC Res. 2111 of 24 July 2013, para. 10 (b) and (c).

[73]UNSC Res. 2093 of 6 March 2013, para. 37.

[74]UNSC Res. 2111 of 24 July 2013, para. 10 (d).

[75]UNSC Res. 2093 of 6 March 2013, para. 33, UNSC Res. 2111 of 24 July 2013, paras. 6 and 14, UNSC Res. 2142 of 5 March 2014, para. 2, UNSC Res. 2182 of 24 October 2014, para. 1, and UNSC Res. 2244 of 23 October 2015, paras. 1 and 2. The resolutions do not explain why these weapons are mentioned and the formulations are not always optimal; see Stockholm International Peace Research Institute, UN arms embargo on Somalia, 14 November 2014, downloadable from http://www.sipri.org/databases/embargoes/un_arms_embargoes/somalia.

[76]UNSC Res. 2111 of 24 July 2013, para. 11 (a). See also UNSC Res. 2142 of 5 March 2014, paras. 2 and 8. The resolutions do not elaborate on what would be subsumed under the notion of "in service of its security forces"; see Stockholm International Peace Research Institute (2014). The system is continued in UNSC Res. 2244 of 23 October 2015, para. 2. The Somali authorities are required to provide information regarding such acquisitions, but these requirements are not always met. See UNSC Res. 2182 of 24 October 2014, para. 2, and UNSC Res. 2244 of 23 October 2015, para. 6.

[77]UNSC Res. 751 of 24 April 1992, para.11.

[78]See The Security Council Committee pursuant to resolutions 751 (1992) and 1907 (2009) concerning Somalia and Eritrea (https://www.un.org/sc/suborg/en/sanctions/751).

[79]UNSC Res. 1425 of 22 July 2002, para. 3.

[80]UNSC Res. 1519 of 16 December 2003, para. 2.

expanded in Res. 2036 to also include the charcoal export, as the UNSC expresses "concern that charcoal exports from Somalia are a significant revenue source for Al Shabaab and also exacerbate the humanitarian crisis."[81]

The current mandate expires on 16 December 2016,[82] and until recently, no separate enforcement powers were provided in relation to this embargo, in contradistinction to the embargos of the Federal Republic of Yugoslavia (Serbia and Montenegro) and Iraq.[83] Here, it might have been argued that the reference in Res. 733 to "immediately implement a general and complete embargo on all deliveries of weapons and military equipment» included the right to stop and search the vessels suspected of transporting weapons to Somalia, as the term «implement" would seemingly have to include this.[84] However, Klein is correct in pointing out that as the UNSC at the time typically explicitly authorized the halting of vessels and inspection of cargo, such an interpretation would be improper.[85] As a consequence, there has never been a lack of weapons.[86] Lately, it might even seem as if Somali pirates have financed violations of this embargo.[87]

3.2 *The Reach of the Enforcement Authority Granted in UNSC Res. 2182*

The absent enforcement powers were rectified in para. 15 of Res. 2182 of 24 October 2014, which establishes a maritime interdiction regime to be enforced by states, acting nationally or through voluntary multinational naval partnerships, in cooperation with the Federal Government of Somalia. That government must also notify the UNSG thereof, and the latter office must subsequently notify all Member States accordingly. So far, the Somali government has considered the following Combined Maritime Forces states as acceptable for the purpose of this authorization: Australia, Bahrain, Canada, Denmark, France, Germany, Italy, Japan, the Netherlands, New Zealand, Norway, Portugal, Saudi Arabia, Singapore, Spain, Turkey, UAE, the UK and the USA.[88]

The aim of the mandate is to ensure strict implementation of the arms embargo on Somalia and the charcoal ban. For this purpose, the vessels and aircraft of these states may inspect, without undue delay, in Somali territorial waters and on the high

[81]UNSC Res. 2036 of 22 February 2012, p. 3. See also paras. 22 and 23.

[82]UNSC Res. 2244 of 23 October 2015, para. 31.

[83]Brown (2011), p. 75.

[84]Klein (2010), p. 278.

[85]Id.

[86]UNSC Res. 1851 of 16 December 2008, para. 9.

[87]Ibid., p. 2, and UNSC Res 2020 of 22 November 2011, para. 6.

[88]Report of the Monitoring Group on Somalia and Eritrea pursuant to Security Council resolution 2182 (2014): Somalia, UN Doc. S/2015/801, 19 October 2015, p. 294.

seas off the coast of Somalia extending to and including the Arabian Sea and Persian Gulf, certain types of vessels.[89]

Although commentators have held earlier that UN-authorized maritime interdiction may be undertaken in the territorial sea of third states even when this geographic extension is not explicitly mentioned in the mandate,[90] the very reference to this power only in relation to "Somali territorial waters and on the high seas off the coast of Somalia extending to and including the Arabian sea and Persian Gulf"[91] would seem to argue against the authorization applying to, e.g., the territorial sea of Djibouti.[92]

The vessels covered by the authorization are described in para. 15 as bound to or from Somalia, and which the enforcers have reasonable grounds to believe are

(a) carrying charcoal from Somalia in violation of the charcoal ban;
(b) carrying weapons or military equipment to Somalia, directly or indirectly, in violation of the arms embargo on Somalia; and
(c) carrying weapons or military equipment to individuals or entities designated by the Committee established pursuant to Resolutions 751 (1992) and 1907 (2009).

That a specific vessel has undertaken such a transport before is not in itself sufficient to establish such grounds, but it would definitely help in establishing that threshold as long as corroborating evidence exists. As regards weapons and military equipment onboard vessels, the UNSC clarifies in Res. 2244 (2015) para. 3 that "the entry into Somali ports for temporary visits of vessels carrying arms and related materiel for defensive purposes does not amount to a delivery of such items in violation of the arms embargo on Somalia, provided that such items remain at all times aboard such vessels."

In para. 16 of Res. 2182 (2014), the UNSC moreover calls upon "all Flag States of such vessels to cooperate with such inspections, requests Member States to make good faith efforts to first seek the consent of the vessel's Flag State prior to any inspections pursuant to paragraph 15, authorizes Member States conducting inspections pursuant to paragraph 15 to use all necessary measures commensurate with the circumstances to carry out such inspections and in full compliance with international humanitarian law and international human rights law, as may be applicable, and urges Member States conducting such inspections to do so without causing undue delay to or undue interference with the exercise of the right of innocent passage or freedom of navigation." As the embargo enforcers are not party to the noninternational armed conflict in Somalia, or in an armed conflict with the weapons and charcoal transporters, the reference to international humanitarian law would seem unnecessary.

[89]Para. 15.

[90]McLaughlin (2009), p. 152.

[91]UNSC Res. 2182 of 24 October 2014, para. 15.

[92]In a similar manner the UNSC explicitly authorized counter-piracy operations in Somali "territorial waters" in UNSC Res. 1816 of 2 June 2008, para. 7 (a).

Furthermore, para. 19 grants authority to seize or dispose of any item covered by the embargo which is found during the inspections. Paragraph 19 clarifies that para. 15 also "includes the authority to divert vessels and their crews, to a suitable port to facilitate such disposal, with the consent of the port State" and "the authority to use all necessary measures to seize items pursuant to paragraph 17 in the course of inspections." If the authorization is put to use, obligations of reporting are incurred by the state whose vessels or aircraft carry out the inspection.[93] Thus, an interdicting state must promptly send a notification and later a more lengthy report to the said committee. Presumably, it is possible to provide the required information without having to divulge more classified information than does the Monitoring Group in its reports.[94]

Here, the UNSC strongly underlines that the authorization follows an invitation from Somali authorities, that it only applies to Somalia, and that it cannot be seen as establishing international customary law.[95]

The abovementioned authorization was renewed in Res. 2244 (2015)[96] but has until recently hardly been acted upon.[97]

4 Challenges

4.1 General Issues

No reports are made of weapons-related interdictions during the period of October 2014 to October 2015,[98] whereas the charcoal ban resulted in "a perceptible deterrent effect."[99]

In relation to the latter, the Monitoring Group observes that "[i]ntelligence shared by the combined Maritime Forces with the Monitoring Group led to the successful interception of the MSV Raj Milan (MMSI 419956307) at Port Rashid in Dubai, United Arab Emirates, with support from United Arab Emirates authorities. The issue of the disposal of seized charcoal, the absence of legal follow—through, and the difficulties in identifying a State willing to accept diverted ships have, however, hampered initial efforts to operationalize maritime interdiction. Minimizing the interaction between the naval forces and the vessels carrying charcoal

[93] UNSC Res. 2182 of 24 October 2014, para. 19.

[94] Information considered strictly confidential is simply left out of the public version of the report.

[95] UNSC Res. 2182 of 24 October 2014, para. 21.

[96] UNSC Res. 2244 of 23 October 2015, para. 20.

[97] Report of the Monitoring Group on Somalia and Eritrea pursuant to Security Council resolution 2182 (2014): Somalia, UN Doc. S/2015/801, 19 October 2015, pp. 41–44 and 293–322.

[98] Ibid., p. 41.

[99] Ibid., p. 43.

remains critical for an effective implementation of the charcoal ban."[100] The Monitoring Group also highlights the problems with identifying potentially embargo-breaking vessels as dhows transporting cargo to and from the southern parts of Somalia tend to not use the automatic identification system.[101] This might at least partially be an attempt to avoid detection by the embargo enforcers.[102]

Basically, the port state acted here within its ordinary authority, after having been tipped off by the Combined Maritime Forces, but it gained a right to confiscate the said cargo through the UNSC resolution. Some of the vessels enforcing the embargo are parties to the 1950 European Convention on Human Rights (ECHR),[103] but to the extent that no interdiction at sea is undertaken by such a state party, only a limited amount of issues would seem to arise in relation to that instrument. Rather, the human rights obligations of the relevant port state will apply.

Here, some of the challenges are largely of a practical nature: how to identify the vessels and how to motivate coastal states to allow the vessel into port for the relevant control. As regards the former, a legal problem admittedly arises in relation to reasonable grounds for believing that a vessel is in breach of the embargo since it might be harder to connect landside intelligence regarding a specific shipment with the very vessel later found at sea or in port.

As regards the weapons embargo, transforming the authorization into action has proved difficult. The Monitoring Group identifies the challenges as "how to interpret and apply certain provisions of the authorisation, particularly with respect to dealing with individuals found on board interdicted vessels, and the documentation and disposal of weaponry, including in the context of European Union legal requirements."[104] In a footnote, the Monitoring Group furthermore highlights issues like "the obligations of flag states, the need for the conduct of inspections to accord with international humanitarian and human rights law, seizure and disposal and the reporting requirements consonant on such inspections."[105]

Of importance is naturally the fact that the UNSC has not authorized a deviation from otherwise applicable human rights obligations, or for that matter international humanitarian law. Thus, human rights will *inter alia* limit the amount of force, armed or otherwise, to be used during the boarding of a suspected vessel, regulate the possibility of detaining individuals and the procedural rights held by these, the detainees' right (should they be prosecuted) to a fair trial, and their right to property.

Some of the pertinent legal issues indicated by the Monitoring Group are considered below.

---

[100]Ibid., pp. 43–44 and 316.

[101]Ibid., p. 44.

[102]Id.

[103]Convention for the Protection of Human Rights and Fundamental Freedoms, CETS No. 5.

[104]Report of the Monitoring Group on Somalia and Eritrea pursuant to Security Council resolution 2182 (2014): Somalia, UN Doc. S/2015/801, 19 October 2015, p. 294.

[105]Ibid., p. 294, n 152.

4.2 *To Whom the Honor?*

A typical issue nowadays is whether the relevant naval vessel may be held to represent another entity than its flag state, thereby possibly passing on the responsibility for the detention etc. to that other entity. The Combined Maritime Forces would not seem to be an entity with a sufficient international legal personality to shoulder such obligations,[106] whereas the states participating on EU naval operations seem to have diverging views on whether the individual is apprehended by a state or the EU itself.[107] The looseness of the UNSC control with the enforcers of this operation also argues against placing sole responsibility on the UN. It is to be presumed that any detention will be considered as having been undertaken by the state to which the interdicting vessel belongs. As regards naval vessels and the extraterritorial application of the ECHR, it would seem clear that they bring persons apprehended by such vessels within the "jurisdiction" of ECHR Art. 1.[108]

4.3 *Flag State Consent*

As regards the UNSC requesting the states taking upon themselves to enforce the embargo that they "make good faith efforts to first seek the consent of the vessel's Flag State prior to any inspections" (para. 16), it must be stressed that this is not an obligation. If time allows, it would nevertheless be proper to seek such an authorization, and in order to make this process efficient, contact should be made in advance with the authorities of the neighboring states for this purpose, as well as with other states where the relevant vessels might be flagged. Just as important is nevertheless agreeing in advance with the neighboring states of Somalia for the bringing into their ports of vessels following boarding.

4.4 *Detention of Persons*

As mentioned above, this issue was raised in relation to the mercenary embargo of Libya, although to the knowledge of this author no such detentions took place. Admittedly, the authorization to detain is not as explicit in relation to the charcoal/weapons embargo of Somalia as it was with Libya, but it is submitted that detention

[106]For a general discussion of whether another entity is to be held responsible, see Serdar Mohammed and Others v. Secretary of State for Defence, Court of Appeal (Civil Division), 30 July 2015, [2015] EWCA Civ 843, paras. 50–66.

[107]Petrig (2014), pp. 120 and 121, n 434.

[108]For a good overview of the case law in general, see Serdar Mohammed and Others v. Secretary of State for Defence, Court of Appeal (Civil Division), 30 July 2015, [2015] EWCA Civ 843, paras. 83–106.

is nevertheless authorized by the terms "to use all necessary measures commensurate with the circumstances."[109] Thus, an ECHR member state detaining an individual in these circumstances will not be required to identify one of the alternatives in ECHR Art. 5, para. 1 as applicable. However, nothing is *inter alia* provided in relation to how long individuals found onboard interdicted vessels may be detained, and therefore the other parts of Art. 5 will largely apply without abridgement.

As they have been acting in violation of a UNSC-authorized embargo, it might seem proper to use any valid criminal option available. Violating the embargo is nevertheless no international crime,[110] and even if it was, the prosecution would have to take place before national courts as no international court exists with jurisdiction to try such cases. A successful prosecution would then depend on *inter alia* domestic provisions, making it a crime to violate the embargo, and it might be expected that such provisions are lacking in many national legal systems. If relevant domestic provisions do exist, it is likely that the individuals will be handed over to the flag state of the vessel they sailed on, their state of nationality, or the relevant port state.[111] However, few states will presumably be interested in prosecuting these individuals, and it would therefore be natural for them to be released by the port state authorities after the necessary statements etc. have been taken or by the naval vessel itself if this can take place at an earlier point in time and in a way that would respect the human rights of the detainees. As the list of information that the interdicting party is obliged to report under para. 20 is rather long, it is to be expected that the detention of the said individuals will last more than a few hours at the very least, and thus that the persons will seldom be released before the vessel has made port.

An outstanding question is, typically where the vessel is sold off at an auction, how to bring these individuals back to where they belong.

4.5 *Seizure and Disposal*

An important aspect of inspections is provided for in Res. 2182 (2014) para. 19, where it follows that the state may "seize and dispose of (such as through

[109]A similar reasoning is applied in relation to detention in Afghanistan by ISAF where the authorization is largely similarly phrased, see ibid., paras. 146–148.

[110]The obligations undertaken by the 2013 Arms Trade Treaty, UN Doc. A/RES/67/234B of 11 June 2013, Art. 6, para. 1 at least point is that direction: "A State Party shall not authorize any transfer of conventional arms covered under Article 2 (1) or of items covered under Article 3 or Article 4, if the transfer would violate its obligations under measures adopted by the United Nations Security Council acting under Chapter VII of the Charter of the United Nations, in particular arms embargoes".

[111]Even if the flag state of the naval vessel undertaking the interdiction has such a domestic provision, the experience of state's lack of enthusiasm in prosecuting pirates does not argue for them easily using such an authorization.

destruction, rendering inoperable or unusable, storage, or transferring to a State other than the originating or destination States for disposal) any items identified in inspections pursuant to paragraph 15, the delivery, import or export of which is prohibited by the arms embargo on Somalia or the charcoal ban, [that the UNSC also] authorizes Member States to collect evidence directly related to the carriage of such items in the course of such inspections, and decides that charcoal seized in accordance with this paragraph may be disposed of through resale which shall be monitored by the Somalia and Eritrea Monitoring Group."

This authorization resembles to a certain extent the right granted for the purpose of counterpiracy operations against the Somali piracy plague.[112] The wording from Res. 2020 (2011) is illustrating. Besides calling on states to deploy naval vessels, arms, and military aircraft to fight piracy and armed robbery at sea, the UNSC here also refers to another mode of fighting: "through seizures and disposition of boats, vessels, arms and other related equipment used in the commission of piracy and armed robbery at sea off the coast of Somalia, or for which there are reasonable grounds for suspecting such use."[113]

Fink holds in relation to the piracy authorization that "[i]t might be clear from the purposes of the mandate that certain goods may not pass, which thus implies that those goods can be seized, but it is questionable whether a UN resolution is also sufficient basis to form a title that allows for either immediate change of legal possession or for the goods to become *res nullius* immediately after the capture. If this is not the case, seizing and destroying captured goods, considering it as «booty» or anything else before any further judicial process has taken place, is, although a good military practical solution, done on rather shaky legal grounds."[114]

Actually, the formulation in Res. 2020 (2011) above would seem to grant a right to preventively seize and dispose over vessels and weapons also were only reasonable suspicion of piracy can be established. This would then be in contrast to the general duty to compensate if the suspicion has been unfounded following UNCLOS Art. 110.[115] The same regime would apply to the weapons and charcoal embargo. This threshold of suspicion is probably lower that what is normal in many national legal systems in relation to a permanent loss of items following suspicion of criminal acts.[116] As breaching the said embargo might not even be illegal under domestic law, this gives grounds to pause.

It should also be noted that the new charcoal/weapons authorization diverges in a number of ways from the one used in relation to the Somali piracy threat. Firstly,

[112]UNSC Res. 1846 of 2 December 2008, para. 9, UNSC Res. 1851 of 16 December 2008, para. 2, UNSC Res. 1897 of 30 November 2009, para. 3, UNSC Res. 1950 of 23 November 2010, para. 4, and UNSC Res. 2020 of 22 November 2011, para. 7.

[113]UNSC Res. 2020 of 22 November 2011, para. 7. Italics in original have not been reproduced. Variations of this formulation are used in UNSC Res. 2077 of 21 November 2012, para. 10, UNSC Res. 2125 of 18 November 2013, para. 10, UNSC Res. 2184 of 12 November 2014, para. 11, and UNSC Res. 2246 of 10 November 2015, para. 12.

[114]Fink (2010), p. 21.

[115]Guilfoyle (2009), pp. 68–69.

[116]Friman and Lindborg (2013), p. 184.

the charcoal/weapons authorization exemplifies disposal "such as through destruction, rendering inoperable or unusable, storage, or transferring to a State other than the originating or destination States for disposal." This largely reflects the wording used in UNSC Res. 1970 (2011) para. 12 regarding the arms embargo of Libya. Secondly, Res. 2182 (2014) emphasizes "the importance of all Member States [. . .] taking the necessary measures to ensure that no claim shall lie at the instance of Somalia [etc.] [. . .] in connection with any contract or other transaction where its performance was prevented by reason of the measures imposed by this resolution or previous resolutions."[117] Thirdly, the right to collect evidence is explicitly mentioned. Fourthly, in relation to charcoal, the right to resale under certain conditions is specifically catered for, although that mode of disposal is not obligatory.

One question is here whether the phrases from the UNSC resolutions regarding the charcoal/weapons embargo are sufficient to establish an exception for the right to property under the ECHR Protocol 1 Art. 1 should the relevant acts be considered as falling within the jurisdiction of the state party under ECHR Art. 1.

As this would be a case of potential loss of property, the situation would seem to be covered by the rule found in Art. 1, No. 1. Central questions are then if a limitation of the right to property may be found in international law or whether it instead must be found in domestic law. Additionally, are the words used in the resolutions sufficiently clear if the limitation as such may be found in international law? The answer to both questions is probably yes. In relation to the similarly phrased piracy resolutions, Petrig nevertheless points out that the general reference in the resolutions to carry out the use of force "consistent with this resolution and international law" prohibits a state from dealing with these cases in a summary way.[118]

However, it would seem as if the European Court of Human Rights has instead used the rule in Art.1 No. 2 when dealing with seizure of goods being smuggled, fines, etc.[119] Since the seizure does not necessarily take place after a decision by a court, the criminal alternative ("penalties") in No. 2 would not necessarily be applicable.[120] Should this rule nevertheless be applied, the questions raised above would have to be addressed, as well as procedural requirements like the ability to refer the case to a court.[121] The right to use such procedural rights is often illusory

[117]Para. 18.

[118]Petrig (2015), p. 853.

[119]See as an example Handyside v. the United Kingdom, Appl. No. 5493/72, Plenary court judgment, 7 December 1976, para. 63 (confiscation and destruction of books), Agosi v. the United Kingdom, Appl. No. 9118/80, Chamber judgment, 24 October 1986, para. 51 (confiscation of gold coins smuggled into the United Kingdom), and Air Canada v. the United Kingdom, Appl. No. 9/1994/456/537, Chamber judgment, 26 April 1995, paras. 33–34 (which admittedly dealt with limitations on use until a penalty had been paid).

[120]Phillips v. the United Kingdom, Appl. No. 41087/98, Chamber judgment, 5 July 2001, para. 51, and Friman and Lindborg (2013), p. 185, n. 55.

[121]Agosi v. the United Kingdom, Appl. No. 9118/80, Chamber judgment, 24 October 1986, paras. 55, 59–60, and Air Canada v. the United Kingdom, Appl. No. 9/1994/456/537, Chamber judgment, 26 April 1995, para. 44.

for apprehended pirates,[122] and this would be similar for those apprehended in relation to the charcoal/weapons ban. Moreover, problems may arise in relation to the assessment of proportionality if items are damaged or destroyed during an interdiction operation, either because more force is used than necessary or because the item is later destroyed and thereby not returned to an innocent third party from whom it was stolen.[123] This is especially troublesome where the individual is later found not to have been guilty of violating the ban on charcoal/weapons.[124] Although this right to "summary" seizure and loss is used quite extensively in counterpiracy operations,[125] it would seem as if EU forces on such operations for these reasons apply a higher threshold than the one that follows from the UNSC resolutions,[126] and this they will presumably do also in relation to the charcoal/weapons ban.

As indicated, the charcoal, weaponry, etc. found would have to be documented and disposed of in a satisfactory way. The better way of handling the issue of charcoal is probably, as indicated in para. 19, to auction it off.[127] After having deducted the costs of the state auctioning it away, it would be proper to transfer the remaining sum to the Somali Federal Government, but no such obligation exists. As regards weapons, such cargo could presumably be granted the Federal Government of Somalia, and since the said weapons and ammunition is probably rather new and in larger quantity than what has so far been encountered in pirate skiffs, this might be a natural disposal of the equipment. However, there is no obligation on the interdicting states to do so, and they might just as well decide to destroy the cargo. In both instances, the UNSC highlights in para. 19 the importance of disposing equipment etc. in "an environmentally responsible manner."

5 Conclusions

There are probably many places around the world where additional weapons are what is least needed and where correspondingly an effectively enforced weapons embargo should be welcomed by the world at large. And from a practical point of view, this will often be the only measure that casualty-avoiding wealthy nations are motivated to participate in the enforcement of.[128] Among all the challenges that

[122] Friman and Lindborg (2013), p. 185.

[123] Bodini (2011), pp. 842–843 (in relation to piracy).

[124] Id.

[125] Friman and Lindborg (2013), p. 185.

[126] Id.

[127] This is what happened with the cargo of MSV Raj Milan—a vessel found to have violated the charcoal embargo, see Report of the Monitoring Group on Somalia and Eritrea pursuant to Security Council resolution 2182 (2014): Somalia, UN Doc. S/2015/801, 19 October 2015, p. 316.

[128] The problem is largely that an effective embargo will freeze the relationship of strength between the contenders, leaving an edge to the major party, which might be seen as action in violation of relevant rules meant to protect individuals and groups from oppression. Also, the arms embargo of

must be overcome in order to achieve such enforcement, the legal issues deserve more attention than they have been given so far.

In relation to the implementation hurdles identified in its report on Somalia, the Monitoring Group states that "development of a specific implementation assistance notice should be considered. In the interim the sharing of real time information with the Monitoring Group on vessels which may be operating in violation of the arms embargo should be encouraged."[129] Some suggestions are also given in relation to the charcoal ban,[130] but the Group did not suggest new formulations to be used in upcoming UNSC resolutions for the clarification of the issues raised in chapter 4, and the wording used in Res. 2244 of 23 October 2015 offers little help.

For member states to the ECHR inclined on limiting the scrutiny of human rights institutions, one option would be to seek a transformation of the operation into something resembling a maritime KFOR or ISAF, as these operations have or may presume to be left largely outside the assessment of at least the European Court of Human Rights, as it seemingly will be holding the UN responsible instead. Note, however, that the UN Human Rights Committee may often be an alternative avenue for applicants as it states in General Comment No. 31 that a troop-contributing nation will still be held responsible in such operations.[131]

In relation to seizure and disposal, especially when undertaken in a UN operation as opposed to a coalition of the willing, the UNSC should consider authorizing a sanction panel,[132] which would have the final say in whether detention and seizure were legal; undertake the disposal of the seized items in a prize court fashion; and offer compensation where the embargo enforces were found to be at fault. The panel could be a sub-body of the sanction committee, be located in Djibouti where many of the vessels dock, but deploy (under prearrangements with the neighboring port states) to where the relevant vessel has been brought for adjudication.

Be that as it may, although the current enforcement mandate is welcome, weapons will probably still cross into Somali territory over its borders with Kenya, Ethiopia, and—to a lesser extent—Djibouti, which are presumably not sufficiently policed. The key to avoiding this, as well as piracy and other kinds of organized crime, is achieving a well working state on the Horn of Africa.

Postscript

Since this chapter was finalized, the UNSC has anew authorized enforcement of the weapons embargo of Libya (Res. 2292 (2016)) where much of the terminology of

inter alia Bosnia Herzegovina hindered that state's right to respond in self-defense against the attacks it underwent at the hands of *inter alia* Serb military entities. See Fleck (2013), p. 70.

[129]Report of the Monitoring Group on Somalia and Eritrea pursuant to Security Council resolution 2182 (2014): Somalia, UN Doc. S/2015/801, 19 October 2015, p. 295.

[130]Ibid., p. 50.

[131]Human Rights Committee, General Comment No. 31 [80] The Nature of the General Legal Obligation Imposed on States Parties to the Covenant, UN. Doc. CCPR/C/21/Rev.1/Add. 13, p. 4.

[132]Fink has previously indicated this as a general weakness with the UN embargo regime. See Fink (2010), pp. 20–21.

Res. 1973 is reused. However, good-faith efforts to obtain flag state consent prior to inspection must now be made (para. 3), and the category of vessels able to undertake such inspections is also identified (para. 6), whereas it is explicitly mentioned that the authorization does not apply to vessels benefitting from sovereign immunity (para. 7). Explicit reference is also made of the authority to divert the vessel to a suitable port (para. 8), and this provision may have been imported from Res. 2182 (2014) (para. 19).

Also, the Combined Maritime Forces have on two occasions in March 2016 (the Australian HMAS Darwin and the French FS Provence) seized large amounts of weapons. The vessels used for smuggling were first inspected in order to determine their nationality, and when deemed stateless, the dhows were searched and the weapons then found were seized.[133]

Acknowledgements The author would like to thank the peer reviewer for helpful suggestions.

References

Journals and Articles

Aaron MR, Nauta DRD (2013) Operational challenges of the law on air warfare: the example of operation unified protector. Military Law Law War Rev 52:353–377

Bodini SP (2011) Fighting maritime piracy under the European convention on human rights. Eur J Int Law 22:829–848

Fink MD (2011) UN-mandated maritime arms embargo operations in operation unified protector. Military Law Law War Rev 50:237–260

Fink MD (2013) Maritime embargo operations: naval implementation of UN sanctions at sea under articles 41 and 42 of the UN Charter. Neth Int Law Rev 60:73–92

Song Y-H (2007) The U.S.-led proliferation security initiative and UNCLOS: legality, implementation, and an assessment. Ocean Dev Int Law 38:101–141

Books and Chapters

Brown N (2011) Jurisdictional problems relating to non-flag state boarding of suspected ships in international waters: a practitioner's observations. In: Symmons CR (ed) Selected contemporary issues in the law of the sea. Martinus Nijhoff, Leiden, pp 69–82

Dinstein Y (2012) War, aggression and self-defence, 5th edn. Cambridge University Press, Cambridge

Doswald-Beck L (ed) (1995) San Remo manual on international law applicable to armed conflicts at sea. Cambridge University Press, Cambridge

Fleck D (2013) Arms control and disarmament law: its role in addressing new security threats. Military Law Law War Rev 52:61–80

[133]See https://combinedmaritimeforces.com for details.

Friman H, Lindborg J (2013) Initiating criminal proceedings with military forces: some legal aspects of policing Somali pirates by navies. In: Guilefoyle D (ed) Modern piracy: legal challenges and responses. Edward Elgar, Cheltenham, pp 172–201

Guilfoyle D (2009) Shipping interdiction and the law of the sea. Cambridge University Press, Cambridge

Klein N (2010) Maritime security and the law of the sea. Oxford University Press, Oxford

McLaughlin R (2009) United Nations naval peace operations in the territorial sea. Martinus Nijhoff, Leiden

Petrig A (2014) Human rights and law enforcement at sea: arrest, detention and transfer of piracy suspects. Brill Nijhoff, Leiden

Petrig A (2015) Piracy. In: Rothwell DR, Oude Elferink AG, Scott KN, Stephens T (eds) The Oxford handbook of the law of the sea. Oxford University Press, Oxford, pp 843–865

Rothwell DR, Stephens T (2016) The international law of the sea, 2nd edn. Hart Publishing, Oxford and Portland

Tanaka Y (2015) The international law of the sea, 2nd edn. Cambridge University Press, Cambridge

von Heinegg WH (2010) Maritime interception/interdiction operations. In: Gill TD, Fleck D (eds) The handbook of the international law of military operations. Oxford University Press, Oxford, pp 375–393

Online Publications

USA-USSR Joint Statement on the Uniform Interpretation of Rules of International Law Governing Innocent Passage (http://cil.nus.edu.sg/rp/il/pdf/1989%20USA-USSR%20Joint%20Statement%20on%20the%20Uniform%20Interpretation%20of%20Rules%20of%20International%20Law-pdf.pdf)

U.S. Department of State, Statement of Interdiction Principles (http://www.state.gov/t/isn/c27726.htm)

Stockholm International Peace Research Institute, UN arms embargo on Somalia, 14 November 2014, downloadable from http://www.sipri.org/databases/embargoes/un_arms_embargoes/somalia

The Security Council Committee pursuant to resolutions 751 (1992) and 1907 (2009) concerning Somalia and Eritrea (https://www.un.org/sc/suborg/en/sanctions/751)

Other

Convention for the Protection of Human Rights and Fundamental Freedoms, CETS No. 5

UN Convention on the Law of the Sea 1833 UNTS 3

Arms Trade Treaty, UN Doc. A/RES/67/234B of 11 June 2013

UNSC Res. 221 of 9 April 1966

UNSR Res. 661 of 6 August 1990

UNSC Res. 665 of 25 August 1990

UNSC Res. 733 of 23 January 1992

UNSC Res. 751 of 24 April 1992

UNSC Res. 787 of 16 November 1992

UNSC Res. 820 of 17 April 1993

UNSC Res. 875 of 16 October 1993

UNSC Res. 917 of 6 May 1994

UNSC Res. 1132) of 8 October 1997

UNSC Res. 1425 of 22 July 2002

UNSC Res. 1519 of 16 December 2003

UNSC Res. 1540 of 28 April 2004

UNSC Res. 1701 of 11 August 2006
UNSC Res. 1744 of 21 February 2007
UNSC Res. 1772 of 20 August 2007
UNSC Res. 1816 of 2 June 2008
UNSC Res. 1846 of 2 December 2008
UNSC Res. 1851 of 16 December 2008
UNSC Res. 1897 of 30 November 2009
UNSC Res. 1940 of 29 September 2010
UNSC Res. 1950 of 23 November 2010
UNSC Res. 1970 of 26 February 2011
UNSC Res. 1977 of 20 April 2011
UNSC Res. 2009 of 16 September 2011
UNSC Res. 2016 of 27 October 2011
UNSC Res. 2020 of 22 November 2011
UNSC Res. 2036 of 22 February 2012
UNSC Res. 2077 of 21 November 2012
UNSC Res. 2093 of 6 March 2013
UNSC Res. 2094 of 7 March 2013
UNSC Res. 2111 of 24 July 2013
UNSC Res. 2125 of 18 November 2013
UNSC Res. 2142 of 5 March 2014
UNSC Res. 2146 of 19 March 2014
UNSC Res. 2182 of 24 October 2014
UNSC Res. 2184 of 12 November 2014
UNSC Res. 2240 of 9 October 2015
UNSC Res. 2244 of 23 October 2015
UNSC Res. 2246 of 10 November 2015
UNSC Res. 2292 of 14 June 2016
Report of the Monitoring Group on Somalia and Eritrea pursuant to Security Council Resolution 2182 (2014): Somalia, UN Doc. S/2015/801, 19 October 2015
Human Rights Committee, General Comment No. 31 [80] The Nature of the General Legal Obligation Imposed on States Parties to the Covenant, UN. Doc. CCPR/C/21/Rev.1/Add. 13
Handyside v. the United Kingdom, Appl. No. 5493/72, Plenary court judgment, 7 December 1976
Agosi v. the United Kingdom, Appl. No. 9118/80, Chamber judgment, 24 October 1986
Air Canada v. the United Kingdom, Appl. No. 9/1994/456/537, Chamber judgment, 26 April 1995
Phillips v. the United Kingdom, Appl. No. 41087/98, Chamber judgment, 5 July 2001
Serdar Mohammed and Others v. Secretary of State for Defence, Court of Appeal (Civil Division), 30 July 2015, [2015] EWCA Civ 843

The Right of Innocent Passage: The Challenge of the Proliferation Security Initiative and the Implications for the Territorial Waters of the Åland Islands

Pirjo Kleemola-Juntunen

1 Introduction

Modern law of the sea developments has raised questions relating to the application of treaty-based, non-formalised and non-treaty based regulations of cooperation, as well as norms of international customary law and State practice, which have been introduced after the 1921 Convention relating to the Non-fortification of and Neutralisation of the Åland Islands.[1] The Åland Islands are unique within Europe, and their legal status cannot be compared to anywhere else in the region. The Åland archipelago is located within the entrance to the Gulf of Bothnia, between Finland and Sweden. As a result of the geographical connection to the Finnish mainland, the demilitarised and neutralised sea area of the Åland Islands is located within the territorial sea and internal waters of Finland. The Åland Strait, a narrow stretch of water connecting the Gulf of Bothnia with the Baltic Sea between the Åland Islands

This chapter is written as a part research project "Demilitarisation in an increasingly militarised world. International perspectives in a multilevel framework – the case of the Åland Islands". The research project is a co-operation between the Åland Islands Peace Institute (ÅIPI) and the University of Lapland and its Arctic Center in Rovaniemi (Finland). The Project is funded by the KONE Foundation. The writer is working as a Post-doc researcher in the Project.

[1]Convention relating to the Non-fortification of and Neutralisation of the Aaland Islands, adopted on 20 October 1921 and entered into force 6 April 1922. 9 LNTS 211. Parties to the Convention include: Denmark, Estonia, Finland, France, Germany, Great Britain, Iceland (Union with Denmark in 1921) Italy, Latvia, Poland and Sweden.

P. Kleemola-Juntunen (✉)
Åland Islands Peace Institute, Åland, Finland

Northern Institute for Environmental and Minority Law, Arctic Centre, University of Lapland, Rovaniemi, Finland
e-mail: pirjo.kleemola-juntunen@ulapland.fi

G. Andreone (ed.), *The Future of the Law of the Sea*,
DOI 10.1007/978-3-319-51274-7_12

and Sweden is a particularly important sea route.[2] In terms of the law of the sea, the development of weapon technology has made warships more effective and enhanced their roles in warfare. Coastal States have also become more alert to the threat of nuclear-powered ships and ships carrying nuclear weapons. Furthermore, in the twenty-first century, the number of unlawful activities at sea has become a challenge to maritime trade. Today, the most significant of the unlawful activities is terrorism, and as a result there has been an increase in the interest shown in maritime security by the international community. Traditionally, weapons of mass destruction (WMD) have been in the hands of States. Recently, it has become possible for non-State actors to acquire WMD and related materials, and this has increased the probability of unlawful trafficking of WMD across the world. The demilitarised sea around the Åland Islands is at risk of this unlawful behaviour.

The Proliferation Security Initiative came about following the terrorist attacks of 11 September 2001 and after the So San incident.[3] The Proliferation Security Initiative (PSI) is a cooperation arrangement that aims to prevent the proliferation of weapons of mass destruction, their delivery systems and related materials. It was introduced by the United States as a measure to prevent terrorist attacks in 2003, and the principle has been endorsed by the UN Security Council Resolutions adopted under Chapter VII of the UN Charter,[4] as well as by the adoption of the 2005 Protocol to the Convention for the Suppression of Unlawful Acts Against the Safety of Maritime Navigation.[5] The 2005 SUA Protocol was the first treaty that recognised the trafficking of WMD and related materials as illegal behaviour.[6] As a

[2]According to HELCOM publication Shipping Accidents in the Baltic Sea in 2013 14,433 ships on the Åland West route and 1397 ships on the Åland East route have crossed AIS fixed lines through the Åland Strait during the year 2013, HELCOM (2014), pp. 3–7.

[3]The missiles of North Korean origin were in transit to Yemen by a ship flying under Cambodian flag. The So San was intercepted and boarded by the Spanish Navy relying on U.S. intelligence and subsequently released due to lack of legal support for the seizure. See more Byers (2004), pp. 526–527.

[4]UN Security Council Resolutions 1540 (2004), 1673 (2006), 1810 (2008), 1977 (2011), 2055 (2012).

[5]Protocol to the Convention for the Suppression of Unlawful Acts Against the Safety of Maritime Navigation and Protocol for the Suppression of Unlawful Acts Against the Safety of Fixed Platforms on the Continental Shelf adopted on 1 November 2005 and entered into force on 28 July 2010 ('2005 SUA Protocol').

[6]Durkalec (2012), p. 14, Protocol of 2005 to the Convention for the Suppression of Unlawful Acts Against the Safety of Maritime Navigation, at https://www.unodc.org/tldb/pdf/Protocol_2005_Convention_Maritime_navigation.pdf. The *Achille Lauro* incident of 1985 gave rise to the Suppression of Unlawful Acts Against the Safety of Maritime Navigation and Protocol for the Suppression of Unlawful Acts Against the Safety of Fixed Platforms on the Continental Shelf (the SUA Convention and SUA Protocol), adopted 10 March 1988 and entered into force 1 March 1992, as a measure to prevent unlawful acts which threaten the safety of ships and security of passengers and crew. IMO Convention for the Suppression of Unlawful Acts Against the Safety of Maritime Navigation and Protocol for the Suppression of Unlawful Acts Against the Safety of Fixed Platforms on the Continental Shelf, March 13, 1988, entered into force on March 1, 1992, IMO Doc SUA/CONF/15, ILM 27 (1988), pp. 672–684.

non-treaty-based partnership of States, the Proliferation Security Initiative (PSI) is aimed to complement existing international arms control arrangements such as the Treaty on the Non-Proliferation of Nuclear Weapons (NPT),[7] the Chemical Weapons Convention (CWC)[8] and the Biological Weapons Convention (BWC).[9] The aim of the PSI participants was to stop the illicit transport of WMD on the oceans. The major problem as regards illicit trafficking is not the ready-made WMD but is more often the trafficking of components, technologies and production materials related to WMD. The problem with these items is that majority of them can have civilian as well as military end uses. Such dual-use materials pose a problem because they are mostly used for peaceful purposes, and in these circumstances their transportation is legal. The main concern of the PSI is to prevent States and non-State actors of proliferation concern from acquiring the materials to build WMD.[10]

Although the PSI is a political initiative, after the adoption of the Statement of Interdiction Principles that sets forth the objectives and working methods of the PSI, it has implications for the existing law of the sea.[11] The Statement of Interdiction Principles says that PSI activities will not violate international law. However, subparagraph 4 (d) (1) of the Statement of Interdiction Principles calls on participants to take appropriate actions to stop and/or search in their territorial seas vessels that are 'reasonably suspected' of carrying such cargoes to or from States or non-State actors of proliferation concern and to seize such cargoes that are identified.

The requirement in subparagraph 4 (d) (1) is problematic because although coastal States have sovereignty over their territorial seas, it is limited by Article 19 of UNCLOS, which regulates the right of innocent passage of foreign vessels. However, today 105 States are participants to the PSI, and when they act unilaterally they give rise to new State practice. If several flag States were to accept boarding by a coastal State of a ship exercising the right of innocent passage because the ship was suspected of the illegal trafficking of WMD without prior

---

[7]Treaty on the Non-Proliferation of Nuclear Weapons, adopted on 1 July 1968 and entered into force on 5 March 1970 at https://www.iaea.org/sites/default/files/publications/documents/infcircs/1970/infcirc140.pdf.

[8]Chemical Weapons Convention, adopted on 13 January 1993 and entered into force on 29 April 1997 at http://disarmament.un.org/treaties/t/cwc/text.

[9]Biological Weapons Convention, adopted on 10 April 1972 and entered into force on 26 March 1975 at http://disarmament.un.org/treaties/t/bwc/text.

[10]US Department of State Proliferation Security Initiative at http://www.state.gov/t/isn/c10390.htm, Logan (2005), p. 255, Prosser and Scoville (2004), Beck (2004), p. 16 at http://www.uga.edu/cits/documents/pdf/monitor/monitor_sp_2004.pdf.

"The PSI participants are committed to the following interdiction principles to establish a more coordinated and effective basis through which to impede and stop shipments of WMD, delivery systems, and related materials flowing to and from states and non-state actors of proliferation concern, consistent with national legal authorities and relevant international law and frameworks, including the UN Security Council." Fact Sheet The White House, Office of the Press Secretary (2003).

[11]Winner (2005), p. 130.

permission of the flag State, it would weaken the existing legal norm relating to right of innocent passage that has its roots far back in history. This is especially true because the PSI is targeted to merchant ships, and the origin of the concept is freedom of navigation of merchant ships over the oceans.[12] Thus, at least in theory, the new State practice of the provision regulating innocent passage would narrow the scope of the right of innocent passage.

Although the sovereignty of coastal States extends to the territorial seas, such sovereignty includes some restrictions that do not exist for a State's land-based territory or its internal waters. On the territorial sea, all foreign ships enjoy the right of innocent passage, an old principle concept of the law of the sea, today codified in the UN Law of the Sea Convention. As long as the passage is innocent, the coastal State has restricted jurisdiction to interfere with the passage. It seems that the PSI's main output—the Statement of Interdiction Principles (SOP)—includes elements that contradict with UNCLOS.

This chapter discusses the relationship of the PSI and the resolutions of the UN Security Council, whose aim is to stop the illicit transport of weapons of mass destruction (WMD) on the oceans, to the UN Convention on the Law of the Sea (UNCLOS) and to the 1921 Åland Convention. Today, all parties to the 1921 Åland Convention, as well as Russia, are participants in the PSI, and they are also parties to the UNCLOS and member States of the United Nations.

2 The Demilitarisation of the Åland Islands

The Åland Islands' international legal regime was confirmed in the aftermath of the First World War. The Åland Islands and its surrounding sea area was demilitarised and neutralised by the 1921 Åland Convention. In 1921, the Åland Convention stated that it is complementary to the 1856 Convention between France, Great Britain and Russia, which resulted in the demilitarisation of the Åland Islands after the Crimean War. The 1921 Åland Convention has since been supplemented by the 1940 Treaty between Finland and the Soviet Union. Following the Second World War, the 1947 Peace Treaty confirmed the status quo by declaring that 'the Åland Islands will remain demilitarised according to the present situation' Furthermore, the position of the Åland Islands was also mentioned in the EU Treaty of Accession when Finland joined the EU in 1995.[13]

[12]Thomas (2009), p. 657.

[13]Convention relating to the Non-Fortification and Neutralisation of the Åland Islands, Finnish Treaty Series 1/1922, English translation available in 17 AJIL 1923, Supplement: Official Documents, pp. 1–6. Hereinafter the 1921 Åland Convention, Treaty concerning the Åland Islands between Finland and the Union of Soviet Socialist Republics, Finnish Treaty Series 24/1940. By the Armistice Agreement 19.10.1944, the bilateral treaty between Finland and the Soviet Union

The location of the Åland Islands indicates that they may be of military strategic significance. A military power in control of Åland and with aggressive intentions could use the islands as a base for military operations. Sweden has always been the most active proponent of the demilitarised and neutralised status of the Åland Islands because any power controlling the Åland Islands would be able to threaten Sweden' s east coast and the capital city of Stockholm. In addition, the former Soviet Union was in favour of the demilitarisation of the Åland Islands, and a bilateral treaty between Finland and the Soviet Union was concluded on the initiative of the Soviet Union to this effect.[14]

Security has been the most important question in negotiations concerning the area of the Åland Islands. When Russia conquered Finland during the 1808–1809 war, the strategic situation changed in the Baltic Sea region. According to the 1809 Peace Treaty, the Torneå and Muonio rivers and the Gulf of Bothnia combined to form the Russia–Sweden border, and the Åland Islands belonged to Russia.[15] This situation was worrying from a Swedish perspective because the Åland Islands were a strategic stronghold. In this sense, the situation became a serious security issue for Sweden as the Åland Islands becoming a part of Russia created new localised threats. It is easy to understand therefore that the peace negotiations following the 1808–1809 war and any other matters regarding the Åland Islands were of vital importance to Sweden. During peace negotiations, Sweden's primary goal was to retain the Åland Islands, but efforts to secure this objective failed. A second option for Sweden was the non-fortification of the islands, but this proposal was also rejected.

This notion of the non-fortification of the Åland Islands emerged again during peace negotiations following the Crimean War. Sweden had remained neutral during the war and had not taken part in the negotiations. However, Sweden used its diplomatic influence to convince Britain and France to voice concerns on its behalf. Sweden was thus presented with a new opportunity to regain the Åland Islands. If successful, the Åland Islands would belong to Sweden and the special demilitarised status of the islands would no longer be necessary. Efforts made to accomplish this, however, did not bear fruit, and instead in an effort to reach a compromise, the Åland Islands were demilitarised.

---

concerning the demilitarisation of the Åland Islands was re-confirmed. This meant that fortifications on the Åland Islands had to be destroyed, Reactivation of the Treaty concerning the Åland Islands between Finland and the Union of Soviet Socialist Republics, Finnish Treaty Series 9/1948, Peace Treaty with Finland, Finnish Treaty Series 20/1947: English translation available in 42 AJIL 1948, Supplement: Official Documents, pp. 203–223, Commission opinion on Finland's application for membership on 4th November 1992. The 1940 Treaty was confirmed by the 1992 Protocol between Finland and the Russian Federation.

[14]Hannikainen (1994), p. 615.

[15]The Treaty of Fredrikshamn. www.histdoc.net/history/fr/frhamn.html. Accessed 19 Jan 2016.

The 1856 Convention is short and only covers demilitarisation. There is no clear definition in the Convention regarding the territory to which it applies. However, Article I of the Convention refers to the islands of Åland and has therefore only been applied to this specific land area. Thus, it was possible to carry out military operations in the seas surrounding the Åland Islands without infringing on the principle of demilitarisation. The 1856 Convention is also silent regarding defence arrangements permitted during times of war. During the First World War, the demilitarisation of the Åland Islands was not respected. Russia fortified the Åland Islands and used them as a base for military operations against Germany. In the late stages of the war in 1918, the Åland Islands were first occupied by Sweden and then by Germany. The fortifications were demolished in 1919 by Finland, which had emerged as a new independent State.[16] In the aftermath of the First World War, the Åland Islands became an object of a territorial dispute between Finland and Sweden.[17] The League of Nations settled the dispute, and the sovereignty of the Åland Islands was recognised as belonging to Finland. As a result of this settlement, the Åland Islands and their surrounding sea area were demilitarised and neutralised by the 1921 Åland Convention.

A huge threat to the demilitarised and neutralised status of the Åland Islands was posed by the outbreak of the Second World War as the Åland Islands were a key strategic focal point for belligerents. During the war, the legal status of the Åland Islands was respected, although both Germany and the Soviet Union had plans to occupy the islands. Finland, however, decided to fortify the islands. When Finland informed the parties to the 1921 Åland Convention of its military preparations, they did not express any criticism.[18] Since the end of the Second World War, the Åland Islands have been spared from any further military operations.

The 1921 Åland Convention offers certain exceptions when considering a military presence within the zone during peacetime and when there is armed conflict, and the exceptions are different depending on whether the vessel is Finnish or not. Furthermore, the passage accorded to ships can be divided into two parts: namely, when passing through territorial seas and when entering internal waters. The 1921 Åland Convention limits Finnish warships and aircraft, as well as Finland's authority to regulate any access granted to foreign warships, either when entering or staying within the zone. But the Åland Convention also refers to the rules of international law and practice, in the event of the innocent passage of warships. When the Åland Convention was concluded, the international law of the sea was based on customary law, which left coastal States wide discretionary powers to determine the nature of passage. At present, the international law of the sea consists of rules that are applied from treaty law and customary international

[16]Hannikainen (1994), p. 617.

[17]O'Brien (2012).

[18]Hannikainen (1994), p. 618.

law. Although the conventions relating to the right of innocent passage sought to codify the customary law, the very act of codification reduces a State's discretion.

The preamble to the 1921 Åland Convention says that the Convention has been made in order to guarantee peace and stability, in the sense that the Åland Islands shall never become a threat from a military perspective. The phrasing of the preamble to the Convention clearly shows that there was a common interest among States to secure the region, with a particular focus on the Åland Islands. It is understandable that the legal scope of the Convention was expanded to cover the islands' surrounding sea areas so as to prevent military activities from occurring in the future. It may also be said the geographical range of the 1921 Åland Convention is connected to various security issues and the ability of States to handle these issues within the limits of the region. The demilitarised and neutralised zone therefore exists as a consequence of localised security threats that were identified by parties to the 1921 Convention.

The 1921 Åland Convention established Åland's three-nautical- mile territorial sea, thereby separating the demilitarised and neutralised sea areas from other parts of Finland's territorial sea. A coastal State has sovereignty over its territorial sea. The right of innocent passage is the main restriction imposed by international law over any coastal State wishing to exercise sovereignty over its territorial sea. Taking the sea within the Convention, it also introduced rules of the international law of the sea to the content of the Convention. Although Article 4 prohibits all kinds of military presence in the area, there are some exceptions to this provision. These exceptions relate to Finland and its right to regulate navigation and the presence of foreign vessels in its territorial waters, which are at the same time also a demilitarised and neutralised area. During the negotiation process of the Convention, States also had to solve the question of the right of innocent passage through the demilitarised and neutralised zone.

Article 5 of the 1921 Åland Convention grants warships the right of innocent passage through areas of the Åland Islands' territorial waters that are part of the neutralised zone. However, the article makes references to international rules and usages in force that therefore sets limits to the application. The Article indicates that Finland's authority to enact rules that would prohibit the innocent passage of warships through the territorial waters of the Åland Islands is restricted. The Article is also applicable to Finnish warships, and as a result the 1921 Åland Convention restricts a coastal State's jurisdiction over its own territorial sea. The possibility of prohibiting innocent passage was never properly addressed during the 1921 Conference discussions, and only Finland had expressed its view on the matter by stating its right to prohibit innocent passage in special circumstances.[19]

---

[19]"... le droit d'interdire le passage inoffensive dans des circonstances spéciales" in Actes de la Conférence, p. 64.

3 An Overview of the Proliferation Security Initiative

The Proliferation Security Initiative (PSI) arose out of the 11 September 2001 terrorist attacks and after the So San incident. The So San incident showed that the United States had no legal authority to seize the missiles of North Korean origin that were in transit to Yemen by a ship flying under the Cambodian flag. The United States lacked a clear legal authority to seize the missiles, but there was also no provision under international law prohibiting Yemen accepting the delivery of the missiles from North Korea.[20] The PSI was originally proposed by the United States in 2003 in Krakow, Poland, by President Bush, who stated that the 'greatest threat to peace is the spread of nuclear, chemical and biological weapons' and announced the PSI.

'When weapons of mass destruction or their components are in transit, we must have the means and authority to seize them. So today I announce a new effort to fight proliferation called the Proliferation Security Initiative. The United States and a number of our close allies, including Poland, have begun working on new agreements to search planes and ships carrying suspect cargo and to seize illegal weapons or missile technologies. Over time, we will extend this partnership as broadly as possible to keep the world's most destructive weapons away from our shores and out of the hands of our common enemies.'[21]

The PSI is not a treaty but is rather a statement of intention to prevent the movement of weapons of mass destruction and related materials at ports and different maritime zones without maintaining any organisational frameworks.[22] The PSI aims to complement existing international arms control arrangements. It refers to the rules of international law but not specifically to the norms of the law of the sea.[23] The PSI represents new forms of international cooperation beyond international treaties and organisations.

Initially this United-States-led initiative co-opted 10 States (Poland, Australia, France, Germany, Italy, Japan, the Netherlands, Portugal, Spain and the United Kingdom). These 11 original participants are the core group of the PSI. They adopted the Statement of Interdiction Principles, which was announced on

---

[20] Ahlström (2005), p. 741, Logan (2005), p. 253, see also Garvey (2005), pp. 128–129.

[21] Remarks by the President to the People of Poland (2003) http://georgewbush-whitehouse.archives.gov/news/releases/2003/05/20030531-3.html.

[22] In the Proliferation Security Initiative meeting in London 9–10 October 2003 the participants to the meeting agreed that "the PSI was a global initiative with an inclusive mission. Successful interdiction of trafficking in WMD, their delivery systems and related materials requires the widest possible co-operation between states. Participation in the PSI, which is an activity not an organisation, should be open to any state or international body that accepts the Paris Statement of Principles and makes an effective contribution." at http://dfat.gov.au/international-relations/security/non-proliferation-disarmament-arms-control/psi/Pages/proliferation-security-initiative-london-9-10-october-2003-2.aspx. Klein (2011), p. 150, Jinyuan (2012), p. 97.

[23] See Ministry for Foreign Affairs of Finland http://formin.fi/public/default.aspx?contentid=325890&contentlan=2&culture=en-US.

4 September 2003 in Paris and which is a significant public output of the initiative.[24] The interdiction principles identify concrete actions to collectively or individually interdict shipments of WMDs, their delivery systems and related materials.[25] However, the Statement of Interdiction Principles does not bind participants to the PSI legally; it is a political commitment and practical cooperation to help impede and stop the flow of WMDs, their delivery systems and related materials to and from States and non-State actors of proliferation concern. The Interdiction Principles set forth the objectives and working methods of the PSI.[26]

The PSI cooperation is operated by exercises and bilateral ship-boarding agreements. The aim of the PSI partnership is to establish a network that impedes and stops the illicit trafficking of WMDs and related materials, as well as their delivery systems.[27] Today, 105 States have publicly endorsed the PSI, and the European Union has given its support to the cooperation.[28] Finland has been a participant in the PSI since 2004. Although the PSI has no permanent institutional structure, it has an Operational Experts Group, which comprises 21 States (Australia, Argentina, Canada, Denmark, France, Germany, Greece, Italy, Japan, Republic of Korea, the Netherlands, New Zealand, Norway, Poland, Portugal, Russia, Singapore, Spain, Turkey, the United Kingdom and the United States). The group meets frequently, and its task is to take care of the planning for the initiative to ensure the PSI's effectiveness by contributing customs, law enforcement, military and other security experts and assets to interdiction exercises, hosting PSI meetings, workshop and exercises with other PSI-endorsing States.[29] Through the PSI partnership, States have established a network that impedes and stops the illicit trafficking of WMDs and related material, as well as their delivery systems.[30]

It is clear that the international community has a negative attitude towards the proliferation of WMDs. Evidence of this is found in the number of participants in the treaties, such as the Treaty on the Non-Proliferation of Nuclear Weapons (NPT), the Biological Weapons Convention (BWC) and the Chemical Weapons Convention (CWC). However, there are number of States that are not party to the NPT Convention, e.g. North Korea. Although these conventions prohibit the proliferation, transport and sale of biological and chemical weapons, they do not grant high

[24]Ahlström (2005), p. 745.

[25]Proliferation Security Initiative: Chairman's Statement at the Third Meeting (2003).

[26]Ahlström (2005), p. 745, Winner (2005), p. 130.

[27]http://www.psi-online.info/Vertretung/psi/en/01-about-psi/0-about-us.html, Tornberg (2009), p. 140.

[28]China, India and Pakistan are not participants of the PSI.

[29]See Operational Experts Group at http://www.psi-online.info/Vertretung/psi/en/04-Operational-Experts-Group/0-operational-experts-group.html (4.2.2016). China is not a participant of the PSI, but it has a joint declaration with the European Union Joint declaration of the People's Republic of China and the European Union on Non-proliferation and Arms Control, C/04/348, Brussels, 8 December 2004, 15854/04 (Presse 348).

[30]http://www.psi-online.info/Vertretung/psi/en/01-about-psi/0-about-us.html, Tornberg (2009), p. 140.

seas interdictions, even *inter partes*. As the Legality of the Threat or Use of Nuclear Weapons Case recognises, these conventions are not sufficient evidence of State practice or *opinio juris* to create a legal obligation or to prohibit the use of certain weapons of mass destruction.[31] The PSI was intended to establish a last means for stopping the transfers of WMDs and related materials, in case the proliferators had managed to load such material aboard a ship. The geographical scope of the initiative focused on the high seas because ships on the high seas are subject to the authority of the State whose flag they fly.[32]

The major problem regarding illicit trafficking is not the ready-made WMDs but components, technologies, production materials and means of delivery associated with WMDs. The problem with these items is that the majority of them are civilian as well as military end uses. The dual-use materials pose a problem because they are mostly used for peaceful purposes, and their trade is legal. The proliferation problem caused by dual-use materials is significant because 95 of the elements for WMDs are dual use. Added to this is the problem that globalisation and technological advancement and the dissemination and accessibility of knowledge and technology necessary to acquire WMD capabilities have increased exponentially since the 1990s. This development has not only increased the ability of States to obtain WMDs but has also enabled non-State actors to obtain them. Thereby, the main concern of the PSI is to prevent States and non-State actors of proliferation concern from acquiring the materials to build WMDs.[33] However, States have to bear in mind that especially in the case of nuclear materials, the legal transfer of nuclear materials is also an issue. Regarding the right of innocent passage, Article 23 of UNCLOS establishes requirements[34] for the trafficking of foreign nuclear-powered ships and ships carrying nuclear or other inherently dangerous or noxious substances. Thus, government transportation is out of the scope of the PSI. It applies only to commercial transportation. Consequently, any unlawful activities undertaken by warships in the exercise of their official duties will be governed by rules of international law.[35] According to UNCLOS, warships are required to comply with the laws and regulations of the coastal State concerning passage through the territorial sea. If a foreign warship disregards a request for compliance made to it, the coastal State may require it to leave the territorial sea immediately.[36]

[31]Legality of the Use by a State of Nuclear Weapons in Armed Conflict, Advisory Opinion, I. C. J. Reports 1996, pp. 66, 226.

[32]Durkalec (2012), p. 2, Ahlström (2005), p. 744, Byers (2004), p. 527.

[33]http://www.state.gov/t/isn/c10390.htm, Fidler (2003), Prosser and Scoville (2004), Beck (2004), p. 16, Logan (2005), p. 255, Jimenez Kwast (2007), pp. 164–167.

[34]Article 23 requires that foreign nuclear-powered ships and ships carrying nuclear or other inherently dangerous or noxious substances provide certain documents upon request and observe special precautionary measures established for them according to international agreements when they exercise their right of innocent passage.

[35]Lehto (2008), p. 57.

[36]UNCLOS art. 30.

Over 90% of international trade is transported by sea; therefore, the marine transport of WMDs and related materials is the core concern of the PSI. Today, maritime shipping is fast and cost-effective, owing to the use of standardised containers that can be directly transferred to and from ground networks at the ports. The effectiveness of the transportation system in ordinary commercial shipping increases the possibility that WMD-related materials are trafficked undetected. At sea, the boarding and inspection of big containerships requires well-resourced and trained forces and is still difficult and dangerous. Furthermore, any kind of delay in shipping results in increased costs.[37] Taking into consideration that the illicit shipment of WMD-related materials is not frequent, the costs of the implementation of the PSI by stopping and searching numerous ships that are not causing any threat would be unreasonable for commercial shipping.[38]

The PSI is an effort to cover the weaknesses of the international non-proliferation regime, and that is also the purpose of Security Council Resolution 1540 (2004) adopted under Chapter VII UN Charter. The resolution endeavours to fill gaps in international non-proliferation efforts by obliging States to refrain from providing any form of support to non-State actors that attempt to develop, acquire, manufacture, possess, transport, transfer or use nuclear, chemical or biological weapons and their means of delivery and encourages States to take effective measures to conform their relevant obligations and responsibilities.[39] However, the resolution does not authorise interdiction. Regarding criminal jurisdiction, the illicit trafficking of WMDs and related materials is not classified as universal crime, and thus it is not subject to universal jurisdiction.

The PSI participants' readiness to conduct interdiction operations is developed by exercises. These exercises have made it possible for different authorities of the participants in the PSI such as armed forces, customs, police and intelligence to meet and create connections with each other. Today, the PSI is increasingly focused on commercial trade in dual-use materials, which has also made the exercises more civilian oriented. Although the exercises involve more civilian law enforcement authorities, most of the exercises still have a strong military aspect.[40]

3.1 Scope of the UNSC Resolutions

The Security Council has linked the proliferation of nuclear, chemical and biological weapons, as well as their means of delivery, with the notion of a threat to

---

[37]Jimenez Kwast (2007), p. 167 see also fn. 24, Kraska (2009), p. 123.

[38]Logan (2005), p. 259.

[39]UN Security Council Resolution 1540 (2004), Jimenez Kwast (2007), p. 169, see also Resolutions 1673 (2006), 1805 (2008) and 1977(2011), extending the mandate of the Committee to April 25 2021.

[40]Durkalec (2012), pp. 15–16.

international peace and security with Resolution 1540 of 28 April 2004 adopted under Chapter VII of the Charter and the others that followed it. The resolutions also define the means of delivery to cover missiles, rockets and other unmanned systems capable of delivering nuclear, chemical or biological weapons that are specially designed for such use.[41]

The resolutions state that the proliferation of WMDs, as well as their means of delivery, constitutes a threat to international peace and security and oblige States to refrain from providing any form of support to non-State actors that attempt to develop, acquire, manufacture, possess, transport, transfer or use WMDs and related materials and to adopt and enforce appropriate effective laws that prohibit any non-State actor to manufacture, acquire, possess, develop, transport, transfer or use these materials. Furthermore, the resolution required States to take and enforce effective measures to establish domestic controls to prevent the proliferation, including (a) accountability, (b) physical protection, (c) border controls and law enforcement efforts and (d) national export and trans-shipment controls.[42] Security Council Resolution 1540 as well as further resolutions do not specifically authorise the non-flag States to board ships and to seize WMD-related cargo or provide any other enforcement authority.[43] Interdiction was not included in the resolution because of China's opposition to the matter, and thus the resolution only refers to international cooperation to prevent illicit trafficking.[44]

The main rule in international law of the sea recognises the exclusive jurisdiction of a flag State on the high seas. Some provisions of UNCLOS are exceptions to this main rule. The exceptions are related universal crimes occurring on the high seas. Today, State practice and treaties after over a decade of adoption of the PSI and UNSC Resolutions have not amended the flag State's exclusive jurisdiction over its ships, although the popularity of the PSI might indicate that there has emerged a norm of customary international law against proliferation allowing States to take certain actions to prevent it.[45] However, during the negotiations for the resolution on non-proliferation of weapons of mass destruction, a representative of the United Kingdom stated that the resolution does not 'authorize enforcement action against States or against non-State actors in the territory of another country. The draft resolution makes clear that it will be the Council that will monitor its implementation. Any enforcement action would require a new Council decision.' The representative of the United States also stated that the resolution is 'not about enforcement'.[46]

---

[41]UN Security Council Resolutions 1540(2004), 1810(2008), 1977(2011).

[42]UN Security Council Resolution 1540 (2004), Durkalec (2012), p. 13.

[43]Rayfuse (2005), p. 198, Logan (2005), p. 270, Durkalec (2012), p. 13.

[44]Winner (2005), p. 136.

[45]Logan (2005), p. 271.

[46]UN Security Council meeting of April 22 S/PV.4950 (2004), p. 12, 17. The Council has adopted enforcement actions against Iran (Resolution 1929 (2010)) and North Korea (Resolution 1874 (2009)). These Resolutions 'call on all states to inspect all cargo to and from Iran and North Korea

UN Security Council Resolutions are a step towards a universal global non-proliferation regime. Today, they supplement existing non-proliferation and disarmament laws and regulations, but the resolutions do not grant any new authority or jurisdiction to States.[47]

4 Legal Problems with the PSI and the Right of Innocent Passage

Security Council Resolutions, the 2005 SUA Protocol being the first international convention recognising the trafficking of WMDs and related materials as illegal, and the right of self-defence do not provide any enforcement power, or if they do, the power is limited in certain circumstances on the high seas where a flag State has exclusive jurisdiction over the ship and crew.[48] Thus, the use of national military and law enforcement power is regulated by the rules of the law of the sea. Even though the freedom of navigation, one of the oldest principles of the customary international law, is limited in certain circumstances, even on the high seas, where a warship has a right to board vessels, regarding the illicit trafficking of WMDs and related materials, the problem is that the illicit trafficking of WMDs and related materials is not universally condemned in the same terms in the law of the sea as the slave trade.[49] On the territorial seas, the principle of freedom of navigation is exercised through the concept of the right of innocent passage. Regarding the illicit trafficking of WMDs and related materials, it would be wrong if the right of ships to exercise freedom of navigation on the high seas were to be more limited than the right to innocent passage within a coastal State's territorial waters.

The United Nations Convention on the Law of the Sea authorises the boarding of a foreign ship on the high seas in cases of piracy, slave trade, unauthorised broadcasting or when a ship is stateless.[50] Foreign warships or coast guard authorities may interdict and search a ship only in the aforementioned cases. A coastal State has power under international law to stop and seize cargo on its internal waters and territorial sea, except when a foreign ship is exercising the right of innocent passage. The 2005 SUA Protocol criminalised and created new enforcement procedures to prevent maritime terrorism and the use of ships by terrorists and for

---

that is in their territory, including seaports and airports, if there are "reasonable grounds" to believe the cargo contains items of which the supply, sale, transfer or export is prohibited. Both resolutions also call on states to cooperate in inspections and, more significantly, they authorize all UN members to seize and dispose of prohibited cargo'. Durkalec (2012), p. 13. The Resolution 1929 (2010) was terminated by the Resolution 2231(2015) see http://www.un.org/en/sc/2231/.

[47] Allen (2007), p. 59.

[48] Jinyuan (2012), p. 98, Dixon (2006), p. 23, Durkalec (2012), p. 14.

[49] Cirincione and Williams (2005).

[50] UNCLOS art. 110.

terrorist purposes. Article 8*bis* created a new procedure for boarding a ship on the high seas, which is suspected of being involved in offences under the SUA Convention.[51] However, the Convention does not contain any change for the flag State's exclusive jurisdiction on the high seas; thus, a State party to the Convention has to ask the flag State's authorisation to board and to take appropriate measures.

The UN Law of the Sea Convention does not directly speak about security issues; although some provisions regulate warships, they do not deal with naval warfare, disarmament, demilitarisation or denuclearisation. The lack of discussions of military operations in the UN Law of the Sea Conference was not accidental; they were deliberately left out of the discussions.[52] The Convention refers to security in the context of the right of innocent passage.[53] The indirect references to the security issues indicate that the Convention's intention is to regulate the uses of the seas in times of peace. A coastal State may temporarily suspend the right of innocent passage if it deems such suspension essential for the protection of its security (Article 25(3)). Article 21 grants a coastal State the possibility to regulate the passage of ships exercising their right of innocent passage; however, those laws and regulations may focus on the safety of navigation and protection of the marine environment, not security matters.

5 Interdictions by Participants in Their Territorial Sea

5.1 Sovereignty and Jurisdiction of Coastal States

Aside from territorial sea claims, States are primarily concerned about the rights of access and resource exploitation within their territorial waters. The most important topics concerning legislative jurisdiction have been navigation, customs, fishing, sanitation and security. Oceans have always served as the most convenient highway for launching attacks, and because of this, coastal State security interests are grounded by a crucial understanding that territorial seas provide important routes to follow when reaching shores. Coastal State claims to authority over territorial sea areas are commonly described as an assertion of sovereignty over a part of coastal State's land territory.[54]

Sovereignty includes territorial sea claims made by States as they seek to control access to their waters. In aiming to secure comprehensive and continuous authority to deny passage through their territorial seas, the focus of coastal States has mainly centred upon the concept of innocent passage. Moreover, States have sought a

[51]Bergin (2005), pp. 89–90.

[52]O'Connell (1984), p. 825, Hakapää (1988), pp. 69–70, Vukas (2004), pp. 4–5, Rayfuse (2005), p. 189.

[53]UNCLOS art. 19.

[54]McDougal and Burke (1987), p. 179.

number of claims that have included occasional exclusive competence to deny passage in regard to specific cases, a right to prescribe policy for territorial sea cases, a right to prescribe and apply policies to solve problems aboard vessels and a right to the exclusive appropriation of resources.[55]

Sovereignty over territorial sea areas grants coastal States the following rights:

- Coastal States have an exclusive right to fish and to exploit the resources of the seabed and subsoil.
- They have exclusive enjoyment of the air space above the territorial sea area as foreign aircraft does not enjoy the same rights of innocent passage as foreign vessels do.
- A coastal State has an exclusive right to transport goods and passengers from one part of its territory to another part.
- During times of war when a coastal State is neutral, belligerent States are not allowed to engage in combat, or capture merchant vessels, within the coastal State's territorial sea.
- Foreign vessels must obey regulations concerning navigation, health, customs duties and immigration that are enacted by a coastal State.[56]

In addition to these rights, a coastal State has both civil and criminal jurisdiction over merchant vessels exercising the right of innocent passage, as well as persons on board such vessels.[57] Regarding warships, however, a coastal State does not have this kind of jurisdiction and may only demand that the warship leave its territorial sea if it does not comply with persistent requests to adhere to coastal State regulations.[58]

The Statement of Interdiction Principles says that the PSI activities will not violate international law. However, subparagraph 4 (d) of the Statement of Interdiction Principles calls participants to take appropriate actions to do the following:

(1) stop and/or search in their internal waters, territorial seas or contiguous zones (when declared) vessels that are reasonably suspected of carrying such cargoes to or from States or non-State actors of proliferation concern and to seize such cargoes that are identified; and
(2) to enforce conditions on vessels entering or leaving their ports, internal waters or territorial seas that are reasonably suspected of carrying such cargoes, such as requiring that such vessels be subject to boarding, search and seizure of such cargoes prior to entry.

[55] McDougal and Burke (1987), p. 179.

[56] UNCLOS arts. 2, 19, 21, Convention (XIII) concerning the Rights and Duties of Neutral Powers in Naval War, adopted on 18 October 1907 and entered into force on 26 January 1910 art. 1, Malanczuk (1997), pp. 177–178.

[57] TSC arts. 19, 20, UNCLOS arts. 27, 28, Malanczuk (1997), p. 178.

[58] TSC art 23, UNLOSC art. 30.

The requirement in subparagraph 4 (d) (1) is problematic because although coastal States have sovereignty over their territorial seas, it is limited by the right of innocent passage of foreign vessels.[59] A coastal State may not hamper the passage of foreign ships through the State's territorial sea if not being prejudicial to the peace, good order or security of the coastal State.[60] Instead, if the interdiction of a ship under a flag different from the coastal State takes place in the internal waters by the coastal State authorities, the act is in accordance with the law of the sea.[61]

However, in the territorial sea, the right of innocent passage makes the situation complicated. Regarding the coastal State's legislative competences, the 1982 UN Law of the Sea Convention contains specific provisions relating to innocent passage. According to Article 21 (1):

> [a] a coastal State may adopt laws and regulations, in conformity with the provisions of this Convention and other rules of international law, relating to innocent passage through the territorial sea, in respect of all or any of the following:
> (a) the safety of navigation and the regulation of maritime traffic;
> (b) the protection of navigational aids and facilities and other facilities or installations;
> (c) the protection of cables and pipelines;
> (d) the conservation of the living resources of the sea;
> (e) the prevention of infringement of the fisheries laws and regulations of the coastal State;
> (f) the preservation of the environment of the coastal State and the prevention, reduction and control of pollution thereof;
> (g) marine scientific research and hydrographic surveys;
> (h) the prevention of infringement of the customs, fiscal, immigration or sanitary laws and regulations of the coastal State.

In addition, coastal States must give due publicity to their laws. Moreover, such laws may not affect the design, construction, manning or equipment of foreign vessels unless they conform to generally accepted international standards.[62]

Article 21 limits a coastal State's prior legislative competences and therefore provides a jurisdictional compromise between coastal State and flag State interests. Instead, the article grants coastal States certain legislative competences but eliminates the risk of divergent design, construction, manning and equipment standards that might be hard to accommodate when vessels set out on voyage.[63] Foreign vessels have to comply with coastal State laws that are enacted analogously with the Convention.[64] Furthermore, in accordance with Article 21(4), '[f]oreign ships

[59]UNCLOS art. 17.

[60]Klein (2011), p. 200.

[61]Wolfrum (2009), p. 90.

[62]Yearbook of International Law Commission (1956), p. 274, Churchill and Lowe (1999), p. 94.

[63]Churchill and Lowe (1999), p. 94, Harrison (2013), p. 170.

[64]Churchill and Lowe 1999, pp. 94–95. According to Article 22 a coastal State is not allowed to dismiss recommendations made by the IMO, a competent international organisation, when ordering sea lanes. Harrison (2013) argues, however, that the IMO has only a recommendatory role in this situation (p. 180).

exercising the right of innocent passage through the territorial sea shall comply with all such laws and regulations and all generally accepted international regulations relating to the prevention of collisions at sea'.[65] It is irrelevant whether a flag or coastal State is party to conventions containing such regulations.[66]

Article 21 contains an exhaustive list that clearly restricts the matters that a coastal State may regulate. Thus, a coastal State cannot draw any authorisation from Article 21 to implement the PSI unless the coastal State is the destination of the illegal shipment of WMD materials.

5.2 *The Right of Innocent Passage*

The 1958 Territorial Sea Convention determines that passage is innocent if it is not prejudicial to the peace, good order or security of the coastal State. The Convention mentioned two activities that were dissociated from the character of innocence. The passage of foreign fishing vessels was not considered innocent if vessels did not observe coastal State laws and regulations. Laws and regulations made and published by coastal States had generally been created with the intention of preventing vessels from fishing in territorial sea areas. The second exception to the rule was that submarines had to navigate on the surface and display their national flag. Otherwise, legal competence was left to the broad jurisdiction of the coastal States in question when determining whether passage was innocent or not.[67]

The situation concerning the concept of innocence changed in 1982 after the UN Law of the Sea Convention was adopted. UNCLOS includes more specific definitions concerning innocent passage. In Article 19(2), a list of activities that are considered prejudicial to the peace, good order or security of the coastal State are mentioned as follows:

> 1. Passage is innocent so long as it is not prejudicial to the peace, good order or security of the coastal State. Such passage shall take place in conformity with this Convention and with other rules of international law.
>
> 2. Passage of a foreign ship shall be considered to be prejudicial to the peace, good order or security of the coastal State if in the territorial sea it engages in any of the following activities:
>
> (a) any threat or use of force against the sovereignty, territorial integrity or political independence of the coastal State, or in any other manner in violation of the principles of international law embodied in the Charter of the United Nations;
>
> (b) any exercise or practice with weapons of any kind;
>
> (c) any act aimed at collecting information to the prejudice of the defence or security of the coastal State;

---

[65]For example, Convention on the International Regulations for Preventing Collisions at Sea, London, 20th October 1972, which entered into force on 15th July 1977, 1050 UNTS 16.

[66]Churchill and Lowe (1999), p. 95.

[67]TSC art. 14.

(d) any act of propaganda aimed at affecting the defence or security of the coastal State;
(e) the launching, landing or taking on board of any aircraft;
(f) the launching, landing or taking on board of any military device;
(g) the loading or unloading of any commodity, currency or person contrary to the customs, fiscal, immigration or sanitary laws and regulations of the coastal State;
(h) any act of wilful and serious pollution contrary to this Convention;
(i) any fishing activities;
(j) the carrying out of research or survey activities;
(k) any act aimed at interfering with any systems of communication or any other facilities or installations of the coastal State;
(l) any other activity not having a direct bearing on passage.

However, the list is not a comprehensive one because the last item forbids any other activity that is not actually relevant to passage. Nonetheless, any activity that has no direct bearing on passage will not automatically render passage non-innocent.[68]

The list focuses on vessels' activities and therefore suggests that the nature of a vessel is not sufficient grounds for considering passage non-innocent. In addition, Article 23 goes further and sets obligations for foreign nuclear-powered vessels and vessels carrying nuclear or other inherently dangerous or noxious substances when they are exercising the right of innocent passage through the territorial sea.[69] Consequently, UNCLOS does not prohibit the shipment of WMDs or related materials.[70] Article 23 of the Convention only obliges foreign nuclear-powered vessels and vessels carrying nuclear or other inherently dangerous or noxious substances to carry certain documents and to observe special precautionary measures established for such vessels by international agreements when they are exercising the right of innocent passage through territorial seas.[71] Article 23 clearly limits the authority of coastal States as they take into account certain issues related to nuclear-powered vessels and vessels carrying nuclear materials when a decision must be made in relation to whether passage is deemed to be innocent or not. This provision indicates that the nature of the vessel or its cargo does not influence the right of innocent passage as long as it carries with it the appropriate documents and conforms to precautionary measures established by international law.[72] However, as the So San case[73] shows, it is highly probable that a ship involved in the illicit

[68]Pharand (1977), p. 77, Churchill and Lowe (1999), p. 84.

[69]Art. 23: Foreign nuclear-powered ships and ships carrying nuclear or other inherently dangerous or noxious substances shall, when exercising the right of innocent passage through the territorial sea, carry documents and observe special precautionary measures established for such ships by international agreements.

[70]Rayfuse (2005), p. 190. United States required Article 23 to the Convention.

[71]International agreements, such as the International Convention for the Safety of Life at Sea (SOLAS) and its Annex, as well as IMO recommended codes regarding the construction and equipment of ships carrying dangerous liquid chemicals or liquefied gases in bulk, Nordquist et al. (1993), p. 220.

[72]See International Association of Lawyers Against Nuclear Arms Aotearoa New Zealand Branch http://lcnp.org/disarmament/nwfz/submission%20on%20NWF2.htm.

[73]The freighter So San was transporting according to ship's manifest 2000 pounds of concrete, however, it was also transporting missile parts and an unknown chemical, see Joyner (2005), p. 2.

trafficking of WMD materials will not carry documents required by Article 23, nor will it observe precautionary measures. But the coastal State has the right of non-flag enforcement only if a ship carrying WMDs or related materials engages such activities that render its passage non-innocent.[74] Writers have raised the question of whether the illicit trafficking of WMDs and related materials through the territorial seas can be deemed non-innocent.[75]

Some writers have suggested that the mere passage of a ship carrying illicitly WMDs is a violation of the right of innocent passage. Lehrman states that although the list in Article 19(2) does not explicitly refer to trafficking in WMDs or related materials as prejudicial to the peace, it does not foreclose such an interpretation.[76] Kaye argues that 'Clearly the delivery of WMD to terrorists may well be highly prejudicial to the peace, good order and security of a coastal State, an argument could be made that such a passage is therefore not innocent, and the restrictions on coastal State authority over the passing vessel are removed'.[77] Further, Churchill and Lowe claim that activities seen as posing a threat of force affect third States, as well as coastal States. Because a coastal State's security is seen by Churchill and Lowe as being indirectly linked to a third State's welfare, they also believe that there is no need for links to other legal instruments, such as a mutual defence treaty, when aiming to render threats as incompatible with innocent passage. Thus, paragraph 2 may be interpreted in such a way as to allow coastal States to act on the impression that a third State's security is at stake.[78] Further, Joyner holds the view that the wording of Article 19 (2)(a) is wide enough to include a threat of force against a third State.[79]

Ronzitti has an opposing view, arguing that a ship entering territorial sea at one point from the high seas and leaving at another without any intention to enter internal waters or stop at any port does not violate the right of innocent passage.[80] Similarly, Garvey argues that the mere shipment of WMD materials does not constitute a threat to the coastal State.[81] In addition, Logan states that the shipment of WMD materials does not fit within any of the exceptions listed in Article 19. Logan believes that it would be difficult, first, to prove that the shipping of WMD and related materials constituted a threat of force against the coastal State because 95 percent of the materials for WMDs are dual use in nature. Second, it would also be difficult to prove that the shipping of WMD materials threatened the coastal State's sovereignty, territorial integrity or political independence and that the WMD materials were going to be used against that particular State. Third, a

---

[74]Rayfuse (2005), p. 190.

[75]Ronzitti (1990), p. 5, Churchill and Lowe (1999), p. 85, Lehrman (2004), p. 232, Garvey (2005), p. 131, Joyner (2005), p. 529, Logan 2005), p. 259, Kaye (2006), pp. 147–148.

[76]Lehrman (2004), p. 232.

[77]Kaye (2006), pp. 147–148.

[78]Churchill and Lowe (1999), p. 85.

[79]Joyner (2005), p. 529.

[80]Ronzitti (1990), p. 5.

[81]Garvey (2005), p. 131.

violation of the UN Charter requires that the threat or use of force is made in the territorial sea, and thus a coastal State cannot rely on the future use of the WMDs because the use is unlikely to take place in the territorial sea.[82]

The provisions of the 1982 UN Law of the Sea Convention are more detailed than the simple definitions provided in the 1958 Territorial Sea Convention. It seems obvious that the aim of the 1982 UN Law of the Sea Convention was to produce a more objective definition that would leave coastal States less scope for interpretation, as well as less potential to abuse their rights when suspending non-innocent passage. Within the 1982 UN Law of the Sea Convention text, there are particular references made to activities. Therefore, a vessel's presence or passage alone cannot be interpreted as prejudicial to coastal State interests if the vessel does not engage in some specific actions. Thus, the formulation of the provision regulating innocent passage would narrow the scope of the right of innocent passage by adding the illicit trafficking of WMDs and related materials and their delivery systems to the activities that are prejudicial to the peace, good order or security of the coastal State.

The United States and the former Soviet Union signed the bilateral Treaty on the Uniform Interpretation of Norms of International Law Governing Innocent Passage in 1989.[83] Paragraph 3 of this Treaty states the following:

> Article 19 of the Convention of 1982 sets out in paragraph 2 an exhaustive list of activities that would render passage innocent. A ship passing through the territorial sea that does not engage in any of those activities is in innocent passage.

The Uniform Interpretation does not leave an understanding of innocence open to interpretation. Both States are notable maritime powers,[84] and their interpretation was influential at the time of the agreement. It is noteworthy that these States referred to the 1982 UN Law of the Sea Convention. Their common objectives were obviously to contribute to State practice and to promote their own interpretation in the future evolution of customary international law. The Uniform Interpretation was signed on September 1989, at which time the former Soviet Union had signed the 1982 UN Law of the Sea Convention, but the United States had not. However, the former Soviet Union had not ratified the 1982 Convention,[85] and the Convention had not entered into force. The Uniform Interpretation made between the two States is binding upon the two States parties to it but not applicable to third party States. However, the provisions included in the bilateral treaty may become binding on third party States if they become norms of customary international law.[86] The

[82]Logan (2005), p. 259.

[83]LOSB (1989), p. 12.

[84]Nowadays Russia, after the disintegration of the Soviet Union.

[85]The Russian Federation ratified the United Nations Convention on the Law of the Sea on 12th February 1997. The United States signed the Convention on 29th July 1994 and on 7th October 1994 President Clinton transmitted to the Senate the United Nations Convention on the Law of the Sea. Treaty Document 103–39.

[86]Churchill and Lowe (1999), p. 86. The United Nations Convention on the Law of the Sea entered into force 16.11.1994.

restrictive interpretation of the article was created with the best interests of maritime powers in mind because it limits the discretion of a coastal State and thus benefits foreign navies navigating the world's oceans. Although the United States and the Soviet Union considered the list a comprehensive one, in fact it included the phrase 'any other activity not having a direct bearing on passage', which left some scope for further interpretation by coastal States with regard to the nature of passage.[87] At the time they did not see non-State actors as possible users of WMDs, and therefore there is a strong possibility that the Uniform Interpretation is not intended to restrict the interpretation of Article 19(2) in the case of the illicit trafficking of WMDs and related materials and their devices.

Writers' differing opinions regarding the activity making the passage non-innocent illustrate that there is a need to discuss the balance of new modes of threats to coastal States and maritime security caused by non-State actors and the freedom of navigation for merchant vessels that has been historically linked to world interests.

However, although the list is considered non-exhaustive, any activity that has no direct bearing on passage will not automatically render passage non-innocent. Instead, coastal States have to provide evidence of activities that are deemed prejudicial to coastal States' peace, good order or security.[88] A coastal State has to acquire solid intelligence proving that WMD materials were being shipped on the territorial sea.

5.3 *The Territorial Waters of the Åland Islands*

The essential question to ask in relation to the innocent passage of ships illegally carrying WMD and related materials through the territorial waters of the Åland Islands is whether there is sufficient scope available to consider passage non-innocent on the grounds that passage compromises the principles of demilitarisation and neutralisation. Governmental transportation, such as naval warships, is out of the scope of the PSI as it applies only to commercial transportation. Consequently, any unlawful activities undertaken by warships in the exercise of their official duties will be governed by rules of international law.[89] Article 4 (1) of the 1921 Åland Convention says that 'Except as provided in Article 7, no military, naval or air force of any Power shall enter or remain in the zone described in Article 2; the manufacture, import, transport and re-export of arms and implements of war in this zone are strictly forbidden'. However, the 1921 Åland Convention particularly mentions warships that have a right of innocent passage

[87]Hakapää and Molenaar (1999), p. 132.

[88]Churchill and Lowe (1999), p. 84.

[89]Lehto (2008), s. 57.

according to regulations established by international law. Thus, the right of innocent passage of warships is firmly embedded in a separate article, Article 5.[90] This indicates that the intention of parties to the Convention could have been to exclude any evaluation of the innocent passage of warships from being based on concepts of demilitarisation and neutralisation. With this in mind, then, it seems that Finland is not authorised to declare passage non-innocent on the ground that warships prejudice the peace, good order or security of the Islands because of its demilitarised and neutralised status.

The right of innocent passage of merchant ships has its origin in the customary international law and is codified in the 1958 Territorial Sea Convention and, today, in the UN Convention on the Law of the Sea. Thus, although the 1921 Åland Convention does not mention merchant ships, it is unlikely that the right of innocent passage of merchant ships as such was meant to be restricted. Furthermore, this grey area in the Convention's text means that an interpretation of innocent passage must be primarily founded on the rules of international law and practice. Regarding the Finnish national legislation, there are no detailed accounts of what constitutes an act of non-innocent passage.[91] Under the Finnish Territorial Surveillance Act, the entry, stay and departure of vessels to and from Finnish territorial seas is stated to be governed by any relevant separate provisions or international treaties binding on Finland.[92] Hence, any evaluation of the nature of passage, whether innocent or not, rests on the interpretation of UNCLOS.

According to the preamble of the 1921 Åland Convention, it was concluded that the objective of the Convention was to reduce the islands' potential as a military threat. The purpose of the Convention was to protect the coastal States of the Baltic Sea region and not just Finland. Security was an important motive when States signed the Åland Convention. The general protection of the region was achieved by demilitarising and neutralising the land areas and surrounding waters. Thus, demilitarisation and neutralisation ensured the safety of the region by keeping the area free from military deployments or operations. When discussing the territorial waters of the Åland Islands, therefore, one should always bear in mind the interests of the wider group of countries and not just the principal coastal State involved. Therefore, in the case that the illicit trafficking of WMDs and related materials will be used against a party to the 1921 Åland Convention, the shipment poses a threat of force although affecting the third State, the shipment is not in accordance with the Convention's aim and purpose. The demilitarised sea area is established to guarantee peace and stability in the sense that the Åland Islands shall never become a

[90]Article 5 says: "The prohibition to send warships into the zone described in Article 2 or to station them there shall not prejudice the freedom of innocent passage through the territorial waters. Such passage shall continue to be governed by the international rules and usages in force."

[91]Innocent passage is defined in the Finnish Territorial Surveillance Act (755/2000) Section 2 and includes a specific reference to the 1982 UN Law of the Sea Convention.

[92]The Finnish Territorial Surveillance Act, the Finnish Collection of Decrees 178/1938, 755/2000 Section 3.

threat from a military perspective. However, today, non-State actors, for example an international terrorist network, can also get in their hands on a nuclear device, which could constitute a serious and imminent danger to the parties to the 1921 Åland Convention. In this kind of situation, if Finnish authorities had acquired solid intelligence that proved the illicit trafficking of a nuclear device, even a temporary presence of illicit trafficking of nuclear devices within the demilitarised and neutralised zone would become an obvious threat. Therefore, today, this sort of passage does not seem to conform to the 1921 Åland Convention.

5.4 Article 25 of UNCLOS

According to Article 25 of the 1982 UN Law of the Sea Convention, a coastal State may take necessary steps in its territorial sea to prevent passage that is not innocent. The article mentions the concept of 'innocence', which seems to be the main criterion. Passage is another prerequisite that a vessel must fulfil before innocence can be evaluated. The 1982 UN Law of the Sea Convention also defines the concept of 'passage' but is silent about vessels that do not fulfil the Convention's requirements of passage. Churchill and Lowe claim that the right to exclude passage exists in customary international law. Vessels hovering around territorial seas could be deemed non-innocent and may therefore justifiably be excluded from coastal States' waters. As passage is directly linked to the concept of innocence, any violation of passage will automatically be a violation of innocence. The right of innocent passage applies to vessels as they undertake their voyages through the territorial sea of a foreign coastal State. If a vessel were to lose the right to innocent passage, it would then be subject to coastal State jurisdiction, which could possibly lead to an arrest.[93]

Innocent passage may be suspended temporarily for two reasons in particular. Article 25(2) says that

> [i]n the case of ships proceeding to internal waters or a call at a port facility outside internal waters, the coastal State also has the right to take the necessary steps to prevent any breach of the conditions to which admission of those ships to internal waters or such a call is subject.

Hakapää and Molenaar have also remarked on this kind of interference, as they claim that the prevention of innocent passage could take place when a coastal State suspects a foreign vessel of smuggling alcohol or drugs into its territorial waters.[94] Paragraph 2 might imply that the coastal State could stop inbound ships that it suspected of illicit trafficking of WMD or related materials and their devices.[95] The

[93]Churchill and Lowe (1999), p. 87.

[94]Hakapää and Molenaar (1999), p. 133.

[95]Logan (2005), p. 261.

other reason for suspending innocent passage arises when a coastal State believes that suspension is completely necessary for the protection of itself and its interests.[96] A coastal State has a right to suspend passage through its territorial sea and may determine whether the passage of a vessel prejudices its security. It is noteworthy that the right temporarily to suspend innocent passage covers merchant vessels and warships. Coastal States may exercise this right to exclude foreign vessels from restricted areas, but the suspension has to be non-discriminatory and published before becoming effective.

The illicit trafficking of WMDs and related materials and their devices does not seem to fit within the scope of Article 25(3) because its suspension may not be discriminatory and the PSI interdiction is aimed at a specific ship or actors of concern. Furthermore, the PSI interdiction operations have to occur in a specific area. However, Article 25(1) may establish the legal basis for the PSI interdictions if the illicit trafficking of WMDs and related materials makes the passage non-innocent according to Article 19(2).

According to Article 25(1) of UNCLOS, coastal States are allowed to take necessary steps to prevent non-innocent passage from taking place in their territorial seas. What are, then, 'the necessary steps' that a coastal State may take after the passage is rendered non-innocent? The ship in non-innocent passage is subject to full coastal State authority, and 'the coastal State may use any necessary force, proportionate to the circumstances, to require a delinquent vessel to leave its territorial sea'.[97]

5.5 Criminal Jurisdiction in the Territorial Sea

A coastal State has both civil and criminal jurisdiction over merchant vessels exercising the right of innocent passage, as well as persons on board such vessels.[98] However, a coastal State may exercise criminal jurisdiction over foreign ships in its territorial sea only according to Article 27 of UNCLOS.[99] Article 27 of UNCLOS states the following:

> 1. The criminal jurisdiction of the coastal State should not be exercised on board a foreign ship passing through the territorial sea to arrest any person or to conduct any investigation in connection with any crime committed on board the ship during its passage, save only in the following cases:
>
> (a) if the consequences of the crime extend to the coastal State;
>
> (b) if the crime is of a kind to disturb the peace of the country or the good order of the territorial sea;

[96]For example, when undertaking weapon exercises on its own or with a third State. See UNCLOS art. 25 (3).

[97]Shearer (1986), p. 325.

[98]TSC arts. 19, 20, UNCLOS arts. 27, 28, Malanczuk (1997), p. 178.

[99]Rayfuse (2005), p. 190.

(c) if the assistance of the local authorities has been requested by the master of the ship or by a diplomatic agent or consular officer of the flag State; or

(d) if such measures are necessary for the suppression of illicit traffic in narcotic drugs or psychotropic substances.

2. The above provisions do not affect the right of the coastal State to take any steps authorised by its laws for the purpose of an arrest or investigation on board a foreign ship passing through the territorial sea after leaving internal waters.

3. In the cases provided for in paragraphs 1 and 2, the coastal State shall, if the master so requests, notify a diplomatic agent or consular officer of the flag State before taking any steps and shall facilitate contact between such agent or officer and the ship's crew. In cases of emergency this notification may be communicated while the measures are being taken.

4. In considering whether or in what manner an arrest should be made, the local authorities shall have due regard to the interests of navigation.

5. Except as provided in Part XII or with respect to violations of laws and regulations adopted in accordance with Part V, the coastal State may not take any steps on board a foreign ship passing through the territorial sea to arrest any person or to conduct any investigation in connection with any crime committed before the ship entered the territorial sea, if the ship, proceeding from a foreign port, is only passing through the territorial sea without entering internal waters.

Paragraph 1 uses the phrase 'should not be exercised', while paragraph 5 uses the phrase 'may not take any steps'. The different wording illustrates the different juridical nature of the zones in which the suspected criminal offence took place. In the situation envisaged in Article 27(1), the suspected crime has happened on board a ship during its passage through the territorial sea, and thus the coastal State is entitled to exercise jurisdiction. However, the provision limits the coastal State's authority to four particular cases. It is clear that the interests of the freedom of international trade and navigation are protected unless there are significant causes to supersede them by the demands of criminal justice.[100]

In the situation referred in paragraph 2, it is necessary for the coastal State to have criminalised the illicit passage of WMD and related materials in its domestic legislation in order to allow the coastal State's authorities to interdict or detain ships that are passing through the territorial sea after leaving the internal waters of the coastal State.

Paragraph 5 of Article 27 regulates the situation in which the suspected crime has taken place beyond the territorial sea of the coastal State when a vessel is beyond the reach of the coastal State's criminal law. The wording of paragraph 5 does not seem to give discretion to a coastal State because the phrase 'may not' indicates a clear prohibition regarding the exercise of the coastal State's criminal jurisdiction.[101]

Klein argues that the coastal State's domestic legislation that criminalises the illicit passage of WMDs and related materials would overcome this particular restriction. Thus, the prevention of the proliferation of WMDs and related materials to non-State actors is in the hands of States, depending on their will to use the available legal tools.[102] In addition, Logan holds the view that the protective

---

[100]Brown (1994), p. 64.

[101]O'Connell (1984), p. 962, Brown (1994), p. 64.

[102]Article 27(2), (3), Klein (2011), pp. 201–202.

principle according to which a State has a right to protect itself against threatening acts done outside its territory and Article 27 are legal tools to justify the PSI in the territorial sea. Logan comes to this conclusion based on an analogous interpretation of Article 27(1) (d) relating to the illicit trafficking of drugs.[103] However, although the coastal State has criminalised the illicit passage of WMDs and related materials it would also need to associate this kind of activity with the activities regarded to disturbing the peace of the coastal State or good order of its territorial sea or consequences of the crime extend to the coastal State.[104]

Wolfrum considers that the above-mentioned interpretation of Article 27 is problematic. Application of Article 27 requires that the crime has been committed on board the ship passing through the territorial sea and the crime disturbs the peace of the coastal State or good order of its territorial sea. This kind of interpretation of Article 27 would also make it possible to prohibit the transport of nuclear waste, as well as the transport of dangerous substances.[105] This extensive interpretation of Article 27 would be problematic because it would be inconsistent with Article 23. Therefore, the mere passage of a foreign ship through the territorial sea carrying illicitly WMDs or related materials does not meet the requirements for the exercise of the criminal jurisdiction of the coastal State.

Finland is party to the most significant WMD treaties and political arrangements, as well as the SUA Conventions and the 2005 SUA Protocols. Finland has implemented them in its domestic legislation and criminalised the illicit trafficking of WMD and related materials.

In Finland, the responsibility for criminal investigation rests with the police, Customs, the Border Guard and the Defence Forces. The management and organisation of the Finnish Border Guard is within the Ministry of the Interior, from which it follows that the Border Guard's vessels and aircraft are not treated as warships. The demilitarisation regime is regulated directly by a multilevel legal framework, and Finland's sovereign rights as a coastal State are significantly restricted by the 1921 Åland Convention. These restrictions focus on the military presence in the zone.

The responsibility to conduct a criminal investigation in offences made with terrorist intent rests with the police, and they have a right to receive executive assistance, which includes also the use of military force, in the territorial waters and EEZ of Finland from the Border Guard and the Defence Forces.[106] The police have the main responsibility because the use of force against the illicit trafficking of WMDs and related materials is not the use of force against the enemy according to

[103]Logan (2005), p. 263, Klein (2011), p. 202.

[104]Klein (2011), p. 76.

[105]Wolfrum (2009), p. 91, Hakapää (1981), p. 198 refers e.g. murder on board as "other" other activities on board the vessel "which may have "external" effects".

[106]Border Guard Act the Finnish Collection of Decrees 178/1938 578/2005 Section 77a, 79, Laki puolustusvoimien virka-avusta poliisille the Finnish Collection of Decrees 178/1938.

781/1980 Section 1.

the law of armed conflict, as there is no armed conflict, international or national.[107] The police will decide case by case whether the executive assistance is requested from the Border Guard or the Defence Forces.[108]

The Finnish Defence Forces do not have any police powers. Regarding the area of the Åland Islands, the Finnish navy thus has no authority to board a suspected ship, to inspect the ship, to arrest the crew or to take control of any kind over the crew in the maritime zones of Finland. According to the Act on the Defence Forces Section 2 (2)(a), Defence Forces provide 'support for other authorities, including the following:

a) executive assistance to maintain public order and security, to prevent and interrupt terrorist acts, and otherwise to protect society at large'.

According to Section 79 of the Border Guard Act (578/2005), the Border Guard has the right to receive executive assistance from the Defence Forces, among other protective equipment necessary for the safe performance of a dangerous Border Guard function and equipment and the special expertise necessary to combat a security threat to a ship at sea or to passengers on board. However, the assistance does not include the use of firearms or military force.

However, the police have to take into account the international treaty arrangements related to the demilitarisation of the Åland Islands. These treaty arrangements oblige Finland to guarantee the security of the demilitarised Åland Islands. There are three different opinions concerning the interpretation of Article 4 of the 1921 Åland Convention and the presence of Defence Forces in the demilitarised zone in the case of executive assistance.[109] First, the executive assistance of the Defence Forces for the operation requested by the police is under the command of the civil authority, and therefore the troop of the Defence Forces is regarded as civilian, and thus its presence is not regulated by the 1921 Åland Convention. Second, the troop of the Defence Forces is regarded as military, but the 1921 Åland Convention offers certain exceptions when considering a military presence within the zone during peacetime. Thus, the executive assistance of the Defence Forces is based on Article 4 (2)(a) of the 1921 Åland Convention, which says:

> (a) In addition to the regular police force necessary to maintain public order and security in the zone, in conformity with the general provisions in force in the Finnish Republic, Finland may, if exceptional circumstances demand, send into the zone and keep there temporarily such other armed forces as shall be strictly necessary for the maintenance of order.

[107] Treves (2009), p. 412.

[108] Government Proposal HE 220/2013 vp., Laki puolustusvoimien virka-avusta poliisille the Finnish Collection of Decrees 178/1938.
781/1980.

[109] Ministry of Defence (2014), p. 7.

Thereby, the military presence does not in this kind of exceptional situation violate the limitations set on Finnish naval visits by the 1921 Åland Convention. The responsibility for the provision of executive assistance in the area of the Åland Islands regarding the Defence Forces rests mainly with the Finnish navy. The third interpretation considers the restrictions of the 1921 Åland Convention as covering the troop of the Defence Forces as well in the case of executive assistance requests by the police.[110] Thus, military presence would not be allowed in the zone, even in exceptional situations. The last interpretation would mean that the police and the Border Guard could not ask for executive assistance from the Finnish navy, even when the activity that renders passage non-innocent occurs in the demilitarised zone. Regarding the illicit trafficking of WMD and related materials, the second option seems plausible in the context of the coastal State authority to enforce protective rules.

6 Concluding Observations

Boarding a foreign ship without permission or other authorisation is in contravention of international law. This kind of activity on the territorial waters of the Åland Islands by the Finnish military authorities, when directed at governmental ships or civilian ships believed to be carrying WMD or related materials, could be interpreted to be against the provisions of the treaty arrangements that demilitarise the sea area around the Åland Islands.

Participants of the PSI are committed to taking appropriate actions to stop and/or search, in their internal waters, territorial seas or contiguous zones, vessels that are reasonably suspected of carrying cargoes of WMDs, their delivery systems or related materials to or from States or non-State actors of proliferation concern and to seize such cargoes that are identified. The State always has a right to take interdiction operations against its own vessels. However, in the demilitarised zone of the Åland Islands, this might be problematic, even against ships flying the Finnish flag.

The 3-nautical-mile demilitarised sea area around the Åland Islands belongs to Finland's internal waters and territorial sea. Thus, Finland's authority to regulate innocent passage through the Åland Islands' territorial sea depends on the current legal framework. In the territorial sea, the enforcement of the requirements of the PSI rests on the interpretation of Article 19 (2) of UNCLOS. In spite of claims for an independent nature of the right of innocent passage, coastal States have the authority to prevent passage that is not innocent and to adopt new laws and regulations relating to passage. Taking into consideration the objective and purpose of demilitarisation and neutralisation, the Åland Islands' surrounding sea areas

[110]Ministry of Defence (2014), p. 7.

might differ from other sea areas when it comes to the nature of peace, good order or security. When discussing the territorial waters of the Åland Islands, therefore, one should always bear in mind the interests of the wider group of countries and not just Finland. In a case concerning the illicit trafficking of WMDs and related materials through the demilitarised territorial sea area where there is solid intelligence that the intentions are threatening a party to the 1921 Åland Convention, the shipment would pose a threat of force that is not in accordance with the Convention's aim and purpose. When the passage through the territorial waters of the Åland Islands is rendered non-innocent, any enforcement measures undertaken must meet the provisions of the 1921 Åland Convention.

References

Actes de la Conférence. relative à la non-fortification et à la neutralisation des îles d'Aland, tenue à Genève, du 10 au 20 octobre 1921. (1922). Publie par les soins du Secretariat Permanent de la Société des Nations. Genève

Ahlström C (2005) The proliferation security initiative: international law aspects of the statement of interdiction principles. In: SIPRI Yearbook 2005: armaments, disarmament and international security, pp 741–765

Allen CH (2007) Matitime counterproliferation operations and the rule of law. Praeger Security International

Beck ME (2004) The promise and limits of the PSI. The Monitor 10(1):16–17. Available via, Center for International Trade and Security. http://www.uga.edu/cits/documents/pdf/monitor/monitor_sp_2004.pdf. Accessed 3 Feb 2016

Bergin A (2005) The proliferation security initiative-implications for the Indian Ocean. Int J Mar Coast Law 20(1):85–95

Brown ED (1994) The international law of the sea, volume introductory manual

Byers M (2004) Policing the high seas: the proliferation security initiative. Am J Int Law 98 (3):526–545

Churchill RR, Lowe AV (1999) The law of the sea. Manchester University Press

Cirincione J, Williams J (2005) Putting PSI into Perspective, April 27 2005. In: Carnegie endowment for international peace. http://carnegieendowment.org/2005/04/27/putting-psi-into-perspective. Accessed 18 Feb 2016

Dixon DB (2006) Transnational shipments of nuclear materials by sea: do current safeguards provide coastal states a right to deny innocent passage? The Berkeley Electronic Press Paper 1794, pp 1–34. http://law.bepress.com/expresso/eps/1794. Accessed 9 Feb 2016

Durkalec J (2012) The proliferation security initiative: evolution and future prospects. EU Non-Proliferation Consortium Non-Proliferation Papers, pp 1–21

Fact Sheet The White House, Office of the Press Secretary 2003 September 4. http://www.state.gov/t/isn/c27726.htm. Accessed 16 Feb 2016

Fidler DP (2003) Weapons of mass destruction and international law. ASIL Insights 8(3). https://www.asil.org/insights/volume/8/issue/3/weapons-mass-destruction-and-international-law. Accessed 25 July 2016

Garvey JI (2005) The international institutional imperative for countering the spread of weapons of mass destruction: assessing the proliferation security initiative. J Conflict Secur Law 10 (2):125–147

Government Proposal. HE 220/2013 vp. Hallituksen esitys eduskunnalle laiksi rajavartiolain muuttamisesta sekä eräiksi siihen liittyviksi laeiksi

Hakapää K (1981) Marine pollution in international law, material obligations and jurisdiction with special reference to the Third United Nations Conference on the law of the sea. Suomalainen Tiedeakatemia, Helsinki

Hakapää K (1988) Uusi kansainvälinen merioikeus. Lakimiesliiton Kustannus, Helsinki

Hakapää K, Molenaar EJ (1999) Innocent passage - past and present. Mar Policy 23(2):131–145

Hannikainen L (1994) The continued validity of the demilitarised and neutralised status of the Åland Islands. Zeitschrift für ausländisches öffentliches Recht und Völkerrecht 54:614–651

Harrison J (2013) Making the law of the sea, a study in the development of international. Cambridge University Press

HELCOM (2014) Annual report on shipping accidents in the Baltic Sea in 2013. HELCOM-Baltic Marine Environment Protection Commission

International Association of Lawyers Against Nuclear Arms Aotearoa New Zealand Branch (n.d.) Submission to the Select Committee on Foreign Affairs, Defence and Trade on the New Zealand Nuclear Free Zone Extension Bill. http://lcnp.org/disarmament/nwfz/submission%20on%20NWF2.htm. Accessed 21 March 2016

Jimenez Kwast P (2007) Maritime interdiction of weapons of mass destruction in an international perspective. Neth Yearb Int Law 38:163–241

Jinyuan S (2012) The proliferation security initiative (PSI) and interdiction at sea: a Chinese perspective. Ocean Dev Int Law 43(1):96–118. doi:10.1080/00908320.2012.647515

Joint Statement by the United States of America and the Union of Soviet Socialist Republics, Uniform Interpretation of Norms of International Law Governing Innocent Passage (1989). Law of the Sea Bulletin 1989(14):12–13

Joyner DH (2005) The proliferation security initiative: nonproliferation, counterproliferation, and international law. Yale J Int Law 30:507–548

Kaye S (2006) The proliferation security initiative in the maritime domain. Int Law Stud 81:141–164

Klein N (2011) Maritime security and the law of the sea. Oxford University Press

Kraska J (2009) Grasping "The Influence of Law on Sea Power". Naval War Coll Rev 62 (3):113–135

Lehrman TD (2004) Enhancing the proliferation security initiative: the case for a decentralized nonproliferation architecture. Virginia J Int Law 45:223–276

Lehto M (2008) International responsibility for terrorists acts. A shift towards more indirect forms of responsibility. Acta Universitatis Lapponiensis 139, Lapland University Press, Vaajakoski

Logan SE (2005) The proliferation security initiative: navigating the legal challenges. J Transl Law Policy 14(2):253–274

Malanczuk P (1997) Akehurst's Modern Introduction to International Law. Seventh revised edition. Routledge, London

McDougal MS, Burke WT (1987) The public order of the oceans. A contemporary international law of the sea. New Haven Press, New Haven

Ministry for Foreign Affairs of Finland. http://formin.fi/public/default.aspx?contentid=325890&contentlan=2&culture=en-US. Accessed 16 Feb 2016

Ministry of Defence, Selvitys/Utredning HARE PLM001:00/2014, FI.PLM.2015-2073, 767/00.99.00/2014

Nordquist MH, Nandan SN, Rosenna S, Grandy NR (eds) (1993) United Nations Convention on the Law of the Sea 1982, A Commentary, Volume II. Martinus Nijhoff Publishers

O'Brien P (2012) The Åland Islands Solution. A precedent for successful international disputes settlement. http://legal.un.org/ola/media/info_from_lc/POB%20Aalands%20Islands%20Exhibition%20opening.pdf. Accessed 26 Aug 2014

O'Connell DP (1984) In: Shearer IA (ed) The International Law of the Sea Volume II. Clarendon Press, Oxford

Pharand D (1977) International straits. Thesaurus Acroasium 7:64–100

Proliferation Security Initiative. http://www.psi-online.info/Vertretung/psi/en/01-about-psi/0-about-us.html. Accessed 5 Feb 2016

Prosser A, Scoville HJ (2004) The proliferation security initiative in perspective. Center for Defense Information June 16, 2004. http://www.cdi.org/pdfs/psi.pdf. Accessed 3 Feb 2016
Rayfuse R (2005) Regulation and enforcement in the law of the sea: emerging assertions of a right to non-flag state enforcement in the high seas fisheries and disarmament contexts. Aust Yearb Int Law 24:181–200
Remarks by the President to the People of Poland 2003 May. http://georgewbush-whitehouse.archives.gov/news/releases/2003/05/20030531-3.html. Accessed 16 Feb 2016
Ronzitti N (1990) The law of the sea and the use of force against terrorists activities. In: Ronzitti N (ed) Maritime terrorism and international law. Martinus Nijhoff, Dordrecht, pp 1–14
Shearer IA (1986) Problems of jurisdiction and law enforcement against delinquent vessels. Int Comp Law Q 35(2):320–343
Thomas TV (2009) The proliferation security initiative: towards relegation of navigational freedoms in UNCLOS? An Indian perspective. Chin J Int Law 8(3):657–680
Tornberg R (2009) Folkrättsliga aspekter på Proliferation Security Initiative. Inrädesanförande hållet vid Kungl. Örlogsmannasällskapets ordinarie sammanträde i Stockholm, pp 137–149
Treves T (2009) Piracy, law of the sea, and use of force: developments off Coast of Somalia. Eur J Int Law 20(2):399–414
US Department of State (2003) Proliferation Security Initiative: Chairman's Statement at the Third Meeting. Third Meeting of the PSI, September 3–4, 2003. http://2001-2009.state.gov/t/isn/rls/other/25425.htm. Accessed 16 Feb 2016
US Department of State Proliferation Security Initiative. http://www.state.gov/t/isn/c10390.htm. Accessed 16 Feb 2016
Vukas B (2004). The law of the sea: selected writings. Martinus Nijhoff Publishers
Winner AC (2005) The proliferation security initiative: the new face of interdiction. Wash Q 28 (2):129–143
Wolfrum R (2009) Freedom of navigation: new challenges. In: Norquist MH, Koh TT, Moore NJ (eds) Freedom of seas, passage rights and the 1982 Law of the Sea Convention. Matinus Nijhoff Publishers, Leiden/Boston, pp 79–94
Yearbook of International Law Commission Volume II (1956)

Case

Legality of the Use by a State of Nuclear Weapons in Armed Conflict, Advisory Opinion, I. C. J. Reports 1996, p 66

Printed by Printforce, the Netherlands